CLIMATE DATA AND RESOURCES

CLIMATE DATA AND RESOURCES

A reference and guide

Edward Linacre

London and New York

First published 1992
by Routledge
11 New Fetter Lane, London EC4P 4EE

Simultaneously published in the USA and Canada
by Routledge
a division of Routledge, Chapman and Hall, Inc.
29 West 35th Street, New York, NY 10001

Typeset in Baskerville by
Columns Design and Production Services Ltd, Reading
Printed and bound in Great Britain by
Biddles Ltd, Guildford and King's Lynn

British Library Cataloguing in Publication Data
Linacre, Edward 1925–
Climate data and resources: a reference and guide.
1. Climatology
I. Title
551.6

ISBN 0–415–05702–7
ISBN 0–415–05703–5 pbk

Library of Congress Cataloging-in-Publication Data

Linacre, Edward, 1925–
Climate data and resources: a reference and guide/Edward Linacre.
Includes bibliographical references and index.
ISBN 0–415–05702–7. — ISBN 0–415–05703–5
1. Climatology–Technique. 2. Sydney (N.S.W.)–Climate–Observations. I. Title.
QC981.L55 1991 91-12260
551.6′32–dc20 CIP

CONTENTS

FIGURES

TABLES

PREFACE

This book arose out of a senior undergraduate course on Applied Climatology about the practical use of climate information. It is meant to be helpful to students, meteorologists, engineers, planners, builders, farmers – in fact anyone with a problem involving the climate – and also to teachers who do not have access to a comprehensive, up-to-date library. For them and the others, this is a broad survey of the state-of-the-art, being a systematic digest of over 2000 books and papers as well as new material.

The emphasis is on climate: the relatively constant component of the weather. In particular, attention is focused on the sort of observations which are made at a single climate station, so that most value can be derived from the data which accumulate. Anyone concerned with running a weather station should find the book useful.

References are made to measurements taken at 'Marsfield', which is where Macquarie University is located in Sydney, Australia. The climate station there is fairly ordinary, so the use of the measurements illustrates what can be done anywhere in the world. Amongst other things, you could compare local measurements with the numerous empirical observations that are quoted, to test whether universal generalisations are possible.

There are two parts to the book. The first chapters deal with the specification, obtaining and treatment of climate data, which form the raw material for tackling any climate problem. Temperature and humidity are considered in detail in these chapters, with passing reference to evaporation. The second part of the book outlines what is known about three other important climate elements – solar radiation, wind and precipitation.

Each chapter is related to the current concern with climate change. In addition, the book deals with matters which underlie high-tech aspects of climatology, such as the parameterisation of surface conditions for computer modelling of the atmosphere, the modelling of crop performance, the interpretation of satellite observation of ground conditions, the control of air pollution, and so on.

Works and papers mentioned in the book are detailed in the Bibliography. However, this is a distillation of over six times as many references which have been used, whittled down for economy and easier reading. Complete documentation can be obtained from the author.* I would welcome correspondence.

Edward Linacre
School of Earth Sciences,
Macquarie University,
Sydney, Australia 2113

* Please send an International Reply Coupon (or a stamped addressed envelope within Australia) either for any particular reference, or for details on obtaining complete documentation on a 3·5 inch 72 OK disk for Macintosh (using Microsoft Word).

ACKNOWLEDGEMENTS

I thank Helen Linacre particularly, for her forbearance and support during the years it took to write and rewrite this book.

Grateful acknowledgement is made also of advice from the following on parts of various drafts: Peter Abelson, R. Barge, Keith Bigg, Howard Bridgman, Michael Clarke, Jack Davies, Ann Eyland, Bruce Forgan, Roger Gammon, John Hay, John Hicks, E. Jesson, Malcolm Kennedy, G. Leeper, Susan Linacre, Nick Lomb, Ian McIlroy, Graham McIntyre, Don McNeil, I. McRae, John Oliver, Peter Shaw, I. Strangeways, Peter Valko, Alan Vaughan and Eric Webb. Most of the typing of initial drafts was done by Nancy Michal of the Centre for Evening and External Studies, Macquarie University. The diagrams were drawn by Rod Bashford and his colleagues. Several students have helped in checking references to the literature, and many taking my course on Applied Climatology over the years have contributed items of information and provocation. Anne Prescott assisted in checking. I am grateful to staff of Macquarie University Library, notably Jocelyn Gardner and the Inter-library Loan staff. I also wish to thank Alan Fidler who did an excellent job of copy-editing, and the Desk Editor, Karen Peat.

Part I

CLIMATE DATA

1

INTRODUCTION

We begin by considering certain basic aspects of climate. These include its definition, the history of its measurement and study, the spatial hierarchy of kinds of climate, the units of measurement, and the phenomenon of global climate change.

CLIMATE AND CLIMATOLOGY

'Climate' is an abstraction like 'goodness', easier to recognise than to define precisely. The subtlety of the concept is shown by the variety of definitions in the literature. But a consensus of sixteen published definitions (see Note 1) is as follows: 'Climate is the synthesis of atmospheric conditions characteristic of a particular place in the long-term. It is expressed by means of averages of the various elements of weather, and also the probabilities of other conditions, including extreme values.'

The word 'climatology' has two meanings. In the phrase 'the climatology of a place', the word means the quantitative *description* of the climate there. In other contexts, the word means the scientific *study* of climate, the examination of generalisations from past examples of atmospheric behaviour. Or, to put it differently, climatology is the long-term pattern underlying day-to-day weather.

It is closely connected with meteorology. In English-speaking countries, climatology used to be regarded as merely the statistical aspect of meteorology, in contrast to the French and Russian view that meteorology was a subsection of climatology. Nowadays, the subjects are seen as parallel but different. Climatology is more concerned with long-term features of the atmosphere, tends to be statistical rather than determinist, and relates more to spatial than temporal differences. However, they overlap, and several topics could be labelled either way. One example is the study of climate change which we shall consider later. Fortunately, exact demarcation is not important, and we can regard climatology and meteorology simply as complementary aspects of atmospheric science.

A more practical explanation of climatology is in terms of questions which a climatologist can answer. Examples are as follows:

1 How can measurements of air temperature, humidity, wind or rainfall be made meaningfully?
2 How best can values for such climatic elements be estimated, at times and places without measurements?
3 What are the implications of climate data? For instance, is one wet season likely to be followed by another?
4 How much solar energy is available at a particular place?
5 How much wind energy can be captured there?
6 How does rainfall vary over a region?

Each of these questions, along with many others, will be answered in the chapters that follow.

HISTORY OF CLIMATOLOGY

The study of meteorology probably began with the sky religions of Egypt in about 3000 BC. Bones were engraved with weather details for particular days at about 1500 BC in China, indicating, for instance, that northern China was warmer than now. Written records of weather anomalies in north China have been made since the Zhou dynasty (1111–246 BC), the format being standardised during the Han dynasty (206 BC – AD 220). So it has been possible to compile a record of winter thunderstorms over a span of 2200 years, revealing that the climate of north-west China was just as dry then as it is today.

In ancient Greece, the poetry of Hesiod in 800 BC described signs portending various kinds of weather, and Thales in 600 BC related the occurrence of rain to the rising of certain stars. Later, Aristotle's *Meteorologica* (334 BC) was the single, undisputed authority in Europe on all weather processes until the rise of science in the 16th and 17th centuries.

The climate was first mentioned sometime during the Xia dynasty in China (21st–16th century BC), when there was a 400-word text (*Xia Xiao Zheng*) which included descriptions of the weather to be expected in each month. In other words, the conscious study of climates began later than meteorology – perhaps because it takes time to perceive the abstraction which is climate behind the short-term evidence of weather, or because people did not travel as far or as fast in those days, so that climatic differences between places were experienced less dramatically than the day-to-day vagaries of weather at one place, which is the stuff of meteorology.

More recently, there has been an accelerating development of climatology, as indicated in Table 1.1. Study was first concentrated in Greece, then Rome, and also in Italy during the 16th and 17th centuries. The centre of activity shifted further westwards to France in the 18th century, and to Germany in the 19th, alongside a growing English-language contribution.

Table 1.1 Notable events in the history of climatology

Date (approximate)	*Event*
BC	
450	Empedocles' theory that earth, air, fire and water constitute the universe.
440	Herodotus compared the climates of places.
400	Hippocrates' treatise *Airs, Waters and Places* expressed the idea that weather moulds the life of man.
370	Eudoxus' book on bad-weather predictions claimed there was a periodicity in weather phenomena.
334	Aristotle's *Meteorologica* was the first book on the science of the atmosphere.
300	Theophrastus listed winds and weather portents in his *Book of Signs*.
200	Erastosthenes explained climates in terms of the sun's position in the sky.
175	Aratus' poem *Phaenomena* included 'Weather Signs'.
150	Polybius divided the Earth's climates into six latitudinal belts.
140	Hipparchus zoned the world according to the day-length at the summer solstice.
100	Virgil discussed the effect of weather on agriculture in his *Georgics*.
AD	
50	Senaca wrote *Questions of Nature*.
130	Ptolomy used a climatic classification of seven zones with different day lengths, extending to 62°N.
700	The Venerable Bede completed *De Natura Rerum*.
850	Wind-vanes for observing wind direction were introduced.
1200	Idrisi (an Arabian scientist) used seven climatic zones, each of ten divisions.
1340	William Merle recorded weather conditions at Oxford and Lincoln, England.
1481	William Caxton printed an English translation of the anonymous 13th century *Speculum Vel Imago Mundi*.
1500	Leonardo da Vinci invented the hygrometer and designed a pressure-plate anemometer.
1563	William Fulke published *A Goodly Gallerye with a most pleasant Prospect into the Garden of Naturall Contemplation, to behold the naturall causes of all kynde of Meteors, as well fyery and ayery, as watry and earthly, etc.*
1593	Galilei Galileo (or his pupil Santorio) invented the air thermometer.
1632	Galileo suggested that the Trade winds result from the Earth's rotation.
1638	Francis Bacon attributed Trade winds to thermal expansion of the atmosphere at low latitudes.
1639	Rainfall was measured in Italy, the gauge being invented by Castelli, a pupil of Galileo.
1641	Ferdinand II of Tuscany developed a sealed, evacuated thermometer.
1643	Evangelista Torricelli invented the barometer.
1647	Blaise Pascal demonstrated the reduction of pressure up a mountain.
1653	Ferdinand II created a network of observing stations.
1660	A dewpoint hygrometer was constructed in Florence.
1662	Christopher Wren invented a self-recording, tipping-bucket rain gauge.
1664	Longest continuous series of weather observations were begun in Paris.
1667	Robert Hooke published details of his tilting-plate anemometer.
1677	Richard Townley began regular measurements of rainfall in England.
1683	Edmund Halley published a comprehensive map of.global winds.

Table 1.1 continued

Date (approximate)	*Event*
1686	Halley explained the Trade wind and monsoon circulations.
1693	William Dampier began collecting information on the distribution of winds, e.g. around a tropical storm.
1694	Halley conducted experiments showing the effects of sun and wind on evaporation.
1710	Gabriel Fahrenheit developed a 32° – 212° temperature scale.
1729	Pierre Bouguer explained the attenuation of sunlight by the atmosphere.
1730	Rene-Antoine de Reamur developed a 0°–80° temperature scale.
1735	Reamur proposed the 'degree-day' as a unit of plant growth.
	George Hadley developed a theory linking the Earth's rotation and the trade winds.
1736	Robert Marsham began phenological records near Norwich, England.
1737	Anders Celsius developed a 100-unit temperature scale.
1743	Benjamin Franklin described the migration of storms, etc.
1744	Georges-Louis LeClerc described the Earth's hydrologic cycle.
1750	Carl von Linne established a network of 18 phenological stations in Sweden.
1759	John Smeaton published Samuel Rouse's inland version of a scale of wind speed like Beaufort's.
1760	Joseph Black formulated the concept of latent heat.
1777	Mathew Dobson measured evaporation in Liverpool, England.
1781	John Hemmer of Mannheim used a network of 39 climate stations in 18 countries.
1782	James Six invented his maximum–minimum thermometer.
	Ascent into the atmosphere by the Montgolfier brothers.
1783	Horace de Saussure invented the hair hygrometer.
1787	Richard Kirwan published a paper on the latitudinal distribution of temperature.
	John Dalton's rain-gauge network began in north-west England (till 1844).
1788	Lapse rates of mountain temperature observed by de Saussure.
	Erasmus Darwin explained cloud formation by adiabatic expansion.
1802	Dalton developed his laws of the pressure of vapours, and explained condensation following expansion.
	Luke Howard classified cloud types.
1806	Francis Beaufort developed his wind scale for use at sea.
1814	First correct explanation of dew formation by Charles Wells in England.
1816	Heinrich Brandes made synoptic maps of the weather.
1817	Alexander von Humboldt prepared a diagram of the mean isotherms for much of the Northern Hemisphere.
1818	Howard wrote a description of the climate of London.
1821	Ignatz Venitz put forward his evidence of changes in glaciers.
1823	Joachim Schouw published a treatise on the effect of climate on the distribution of various kinds of vegetation.
1825	Ernest August developed the psychrometer.
1831	William Redfield explained the circular nature of storms.
1835	Explanation of the effect named after him, by Gustave-Gaspard Coriolis.
1836	Angelo Bellani invented an instrument for measuring irradiance.
1837	Claude-Servais Pouilett invented a pyrheliometer and introduced the term;

Table 1.1 continued

Date (approximate)	*Event*
	also measured the solar constant.
	Observations were made of cloudiness at Greenwich, London.
1840	Joseph Atkinson prepared a tentative rainfall map, based on Dalton's work, in 1793.
	Follet Osler published a wind rose.
	Heinrich Dove used the term 'normal' for the mean of a series of observations.
1845	Thomas Robinson revised the cup anemometer, improving on Whewell's model of 1837.
	Henri-Victor Regnault explained the theory of the psychrometer.
1846	Matthew Maury prepared wind charts for ocean areas.
1847	Dove made maps of monthly mean temperature.
1849	Adolphe Quetelet published a description of the climate of Belgium.
1851	Weather maps, using data from 22 stations, were sold at the Great Exhibition in London.
1852	Christoph Buys-Ballot compiled daily weather maps of a large part of Europe.
	John Welsh developed an aspirated psychrometer.
1853	James Coffin constructed a series of wind charts of the Northern Hemisphere, showing three latitudinal belts.
	Jean Houzeau wrote a popular treatise on climatology.
	John Campbell devised a sunshine recorder.
1854	Many European meteorological services were formed, e.g. in London.
1855	Alphonse de Ancolle related vegetation distribution to temperatures
1856	William Ferrel proposed his three-cell model of the general circulation of global winds
1857	Lorin Blodget published *Climatology of the United States*
	Urbain Leverrier organised daily weather bulletins in France.
1859	William Jevons produced a climatology of Australia.
1862	Adolf Muhry made global maps of seasonal rainfall.
1866	Thomas Stevenson introduced his instrument screen at Greenwich, England.
1867	Carl Linser noted relationships between plants and climate, including evaporation.
1868	Alexander Buchan published the first global maps of pressure.
1870	Julius Hann studied the variation of temperature with altitude.
1872	International meteorology conference held in Leipzig.
1875	Coffin prepared wind charts for the world.
1878	Inauguration of the International Meteorological Organization in Utrecht.
1879	George Stokes improved on the sunshine recorder of John Campbell.
	Alexander Supan published world maps of temperature regimes.
1882	Elias Loomis produced world maps of precipitation using annual isohyets.
1883	Hann's data compilation, *Handbook of Climatology*.
1884	Alexsander Voeikov's explanatory *Climates of our Globe and Particularly Russia*; he is sometimes called the 'father' of climatology.
	Vladimir Koeppen prepared a world map of temperature regions.
	Supan suggested 35 climate subdivisions.
1886	Teisserenc de Bort issued maps of monthly cloudiness round the globe.
1890	Establishment of the US Weather Bureau.

Table 1.1 continued

Date (approximate)	*Event*
	Oscar Drude linked vegetation distribution to climates.
1892	Richard Assmann developed an aspirated psychrometer.
1896	Svante Arrhenius pointed out the effect of atmospheric carbon dioxide on ground-level temperatures.
1900	Koeppen introduced a climate classification based on plants.
1902	The terms 'troposphere', 'tropopause' and 'stratosphere' coined by de Bort.
1903	Alexandre Eiffel published a climagram.
1918	Koeppen published his main system of climate classification.
1923	F. Houghten and C. Yaglou introduced the concept of 'effective temperature'.
1924	E.L. Johnson published a 'climograph', a graph of monthly mean temperatures at a place against monthly rainfalls, i.e. a 12-cornered loop
1926	Emmanuel de Martonne suggested a temperature/rainfall aridity index.
1928	Tor Bergeron classified air masses and introduced 'dynamic climatology'.
1930	Molchanov successfully used radiosondes at Pavlovsk.
1931	Warren Thornthwaite evolved a temperature/precipitation classification of climates.
	Fritz Albrecht explained the greenhouse effect.
1940	Rudolph Geiger's book on microclimates, *Climate Near the Ground.*
1945	Paul Siple and C. Passel introduced the concept of the wind-chill factor.
1948	Thornthwaite devised a climate classification based on water budgets of the soil.
	Howard Penman explained the evaporation process.
1951	Foundation of the World Meteorological Organization.
1957–8	International Geophysical Year.
1960	*TIROS* weather satellite.

Note: For more recent work, see the Bibliography and the full documentation available from the author (see Preface).

The developments listed in Table 1.1 meant that climatology had become geographical, static and human-centred by the late 19th century. This is well illustrated by Hann's excellent *Handbook of Climatology*, first published in 1883. It was followed in 1884 by Voeikov's *Climates of Our Globe*, which explained climates in terms of the shapes of continents and the flow of ocean currents. In the 1920s there was the development of 'regional climatology' (the study of the climates of particular regions), followed in the 1930s by 'synoptic climatology', concerned with large-area patterns of pressures and winds, and 'dynamic climatology', which is about the movements of such patterns. The subject was still primarily descriptive.

Climatology became more quantitative with the application of statistics. But there was a simultaneous decline in interest, and the fraction of articles on climatology in the major American geography journals fell from 36 per cent during 1916–1919 to 5 per cent over the period 1928–1931. As a result, climatology had come to be regarded as the dry-as-dust book-keeping end of meteorology by the late 1930s.

A new era began in the 1940s, with a great increase in major publications on climatology. Since then the subject has continued to flourish, partly because of problems related to climate – droughts due to El Nino, the 1972 grain crisis in the USSR, the Sahel drought in the 1970s, urban air pollution, a growing concern for the environment, the need to improve agricultural food production, the possibility of a nuclear winter and increasing awareness of imminent climate change. Also, there has been a stimulus from new opportunities for finding solutions, arising from a rapid increase in the understanding of meteorological processes, better climate-data gathering (e.g. by means of satellites) and the advent of computers. So climatology today is broad and diverse, compounded of meteorology, statistics and geography, as well as economics and biology. The emphasis has shifted towards 'applied climatology': the use of climate data to overcome practical problems.

The post-war revival of the subject has been accompanied by the publication of numerous journals, books and reports. Especially useful sources of information are the technical notes and reports from the World Meteorological Organization (the WMO) in Geneva (see *Bulletin of the American Meteorological Society* 66, 1985: 1599–1605). A bibliography of climate data currently available was compiled by Critchfield (1987: 272).

SCALES OF CLIMATE

At this stage it is useful to clarify the relationship between what is measured at a climate station and what is meant by the climates of areas like a valley, a country or the whole Earth. The word has different meanings according to the size of the domain. For instance, we can speak of the climate of a garden, the climate of a city, or changes of the global climate. It is helpful to label these climates separately, distinguishing the 'microclimate' of a *site* (e.g. a climate station) from the 'topoclimate' of a *locality* (like a valley or city), the 'mesoclimate' of a *region*, the 'synoptic climate' of a *continent*, and the 'global climate' of the whole planet.

The climate within an area of a certain scale results from three main influences: (a) atmospheric processes operating in a space of that size, (b) the averaging of climates of the next smaller scale, and (c) the overriding effect of the environment provided by the next larger scale's climate. As regards the last, any particular scale of climate is a 'forcing variable' influencing the included domains of the next smaller scale, i.e. it is external to them, partly controls them, and is hardly affected by any particular one of them. As an example, the time-averaged topoclimate of a particular

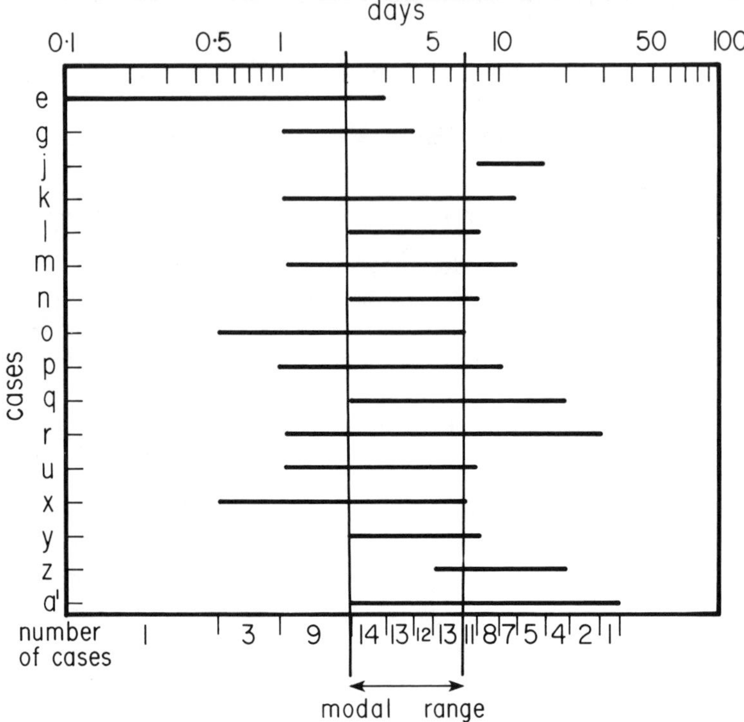

Figure 1.1 An example of the derivation of a characteristic value from data in Table 1.2. In this case the value refers to the time for processes on the synoptic scale. The 'cases' on the left axis concern the rows in Table 1.2. It may be seen that the modal range is 2-7 days, so the characteristic time (i.e. the geometric mean of those extremes) is 3.7 days (about 90 hours). So this value appears in Table 1.3

valley depends on the following: (i) the local rainfall, sunshine, temperature inversions and so on (which are fairly uniform over a valley), (ii) the various microclimates within the valley, and (iii) the regional mesoclimate covering that valley and adjacent localities.

Various proposals have been made about labelling the several kinds of climate. In fact, there have been so many suggestions that yet another would be superfluous. Instead, it seems useful to settle on a consensus of what has already been either proposed for the first time or endorsed by repetition. This can be done by the procedure illustrated in Figure 1.1, which concerns the time-scale of synoptic climate, using data from the last column of Table 1.2. The diagram reveals the most common values, which lie within what is here called the 'modal range', determined graphically. The geometric mean of the extremes of that range is taken as the consensus value. Details of other scales of climate are also given by various authors, the outcome being summarised in Table 1.3.

Table 1.2 Published characteristics of synoptic-scale processes and climate[a]

Case	*Source*	*Horizontal: km*	*Vertical: km*	*Time: days*
b	Tepper 1959: 57	500–2,000	–	–
c	Russell 1960: 401	1,000–10,000	–	–
d	Fukui 1962: 232	50–200	8–30	–
e	Godske 1966: 10	300–4,000	–	0.1–3
f	Geiger 1966	1,000–10,000	–	–
g	Lee 1966: 440	100–1,000	–	1–4
h	Shaw 1967: 296	500–2,000	–	–
i	Wallen 1968	1,000–10,000	–	–
j	Barry 1970: 62	500–5,000	5–20	8–16
k	Mason 1970	400–4,000	–	1–12
l	Fiedler and Panofsky 1970: 1115	250–1,000	–	2–8
m	Munn 1966: 45	100–1,000	–	1–12
n	Pasquill 1971: 448	–	–	2–8
o	Lowry 1972a: 16	100–10,000	1–20	0.5–7
p	Lindner and Nyberg 1973: 86	1,000–50,000	–	1–10
q	Lockwood 1974: 234	500–5,000	–	2–20
r	Orlanski 1975	200–10,000	–	1–30
s	Yoshino 1975	200–10,000	0–200	–
t	Johnson and Ruff 1975	100–1,000	–	–
u	Dutton 1976:5	500–10,000	–	1–8
v	Chandler 1976:1	1,000–4,000	–	–
w	Smith 1979: 142	500–2,000	–	–
x	Gloyne and Lomas 1980: 76	100–10,000	1–20	0.5–7
y	Atkinson 1981: 6	500–2,000	–	2–8
z	WMO 1981a: 7	1,500–6,000	–	5–20
a′	Stull 1985: 432	2,000–5,000	–	2–30

Note: [a] Where a single value has been given in the literature, it has been taken to represent a range from 50% to 200% of that. In some publications the values quoted here were not associated specifically with the synoptic scale so called, but with atmospheric processes of that scale.

The regularity of the variations down columns 3, 8, 10 and 11 in Table 1.3 is remarkable, in view of the wide scatter of the published information from which the values were derived. The regularities point to an inherent self-consistency of the table.

Column 3 shows an increasing randomness of the air's circulations as energy from the sun's particular heating of the equator cascades from the larger processes to the smaller, ending as thermal energy, in conformity with the Second Law of Thermodynamics.

Table 1.3 Scales of atmospheric processes[a]

Scale of climate 1	Rotation Relevant features 2	Rotation axis 3	Geography 4	Domain 5	*Typical extents* horizontal 6	vertical 7	ratio 8	*Deduced characteristics* length[b] 9	time[c] 10	rate[d] 11
Global	Solar heating, general circulation, monsoon	North–south	Whole planet	Earth	20,000 km	20 km	1000	630 km	1 year	0.02 m/s
Synoptic	Front Coriolis effect, anticyclone, Trade wind, ocean, tropical cyclone, latitude, front	Vertical	Continent	Part	1,400 km	13 km	110	135 km	90 h	0.42
Meso	Thunderstorm, gradient wind, sea breeze, mountain, lee wave	Horizontal	Small country or country,	Region	45 km	700 m	64	5.6 km	4.2 h	0.37
Topo	Surface friction cumulus cloud, inversion, thermal	Various	Hill, valley, lake, city	Locality	6.3 km	71 m	89	670 m	1 hour	0.19
Micro	Vertical fluxes, gusts, evaporation, net irradiance, turbulence	Quasi-random	Vegetation, buildings, surface irregularities	Site	14 m	1.4 m	10	4.4 m	1.4 min	0.05
Superficial	molecular diffusion, ground conduction, leaf shape	Random	Roughness	Surface	10 cm	5 mm	20	0.022 m[e]	1.2 sec	0.02

Notes: [a] Based partly on the ideas of Mason (1970) and Linacre (1981).
[b] The geometric mean of the horizontal and vertical extents.
[c] Typical period for a cycle of the processes mentioned in the second column and for the horizontal extents in column 6 (Ludlam 1980: 113).
[d] The dividend of the characteristic length and time.
[e] The product of the thermal resistance of a 5 mm layer of still air (i.e. 0.2 $C°.m^2/W$), and the layer's thermal capacity, i.e. 6 $J/C°.m^2$.

The ratios in column 8 show the relative importance of two of the factors which chiefly control air temperature and atmospheric humidity. These are, firstly, the local vertical fluxes of radiation, rainfall and evaporation, and, secondly, conditions upwind. The latter govern the horizontal fluxes of heat and water vapour in the wind, called 'advection'. The amount of advection at any particular scale of climate depends on the vertical extent of the domain's sides, given in column 7, whilst vertical fluxes depend on the horizontal size. These dimensions are compared by means of the ratios in column 8, showing an increase in the importance of advection at the smaller scales. Strong winds and reduced irradiance also enhance the influence of advection, which smudges the boundaries between adjacent topoclimates especially.

Values in column 10 of Table 1.3 agree with those from an analysis by Van der Hoven of the amounts of energy within wind fluctuations of various frequencies. His measurements showed energy maxima corresponding to fluctuations of 1 minute and 4 days, respectively, which correspond to the 1.4 minutes and 3.7 days in Table 1.3 for microscale and synoptic processes (Houghton 1977: 110). Van der Hoven also found a gap in the energy spectrum for fluctuations of around 20 minutes, implying a sharp distinction between microscale and toposcale processes.

The scales of climate in Table 1.3 relate to those advocated for ecosystem mapping. Bailey (1985) suggested that a 'region' may have a typical horizontal extent of 300 km (instead of 45 km in Table 1.3), and he called the next level in the hierarchy a 'landscape' (instead of 'locality'), with a typical extent of 30 km as against the 6.3 km given in the table. However, the similarities between the climate and ecosystem scales are sufficient to suggest that the scheme in Table 1.3 has a relevance beyond climate itself, to features of the world which are influenced by it.

UNITS

It is unfortunate that most of the literature on climatology is based on superseded units. There is a muddle of Fahrenheit and Celsius, calories and joules and British Thermal Units, miles per hour and knots, and so on – quite apart from special units concocted for particular aspects of climatology such as the clo, acre-foot, Met and langley. Nowadays, these units are all replaced by the Système Internationale (SI), based on the metre, kilogram and second. (Americans use the spelling 'meter', but that is a kind of instrument.) The SI was first proposed by Giorgi in 1901 and has since been adopted universally. The convenience of this decimal system lies in the ease of calculations, of expressing and comparing values, and a general simplification of terms.

Various SI units are tabulated in Table 1.4, along with equivalent older units. Multiples and submultiples of the units are prefixed as follows – tera

Table 1.4 Equivalence of Système Internationale units and others

Item	*Name of unit*	*Relationship to SI unit*	*Approximate conversions* *from SI metric*	*to SI metric*
Length	metre (m)	base unit	1 m = 39.37 inches	1 inch = 0.025 4 m
			= 3.281 feet	1 foot = 0.304 8 m
			= 1.093 6 yards	1 yard = 0.914 4 m
	micrometre (mu m)	0.000 001 m		1 fathom = 1.829 m
	millimetre (mm)	0.001 m		1 chain = 20.12 m
	kilometre (km)	1 000 m	1 km = 0.621 4 mile	1 mile = 1.609 km
			= 0.540 nautical mile[a,b]	1 nautical mile = 1.852 km
Area	square metre	SI unit	1 m^2 = 10.76 square feet	1 square foot = 0.092 90 m^2
			= 1.196 square yards	1 square yard = 0.836 1 m^2
	square centimetre (cm^2)	0.000 1 m^2	1 cm^2 = 0.155 0 sq inch	1 square inch = 6.452 cm^2
	hectare (ha)[a]	10 000 m^2	1 ha = 2.471 acres	1 acre = 0.404 7 ha
	square kilometre	1 000 000 m^2	1 km^2 = 0.386 1 sq mile	1 square mile = 2.590 km^2
			= 247.1 acres	
Volume	cubic metre (m^3)	SI unit	1 m^3 = 35.31 cubic feet	1 cubic foot = 0.028 32 m^3
			= 1.308 cubic yards	1 cubic yard = 0.764 6 m^3
	cubic centimetre (cm^3)[a]	0.000 001 m^3	1 cm^3 = 0.061 02 cu. inch	1 cubic inch = 16.39 cm^3
Fluid volume	litre (L)	0.001 m^3	1 L = 1.760 pints	1 pint = 0.568 3 L
			= 0.220 0 UK gallon	1 UK gallon = 4.546 L
			= 0.264 2 US gallon	1 US gallon = 3.785 L
			= 0.027 50 UK bushel	1 UK bushel = 0.036 37 m^3
			= 0.006 289 barrel	1 US bushel = 0.035 24 m^3
	millilitre (mL)	0.001 L	1 mL = 0.035 20 fluid oz	1 fluid ounce = 28.41 mL
	megalitre (ML)	1 000 000 L (i.e. 1 000 m^3)	1 ML = 0.810 7 acre.foot	1 acre.foot = 1.233 ML

Mass	kilogram (kg)	base unit	1 kg = 2.205 pounds = 0.157 5 stone 1 g = 0.035 27 ounce	1 pound = 0.453 6 kg 1 stone = 6.350 kg 1 ounce avoirdupois = 28.35 g
	milligram (mg)	0.000 001 kg		1 dram = 1.772 g
	gram (g)	0.001 kg		1 grain (Troy) = 64.80 mg
	tonne (t)	1 000 kg	1 t = 19.68 cwt = 1.102 US tons = 0.984 2 ton	1 hundredweight = 50.80 kg 1 US ton = 907.2 kg 1 UK ton = 1 016 kg
Density	kilogram per cubic metre (kg/m^3)		1 kg/m^3 = 0.062 43 pound per cubic foot	1 pound per cubic foot = 16.02 kg/m^3
	gram per cubic (g/cm^3)	1 000 kg/m^3	1 g/cm^3 = 0.036 13 pound per cubic inch	1 lb/cu inch = 27.68 g/cm^3 = 27.68 t/m^3
	tonne per cubic metre (t/m^3)	1 000 kg/m^3	1 t/m^3 = 0.036 13 pound per cubic inch	
Force	newton (N), $kg.m/s^2$	SI unit	1 N = 100 000 dynes = 0.224 8 pound force = 7.233 poundals	1 dyne = 0.01 mN 1 pound force = 4.448 N 1 poundal = 0.138 3 N
	kilonewton (kN)	1 000 N	1 kN = 0.100 4 ton force	1 ton force = 9.964 kN
	meganewton (MN)	1 000 000 N		
Pressure	pascal (Pa) i.e. N/m^2 or $kg/m.s^2$	SI unit	1 Pa = 0.01 millibar = 0.000 01 bar = 0.020 89 pound force per square foot	1 millibar = 100 Pa 1 bar = 100 kPa 1 pound force per square foot = 47.88 Pa
	millibar (mb), for meteorological purposes only[c]	100 Pa	1 mb = 0.750 1 mm mercury 1 kPa = 0.145 0 pound force per square inch = 0.295 3 inch mercury = 7.501 millimetres mercury 1 MPa = 9.871 atmospheres	1 pound force per square inch = 6.895 kPa 1 inch mercury = 3.386 kPa 1 mm mercury = 133.3 Pa = 1.333 hPa 1 atmosphere = 101.3 kPa = 1013 hPa[a]

Table 1.4 continued

Item	Name of unit	Relationship to SI unit	Approximate conversions: from SI metric	Approximate conversions: to SI metric
Velocity	metre per second (m/s)	SI unit	1 m/s = 3.281 feet/sec = 2.237 mph = 1.944 knots = 3.600 km/h[a]	1 foot per second = 0.304 8 m/s 1 mile per hour = 0.447 0 m/s = 1.609 km/h 1 knot = 0.514 4 m/s 1 km/h = 0.277 8 m/s
	kilometre per hour (km/h)		1 km/h = 0.540 knot	1 foot per minute = 5.080 mm/s 1 cusec = 0.028 32 m^3/s
Power	watt (W), i.e. J/s or kg. m^2/s^3	SI unit	1 W = 0.238 8 cal per sec = 0.056 87 BTU/min	1 cal. per sec = 4.187 W 1 BTU per second = 1.055 kW
	milliwatt (mW)	0.001 W	1 mW = 0.014 33 cal per min	1 cal per min = 69.78 mW 1 foot-pound/hour = 0.376 6 mW
	kilowatt (kW)	1,000 W	1 kW = 1.341 horsepower	1 horsepower = 0.745 7 kW
	megawatt (MW)	1,000,000 W		
Energy, work	joule (J) i.e. kg.m^2/s^2	SI unit	1 J = 0.238 8 calorie = 0.000 238 8 Calorie[d] = 10^7 erg	1 calorie = 4.186 J 1 calorie = 4.186 kJ 1 foot-pound = 1.356 J
	kilojoule (kJ)	1,000 J	1 kJ = 0.947 8 British Thermal Unit (BTU)	1 BTU = 1.055 kJ
	megajoule (MJ)	1,000,000 J	1 MJ = 0.277 8 kW h	1 kilowatt-hour = 3.6 MJ
Density of energy flux	watt per square metre (W/m^2), i.e. kg/s^3	SI unit	1 W/m^2 = 0.001 433 cal per centimetre.min = 2.064 cal per square centimetre.day = 0.753 6 kilocal per square centimetre.year = 0.753 6 kilolangley/year	1 cal per square centimetre. min = 69.78 mW/cm^2 1 calorie per square centimetre.day = 0.484 5 W/m^2 1 kilocal/square centimetre.year = 1.327 W/m^2 1 kilolangley/year =1.327 W/m^2

	milliwatt per square centimetre (mW/cm^2)	10 W/m^2	1 mW/cm^2 = 0.317 0 BTU per square foot.hour = 0.014 33 calorie per square centimetre.minute	1 BTU per square foot.hour = 3.155 W/m^2
			1 kJ/m^2 = 0.023 88 langley	1 langley = 1 cal/cm^2 = 41.87 kJ/m^2
Angle	radian (rad)	SI unit	1 radian = 57°18′	100° = 1.745 radians
	milliradian (mrad)	10^{-3} rad	1 radian of latitude = 6 129 km approx.	100 km on Earth = 56 mins lat. = 0.016 29 radian
	degree (1°)[a]	π/180 rad		
	minute (1′)[a]	1/60°	1 rad/s = 2.063×10^5 degrees per hour	
	second (1″)[a]	1/60′		1 degree/h = 4.847×10^{-6} rad/s

Notes: [a] This unit is not strictly part of the Système Internationale but is derived from it.
[b] A nautical mile is approximately the length of one minute of latitude at sea level.
[c] The millibar is now replaced by the hectopascal (hPa, i.e. a unit equal to 100 pascals) to which it is exactly equivalent.
[d] A Calorie equals one kilocalorie.

(T) means 10^{12}, giga (G) means 10^9, mega (M) means 10^6 (i.e. a million times), kilo (k) means 10^3, hecto (h) stands for 10^2 (as in 'hectare', which is 100 m × 100 m), milli (m) means 10^{-3}, micro (Greek μ) 10^{-6}, nano (n) 10^{-9}, pico (p) 10^{-12}. The use of 'centi' or 'deci' is tolerated but not approved, so that 'centimetres' should really be replaced by the equivalent in metres or millimetres. Units such as minutes, litres and hectares are not strictly SI units, but are accepted for convenience.

SI units do not include the Angstrom (equal to 0.1 nanometres), the 'point' of rain (i.e. a depth of 0.254 mm), the dyne (10^{-5} newtons), the 'atmosphere' of pressure (1013 hectopascals), the British Thermal Unit (1055 joules), the calorie (4.186 joules) and the langley (41.86 kilojoules/metre^2). The 'knot' is still tolerated, having the convenient property of representing a minute of latitude per hour, but it is not an SI unit and should be replaced by the metre/second, which is the same as 1.944 knots. The millibar was replaced by its numerical equivalent, the hectopascal, on 1 January 1986, initially by the International Civil Aviation Organization, and subsequently by other institutions.

Temperatures are expressed in 'degrees' because early thermometers were bent into a circle, and the position of the fluid's meniscus was described in terms of its angle around the centre. Nowadays, the Celsius scale replaces the Fahrenheit scale. It is sometimes convenient to remember that 82°F roughly equals 28°C, 61°F is near 16°C, and −40°F equals −40°C. Note 2, p. 312 describes a simple method of converting from one scale to the other.

Conversion of published formulae into their SI equivalents requires care. The chances of mistakes are reduced by the four-stage procedure described in Note 3, pp. 312–13. Some terms have units with several parts, such as the calibration factor of a radiometer (see Chapter 5). This might be given as 15.22 mV/cal/cm^2/min, for example, where mV stands for millivolt. In other words, the instrument reading is 15.22 mV if the irradiance is 1 calorie onto each square centimetre in each minute. In SI units this is 21.8 μV per W/m^2, where μV stands for microvolts. The calibration would be more neatly expressed as 21.8 μV.m^2/W, using a single solidus. Likewise, a unit of cal/cm^2/cm/sec/C° used to describe the thermal conductivity of a material, should be replaced by cal/cm^2.s per C°/cm, or, better, by W/m^2 per C°/m, or, best, by W/m.C°, which is less ambiguous, simpler and standard.

A unit of temperature *difference* is here written as C°. Thus, for example, 11.2°C does not equal 20°F, though 11.2C° does equal 20F°.

Symbols for climatic quantities are a problem because the number to be represented far exceeds the capacity of the Roman and Greek alphabets. Some recommendations have been made by the World Meteorological Organization (WMO 1974b). The symbols in this book are based on common usage and the reader's convenience.

CLIMATE CHANGE

Although the climate is constant in comparison with the weather, it may not be the same from one decade to another, as can be seen in Figure 1.2. Changes in the past have been considerable. Global temperatures were around 5C° lower during the Ice Ages (from 2.5 million to 10,000 years ago) than they are now. Then there was warming until the Altithermal Period around 7000 years ago, when the Earth was about 2C° warmer than now. Afterwards, there was a cooling, and then, in Europe at least, the Medieval Warm Epoch around AD 1000. Later came the Litle Ice Age, with cooling by a degree or so until about AD 1650, especially in the winters. Since that time, there has been a general warming, interrupted by the cooling between 1940–65 seen in Figure 1.2. All changes have been small compared with the year-to-year fluctuations, like those shown in Figure 1.3

Changes of temperature during the Ice Ages were accompanied by parallel variations of atmospheric carbon dioxide, down to about half of present amounts at the coldest time. Likewise, the global warming by about 0.5C° this century has been accompanied by an increase from about 280 parts of carbon dioxide per million of air to 350 ppm by 1987, i.e. by 25 per cent. Also, there have been increasing amounts of other gases, notably spray-pack propellants and methane from more ruminants and rice production, grown to cope with the world's population explosion. As recognised by Fourier in 1827, and calculated by John Tyndall and Svante Arrhenius at the turn of this century, such gases trap heat radiated from the

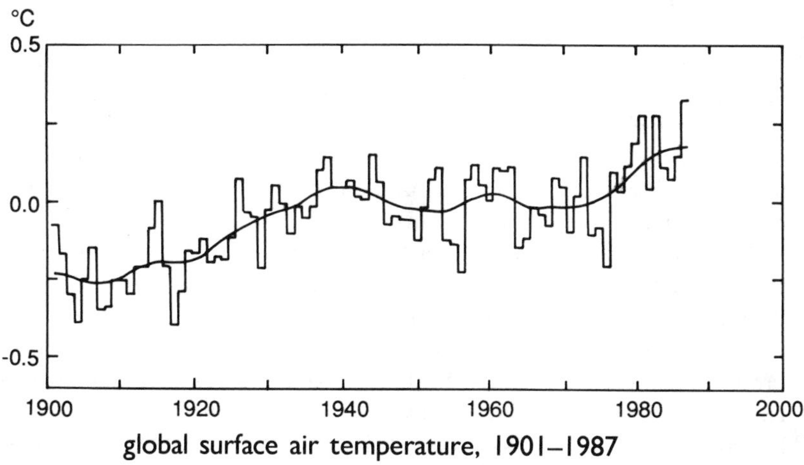

Figure 1.2 Variation of land-based measurements of air temperatures in the southern hemisphere (Jones *et al.* 1988: 790)

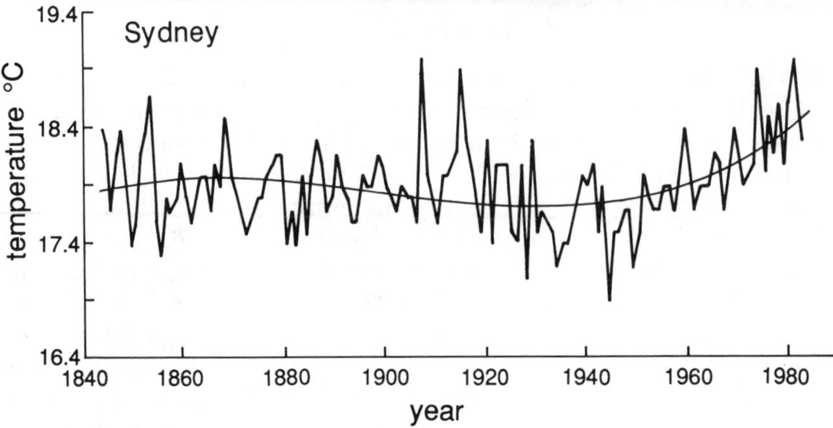

Figure 1.3 Variation of annual mean temperatures at Observatory Hill in Sydney (from data compiled by David Williams)

Earth's surface, just as the glass of a greenhouse was thought to increase temperatures within. The 'greenhouse effect' presently raises global temperatures by 33C°, and without it the seas would be frozen solid. Numerous independent and increasingly sophisticated computer studies during the last decade all point to the probability that present warming will continue as long as the world's population grows and energy consumption per capita increases, as we burn ever more oil, coal, wood and natural gas, creating more and more carbon dioxide (Schneider 1990). Emissions increased fivefold during 1940–80, and continue to multiply (Mitchell 1989).

In addition, there are aggravating factors, e.g. the increased water vapour (another 'greenhouse gas') resulting from greater evaporation from the oceans as they warm: warmer oceans mean more water vapour, hence more greenhouse warming, so yet warmer oceans. This is a 'positive feedback', which complicates the calculation of global warming. Another is the diminished capacity of the oceans to absorb carbon dioxide as they warm up, leaving more of the gas in the atmosphere, and hence yet more warming. A third positive feedback, amplifying any change, is due to the transparency of high-level clouds, which are increased by warming of the oceans and themselves act like a greenhouse gas. On the other hand, more low-level clouds create negative feedback, reducing the warming which creates them by reflecting sunshine away from the Earth.

The multiplicity of processes involved make it impossible to forecast future warming with certainty (Singer 1989). There is still much confusing evidence. The low concentrations of carbon dioxide one would expect during the Ice Ages did occur, but *followed* them, contrary to the

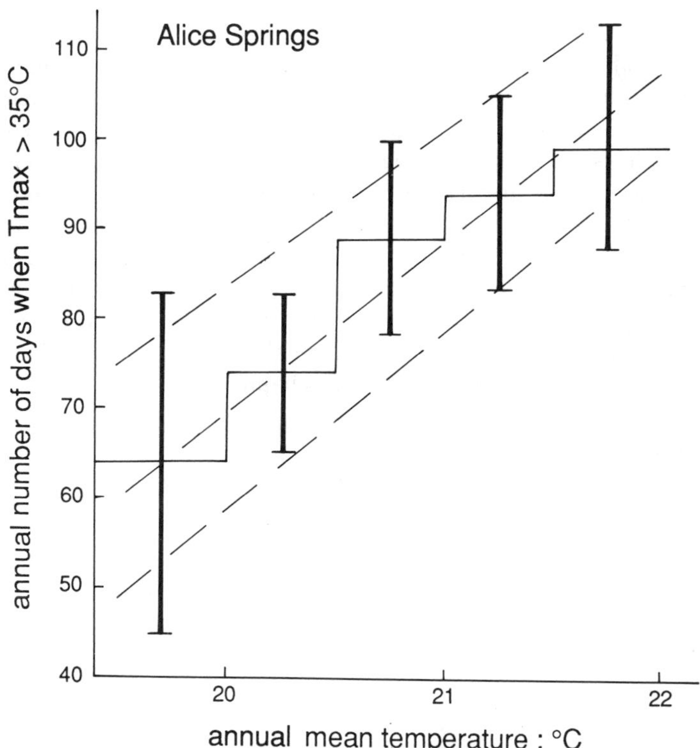

Figure 1.4 Variation of the annual number of days when the maximum temperature exceeded 35°C at Alice Springs (in the middle of Australia) during 1859–87, according to the annual mean temperature. Each vertical bar shows the range of values within one standard deviation of the mean number, for each span of annual mean temperature. The sloping lines imply that a warming by 3C° will increase the number of hot days by about 60 each year, on average. (Similar diagrams for Brisbane, Sydney and Hobart show increases of only 3–6 days/year, presumably because sea-breezes there limit the daily maximum temperatures)

greenhouse hypothesis (Parker 1989). Likewise, much of the warming in the first half of this century preceded the main increase of carbon-dioxide emissions. Then there is the cooling during 1940–65 (Figure 1.2) to be properly explained, and it is uncertain how much of the most recent warming may be due to either natural fluctuations or the El Niño events (see Chapter 7) in 1982–3 and 1986–7. Also, more needs to be known about the role of the deep oceans in sequestering carbon dioxide, and the extent to which more clouds will offset the warming (Ingram 1989).

At present, there is general agreement about the fact of warming but not its rate. Most scientific opinion currently favours a best estimate of between 1.5–4.5C°, say 3C°, by the time that greenhouse gases amount to twice their 1900 value. That might occur within 40 years or so from now, depending on how much fossil fuel is used in the coming decades.

So considerable a warming, comparable with the Ice Age cooling, and of such unprecedented rapidity, would have enormous implications for mankind. For instance, some places would have an increased frequency of the very hot days which affect human health and injure crops, as shown in Figure 1.4. Also, whilst an increase of carbon dioxide tends to promote crop production, harvests inland would be reduced by drier soil due to higher evaporation rates caused by the extra warmth. The impact on grain production would be particularly alarming as the world's population swells.

How to respond to such an enormous but not quite certain threat is a political matter. There is a clear need for much more research, and aspects of this will be mentioned in the following chapters.

2

MEASURING CLIMATE DATA

Climatology needs reliable, long records of daily values which have been obtained with standard equipment. The instruments must be properly installed in suitable places, carefully maintained and conscientiously observed. Even then it is necessary to consider likely errors of the measurements. These matters are dealt with in the present chapter.

We begin by discussing the location and operation of a 'climate station'. This is a set of standard instruments, installed together to measure a particular microclimate usually selected as representative of the surrounding topoclimate or mesoclimate. The next sections of the chapter deal with measurement procedures and, in particular, the problems of measuring air temperature, atmospheric humidity, rainfall and the wind near ground level. These depend partly on the housing of the instruments at the climate station. Lastly, we consider archiving the measured data.

CLIMATE STATIONS

The costs of setting up and operating a climate station are so great that it is necessary first to consider carefully the purpose of the station, to select an appropriate site, equipment, frequency of observation, and so on. There are at least six possible reasons for having a climate station:

1 As a hobby.
2 To provide a description of the environment for other research.
3 As an educational exercise (see Note 1, p. 313).
4 As an aid to agriculture.
5 In anticipation of some unknown future need.
6 To detect climate changes.

The purpose governs the accuracy to be achieved in the measurements.

A climate station resembles a 'weather station' used chiefly for collecting data for forecasting (see Figure 2.1). The main difference is that measurements at a weather station are transmitted to a central forecasting office within an hour or so of the observation, whereas the readings taken

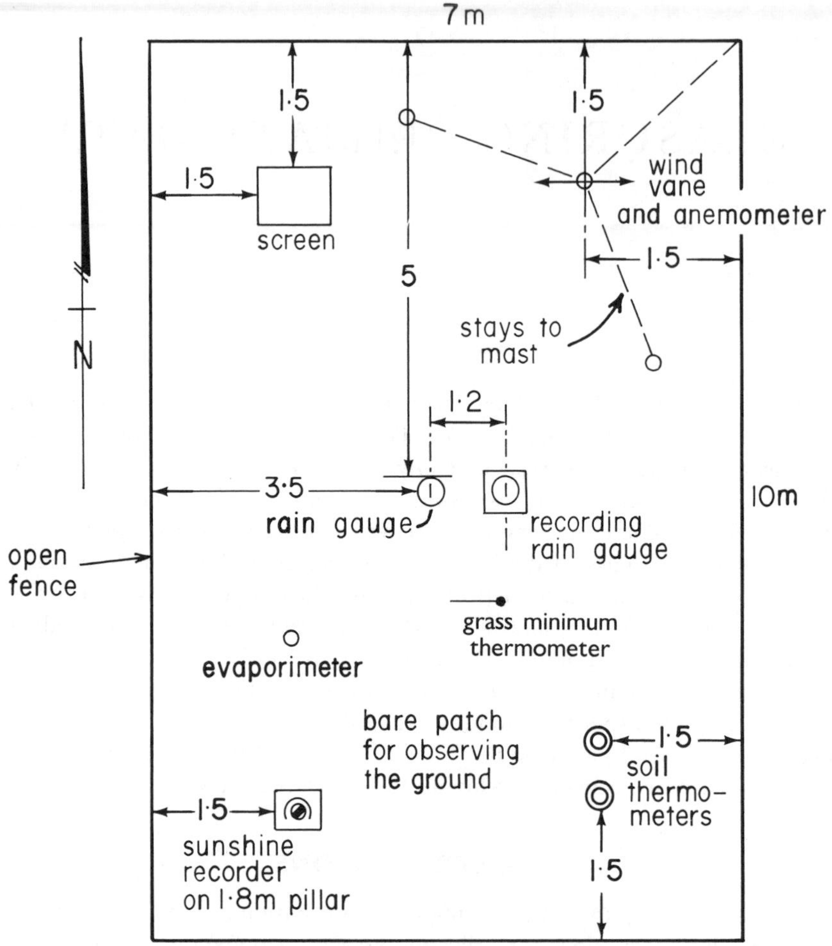

Figure 2.1 Suggested layout of a meteorological enclosure in the southern hemisphere, the sunshine recorder would be in the north-west corner so that no shadow falls on it. (By permission of HM Stationery Office)

daily at a climate station are recorded in a book which is sent to the national weather bureau at the end of each month for analysis and archiving.

There are three kinds of weather station:

1 'First-order' stations (also called 'principal', or 'synoptic' stations). These involve readings, taken either continuously or every three hours, of pressure, dry-bulb and wet-bulb temperatures, relative humidity, wind, sunshine duration and rain. Also, observations are made at fixed times by eye, of the cloud, visibility, and the state of the weather. Preferably, there are measurements of rainfall intensity, solar irradiance and the

evaporation from a pan of water, especially in arid areas.

2 'Second-order' stations. These are more common. At least twice daily, readings are taken of pressure, dry-bulb and wet-bulb temperatures (these are discussed later), wind, cloud and state of the weather, along with the daily maximum and minimum temperatures, daily rainfall and remarks on any fog, frost or thunderstorms, etc.
3 'Third-order' stations are like second-order stations except that readings are less complete or are taken only once daily or even less frequently, or at non-standard times.

A climate station provides less information than a weather station – only daily figures on dry-bulb and wet-bulb temperatures, maximum and minimum temperatures, rainfall, amount of cloud, snow depth, current weather type and visibility. Optional measurements include the number of hours of sunshine daily, the grass minimum temperature, the wind direction and speed, and atmospheric pressure.

'Ancillary' or 'co-operative' stations involve a yet smaller range of measurements, primarily for agriculture. They include thermometers to measure soil temperature, maximum and minimum thermometers for screen temperatures, a sunshine-duration recorder, a humidity meter, a rain gauge and an evaporation pan. (Most of these instruments are described later in this book.) Other equipment desirable at an agro-climate station include a dew-meter, a recorder of wind speed and direction, a rainfall recorder and some instrument to measure solar irradiance.

A network of stations

Although ideally, every plain, coast, island and valley would have a separate climate station – each at a typical site. In practice, it is too costly to have so dense a network.

The proper way to design a climate-station network would be to install a large number of temporary stations and then compare the results from adjacent places. If the correlation between simultaneous measurements from adjacent stations proves to be 'too high' (i.e. an increase at one station is almost always accompanied by a similar increase at the other), or the difference between measured values is 'too low', one of the pair is redundant and should be removed, for economy. In this way, stations would be reduced to a number sufficient to sample all the various topoclimates.

The difficulty with this procedure comes in deciding what correlation is 'too high'. Stringer (1972: 18) suggested that the spacing between adjacent climate stations A and C should be such that the error in interpolating climatic values for an intermediate place B is comparable with the instrumental error at any single station. (The interpolation error would be found by comparing the estimate with actual measurements at B.) Godske

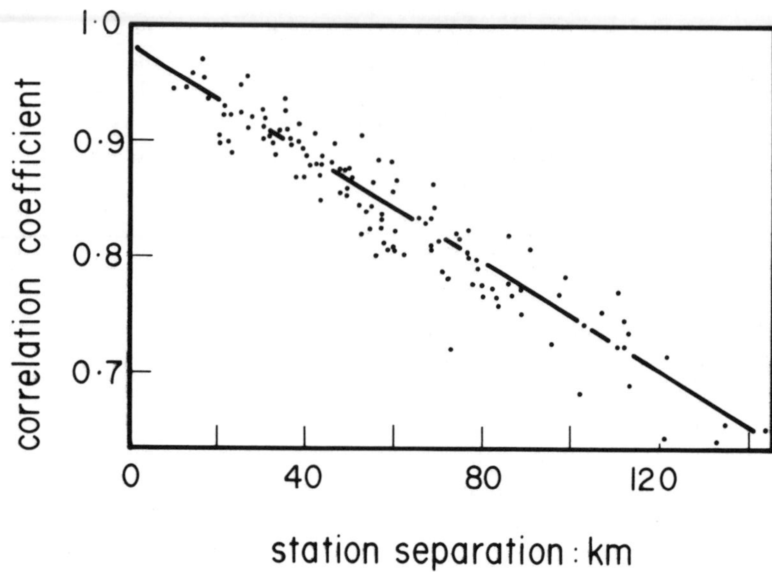

Figure 2.2 Effect of the distance of station separation on the coefficient of correlation of daily sunshine durations at pairs of stations in East Anglia, an extensive flat region of south-east England (Hopkins 1977: 289, by permission of HM Stationery Office)

(1969: 62) arbitrarily and tentatively proposed that satisfactory correlation coefficients (which are discussed in Chapter 4) are between 0.8–0.95 in 90 per cent of synoptic situations. Other writers have recommended a minimum coefficient of 0.7. Coefficients less than that would imply significantly different unsampled topoclimates between the stations.

The appropriate spacing depends on what climate element is being considered. An examination of values from two places 67 km apart in Michigan showed that the elements worst correlated were hourly wind speed and atmospheric stability (the latter being indicated by the difference between temperatures at two heights above the ground). The elements best correlated were hourly temperature and dewpoint. This implies that a network of equipment measuring wind and stability should be more closely spaced than one for temperature and humidity.

Similar indications of the proper spacing of different measurements emerge from other evidence. In the case of measurements of daily sunshine duration in East Anglia (England), a correlation coefficient of 0.7 corresponds to a spacing of instruments by about 100 km (see Figure 2.2). On the other hand, a coefficient of 0.7 for solar irradiance values in summer at 48 places in the USSR implies a spacing of about 500 km, but around 320 km for winter-time daily minimum temperatures in East Anglia. Monthly mean daily minimum temperatures correlate worse than maximum temperatures, necessitating closer spacing of the network.

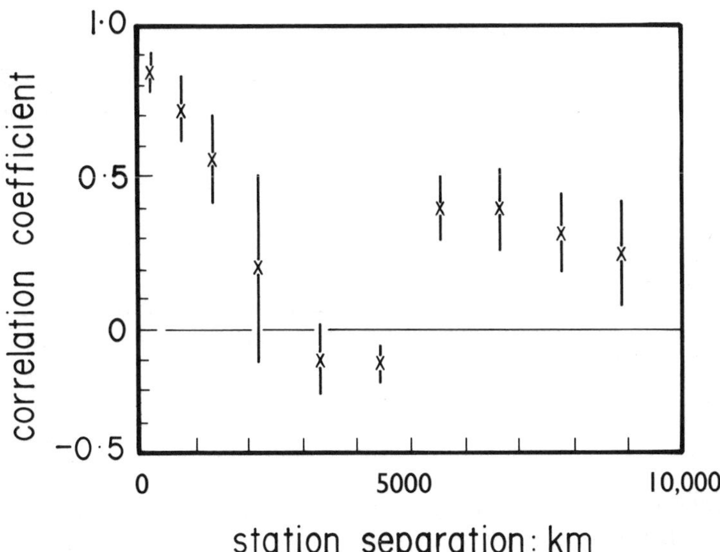

Figure 2.3 Effect of the distance of station separation on the coefficient of correlation of annual mean temperatures during 1943–72 at 47 stations between 47–68°N, taken in pairs (Yamamoto 1980: 322). The bars show the scatter of values for each distance in terms of the standard deviation, discussed in Chapter 4. (By permission of Elsevier Science Publishers)

The correlation depends partly on the period of averaging the measurements. The longer the period, the better the correlation between values from different places. For example, correlations of *daily* mean temperatures in Western Siberia are 0.98 for adjacent thermometers, 0.90 if they are separated by 100 km, and 0.78 for 200 km separation, implying less than 0.7 for 300 km. Whereas Figure 2.3 shows a coefficient more than 0.8 for *annual* mean temperatures 300 km apart.

An alternative to using the correlation coefficient as a measure of similarity between microclimates is calculating the 'association coefficient' (Ac). This is especially suited to discrete climatic events, e.g. the occurrence of hail, frost or fog. The definition of Ac is this:

$$Ac = (a.d - b.c)/(a.d + b.c) \quad (2.1)$$

where a is the chance of the simultaneous occurrence of an event at both of a pair of stations, d the chance of the event's occurrence at neither, and b and c are the chances of occurrence at either one or the other.

Location

So far, we have considered the effects of the various climate elements and of the period of averaging on the desirable spacing between instruments. Another factor is the kind of terrain, as shown in Table 2.1. Irregularities of

Table 2.1 Typical recommended spacings for measurements of various climatic elements.

	Uniform terrain		*Non-uniform terrain*	
	Rural	*City*	*Coast*	*Mountain*
Global irradiance	560 km	–	10 km	160 km
Temperature	160	2–3	15	–
Wind	160	–	15	160
Rainfall[a]	30[b]	15	5[c]	15

Notes: [a] See the section 'Estimating the total rainfall within a catchment' in Chapter 7.
[b] But 60 km in arid and polar zones.
[c] For mountainous islands.

landform cause a regular grid of climate stations be relatively inefficient. In general, the pattern of stations should be most dense in the direction at right angles to 'isopleths' of contour, rainfall, temperature, etc. (An 'isopleth' is a line on a map through points having the same value of some specified element.) For instance, it is recommended that the spacings should be 2–10 km perpendicular to any natural boundary such as the coast, but 20–50 km in the direction parallel to the boundary (WMO 1960).

The distances in Table 2.1 are comparable with the grid size of the best computer models used for forecasting local weather, which was about 100 km in 1987. Data from stations which are more closely spaced would have to be combined for use in the models, reducing the available spatial resolution. This applies also to models used for calculating climate change.

Preferably, climate stations of the various kinds described earlier are arranged in a hierarchical pattern, with the few elaborate stations widely spaced and more numerous stations of lower standard between them. Across areas where the spatial variability of the climate is much greater than the temporal variation over ten years, say, it would be thrifty to move a set of climate-station equipment each few years to obtain data from places between those of the fixed network.

In practice, the spacing of climate stations is often far from ideal, being governed by accidents of history and geography, e.g. the location of towns and cities. Also, their great cost leads to insufficient stations in many areas. This applies especially in developing nations. In Turkey and India, for instance, the mean spacing was about 100 km (in 1958) whilst the average spacing in countries such as Britain, Japan and Germany is about 20 km. In the 270,000 km^2 of New Zealand there were 260 climate stations in 1970, implying an average spacing of 32 km. The Irish network in 1987 consisted of 15 synoptic stations, 85 climatological stations and 650 rainfall stations: the country has an area of 69,000 km^2, so the spacing of climatological stations is 28 km on average. In the USA the spacing of agricultural weather stations is 20–60 km, depending on the crop. Rainfall stations in Japan are about 20 km apart.

The location of a climate station must be specified accurately in terms of longitude, latitude and elevation above sea level. (The word 'altitude' is reserved for the vertical distance above sea level of points which may not be on the ground.)

Exposure of climate stations

The relationship between the surrounding locality and the site of a climate station is called the station's 'exposure'. It is a rather vague concept, an aspect of 'representativeness', the extent to which a measurement accurately reflects what is being measured. It is associated with the problems of spatial and temporal sampling of the environment.

The 'environment' of a domain is the space immediately around it, large enough that its mean properties are unaffected by the domain. Consideration of 'boundary layers' in Chapter 6 indicates that atmospheric conditions unaffected by a surface are found beyond about D/10 from it, where D is the surface's horizontal length. On the other hand, the environment is small enough that variations within it are trivial compared with differences between the domain and the environment.

A climate station should be located on level open ground, with a clear horizon, and not be in a hollow or on a steep slope. One way of checking this, and characterising the exposure numerically, is in terms of the total horizontal angle within which the land less than 8 km away is more than 300 m, say, above the given site.

No obstruction of height H should be within a distance of 10 H from an anemometer, nor 2–4 H from a rain gauge, although a suitable windbreak is required for proper rainfall measurement, as discussed later in this chapter. The ground within 10–25 m of the instruments should be covered with short grass, or, if unavoidable, bare soil, but never a hard surface like concrete, asphalt or crushed rock.

However, these recommendations amount to a counsel of perfection, and bad exposure of climate stations is common. An example is mentioned in Note 2 on p. 314. One sign of poor exposure is an abnormal daily range of temperature, as discussed in Chapter 3.

Poor exposure may arise gradually. It may have been excellent initially, but trees grow, new buildings are erected and nearby roads become busier. If circumstances become too bad, it is necessary to replace the station by one nearby, better exposed. Both the new and the old stations must be operated in parallel for a few years before abandoning the old one, to allow overlapping the long-term records from each so that a single long record can be synthesised.

The location of a climate station should not be changed, nor the equipment or procedure altered, for at least ten years. This reduces discontinuities in the record which might lead to untrue inferences of climate change. For example, a shift of the climate station by only 135 m created a clear and abrupt jump in the temperature record at Observatory Hill in Sydney. Similarly, monthly mean temperatures measured in parts of the USA changed by 2C° or more during the 1960s as a result of improving the technique of temperature measurement by ventilating the thermometers. Likewise, changes of screen design and a gradual alteration of the surroundings of urban climate stations in New Zealand caused a false, apparent warming.

Observing

The usefulness of climate data depends on the skill and conscientiousness of the observer. Observers are responsible for maintaining the instruments, changing the charts on self-recording instruments, taking the readings every day and making weekly or monthly reports. It is not satisfactory to suspend observations during holidays. Observers have to be trained and motivated to detect errors, and need clear, written instructions.

Voluntary observers' measurements are generally of low standard. Hence the move towards automatic weather stations in the USA. A network of 17 automatic climate stations in 1982, for agriculture in the mid-West, had grown to 79 stations by 1990.

The times of observations depend on the status of the climate station. For synoptic weather stations, the daily taking of measurements should be simultaneous throughout the world, with regard to universal time (Tu, or Greenwich Mean Time), not local time. For climatological measurements it is sufficient to use 'official times' which are adopted by each country separately. If daylight saving is introduced in summer, the observations should preferably continue to be made throughout the year at the same Tu. But this is not universal, and users of data have to be wary of discontinuities in any series of temperatures and wind speeds at the beginning and end of the daylight-saving period.

Morning temperature readings are taken at 9 a.m. local time in most temperate countries, but at 6 a.m. in tropical countries. It is important that readings of temperature be taken within the 10 minutes preceding the specified time, because of the rapid warming at that time of day. For instance, an error of about 0.2C° would be caused by reading 10 minutes late, assuming a typical heating of 12C° between 6 a.m. and 2 p.m. The timing is not so critical for observations in the afternoon, when temperatures change less rapidly.

Indicating instruments should be read before recorders. The recommended sequence is this – dry-bulb thermometer, thermograph (whose reading should differ by no more than 1C° from that of the dry-bulb thermometer), wet-bulb thermometer, hygrograph, maximum thermometer, and then the minimum thermometer. Instruments outside the Stevenson screen (discussed on pp. 44–6) are read subsequently, in the following order – grass-minimum thermometer, soil thermometers, rain gauge, wind speed and direction, evaporation and finally sunshine duration.

Where automatic recording equipment is installed at a climate station, observers are needed only for checking the operation of the instruments. This avoids the problems of regular observations at inaccessible places or during holidays. Also, the fact that measurements are taken continuously or at frequent intervals facilitates averaging over periods such as midnight to midnight. Moreover, if the measurements are recorded on to magnetic tape, for instance, it is simple to process the data subsequently by computer. As regards infrequent events like rainfall, it is better that the recorders register only the times of occurrence, rather than running continuously and wasting paper or tape between events. Unfortunately, automatic recording instruments tend to be either expensive or inaccurate. They need periodic checking against indicating instruments and well-trained maintenance staff are required.

Automatic equipment is available for measurements in icing conditions. The sensors are normally enclosed and warmed electrically, and then are exposed each half-hour by the opening of a lid for 2–3 minutes.

Maintenance

Maintenance of equipment is an important consideration. In the USA, for instance, about 9 per cent of the 5000 employees of the US Weather Bureau in 1970 were employed in installing and maintaining the field equipment. Every climate station should be visited by an inspector at least once each two years if it is a principal climate station, or each four years if an ordinary climate station. In the USSR it is recommended that agro-climate stations be inspected at least twice each year. The aim is to check the siting and exposure of instruments, their good order, uniformity in the methods of taking and treatment of observations, and the competence and motivation of the observers.

Some advice on the maintenance of pen recorders is given in Note 3 on p. 314.

So much for the operation of a climate station. Now we will consider measurement accuracy in general, as a preface to examining the problems of observing particular elements.

MEASUREMENT ACCURACY

It is not possible to measure the climate, but only the individual elements which comprise it. A climate *element* is any one of the various properties or conditions of the atmosphere which together specify the physical state of the weather or climate at a given place, for any particular moment or period of time. On the other hand, a climatic *factor* like latitude or shading is a variable which controls a climatic element.

Errors

The art of climate observation consists of reducing the difference between the characteristic of the atmosphere being observed and the instrument reading regarded as equivalent. For example, a normal thermometer actually measures the temperature of the mercury in the bulb and only with care can this be brought close to the temperature of the air, which is what is wanted.

Every measurement has some error, and the climatologist must be aware of it to know how much reliance to put on the observation. Error is often confused with the related concepts of 'precision', 'sensitivity', and 'accuracy'. 'Precision' is exactness, a measure of the reproducibility of the measurement. The 'sensitivity' of an instrument is its response to unit change of input and must at least match the required precision.

The 'error' is the difference from truth (however that is defined), whilst 'accuracy' is the inverse of error. People say 'accuracy' when they mean 'error', e.g. they say 'This instrument has an accuracy of 5 per cent.' Also, the error is commonly expressed as a percentage of the full-scale reading of the instrument, though this understates the error's importance when the reading is only a small fraction of the full-scale value. It is better to express the error either in the reading's units or as a percentage of the reading.

A measurement can include four kinds of error, due respectively to (i) the manner of observation, (ii) the instrument, (iii) the relationship of the instrument to the climatic element being measured (i.e. the 'exposure', discussed already), and (iv) sampling. The importance of the last is indicated by the case of daily mean-temperatures across uniform terrain in Quebec: differences of 1.7°C occur at places only 30 km apart.

A measurement is 'good' if each of the errors is minimised, and if the total error is within a limit appropriate to the problem. Observation errors include blunders such as misreading a thermometer by five or ten degrees, or reading the temperature inaccurately from a thermometer which is not perpendicular to the line of sight, the so-called 'parallax error'. Or the number 3 may be read as an 8. Another problem is what is called the 'personal equation', due to the unconscious bias of some observers to read either too high or too low, or to prefer certain numbers. For instance, observers of the sky's cloudiness may prefer to record 1 or 3 or 7 oktas (discussed in Chapter 5), rather than other values.

There are two sorts of instrument error – systematic and random. Systematic errors (or bias) may be due to faulty calibration or to a lag of the instrument in responding to sudden changes of the measured element. They tend to be consistent and so one can compensate for them. As regards random errors, they include inaccuracies caused by slackness within the instrument. An important feature of random errors is that the average of n readings, each with an error **e**, has an error of only $\mathbf{e}/n^{0.5}$, e.g. the average of four readings has only half the random error of a single measurement.

The size of the error determines how many digits are appropriate in expressing the measurement. It is nonsense to express a temperature as 9.73°C if the error can be 0.1C°, since the actual value can lie anywhere between 9.6°C and 9.8°C. Implications of unreal accuracy are common in the literature. Usually, climatic measurements can be expressed by no more than two significant figures, or at most three. Some consistent convention should be adopted in rounding off (see Note 4, p. 315).

Table 2.2 Tolerable total errors of surface measurements at normal climate stations and automatic weather stations.

	Maximum tolerable error[a]	
Element of weather	*Normal climate station*	*Automatic station*
Sunshine duration	10%	–
Solar irradiance	12 W/m^2	–
Temperature	0.1C°	1C°
Maximum, minimum and ground surface temperatures[b]	0.5C°	–
Dewpoint	0.5C°	2C°
Water-vapour pressure	0.2 hPa	1 hPa
Relative humidity	3%	–
Rainfall and evaporation:		
below 5 mm	0.1 mm	0.5 mm[c]
below 10 mm	0.1 mm	10%[c]
above 10 mm	2%[d]	10%[d]
Rain intensity:		
below 25 mm/h	0.5 mm/h	–
above 25 mm/h	2%	–
Wind speed:		
below 20 m/s	0.5 m/s	2 m/s
above 20 m/s	0.5 m/s	10%
Wind direction	10 degrees	20 degrees

Notes: [a] Note that acceptable *systematic* errors are less than these figures, which are for total errors.
[b] Butler *et al.* (1984: 16).
[c] For rainfall only.
[d] 22 mm error may be acceptable for rainfalls less than 40 mm for the purposes of hydrology (WMO 1983a: 1.24).

Tolerable errors in measuring various climatic elements are shown in Table 2.2. However, the errors may be larger in practice. For example, four identical adjacent thermometers near London gave twice-daily values over a year which individually differed from the average of the four (regarded as the true value) by more than 0.6C° on about 13 per cent of occasions.

Climate station instruments cannot be made highly accurate because they have to withstand all weathers, and the necessary robustness is achieved at the expense of sensitivity. The equipment may be called on to work in temperatures over the range −60°C to +54°C, and winds above 100 m/s. Also, the need for numerous installations and for simplicity of observation and maintenance militates against complex equipment of great accuracy. In addition, piecemeal improvement of a network's equipment is hindered by the requirement that climate stations within a network be uniform, each having an unchanging measuring procedure in order to ensure a homogeneous series of measurements, as discussed in Chapter 4.

Calibration

A factor that determines the price of an instrument like a thermometer is the cost of proper calibration by the maker. Also, regular re-calibration of instruments like a thermo-hygrograph accounts for a significant part of the cost of their maintenance.

Calibration consists of a comparison of readings from the instrument against those from a superior standard equipment alongside. Usually the instruments at the main station of a network are used as a standard for testing and calibrating the instruments from lower grade stations. That standard is in turn validated by comparison against the next highest standard in a hierarchy of increasing accuracy. First there is the working standard, then a reference standard, a national standard, a regional standard, and finally an international 'collective' standard. A 'working standard' is the best spare instrument readily available for the checking of the instrument in use. A 'reference standard' is an instrument of the same kind, kept aside especially for occasional comparison with those in common use, and itself compared with the national standard instrument from time to time. Every few years, the World Meteorological Organisation arranges side-by-side comparisons of national standard instruments from countries within the six WMO regions of the world to define a consensus as a collective standard. Sometimes a superior portable instrument, called a 'travelling standard', is carried around for comparing with reference or national standards.

MEASUREMENT OF TEMPERATURE

Now we will consider the measurement of one particular element: the temperature of the air near the ground. This is normally done with a mercury-in-glass thermometer, which can be calibrated accurately and used down almost to the freezing point of mercury at −39°C. Mercury-in-glass thermometers were used first by Fahrenheit (1686–1736), whose name is associated with a temperature scale. However, that scale was almost universally superseded in 1948 by a scale devised by Anders Celsius in 1742. So temperatures are now given in degrees Celsius (see 'Units' pp. 13–18). The term 'centigrade' is no longer used, because it can apply also to angle measurement and because units are preferably named after people.

With experience, one can assess the temperature reading within 0.1C°, even though the calibration markings are 0.5C° apart. However, it is doubtful that such a degree of resolution is justified, in view of the other errors of temperature measurement. There is an inherent uncertainty of about 0.15C° involved in the calibrating of the thermometers and the total error is often considerable – 'to put it politely the standard of air-temperature measurements falls far below that of other meteorological elements' (Smith 1975: 28). In one reported case, a temperature maximum was probably 10C° too high on account of exposure of the thermometer to sunshine.

A few precautions reduce errors in measuring temperature. For instance, any condensation on the bulb must be lightly wiped off with a tissue before a reading is taken, and the temperature then allowed to settle. The thermometer must be adequately shaded. Also, no time should be lost in taking the reading, especially in inclement weather. Readings should be taken within 30 seconds of the screen being opened, before sky radiation through the doorway affects temperatures within. A second reading should be taken immediately after, to verify the first. The observer must not breathe on the thermometer and care has to be taken to avoid touching the bulb.

Even with the best procedures, adjacent thermometers can differ by almost 2C° on occasion. This is partly due to irregularities within the atmosphere itself. Fluctuations at a climate station near London were up to 1.1C° when there was little wind and it was daytime.

Mercury-in-glass thermometers are used also for soil temperatures at depths of 50 mm, 100 mm, 0.2 m, 0.5 m and 1 metre. The shallow thermometers are made with stems bent at right angles, so that the exposed upper part with the scale faces upwards for easier reading. The exposed part should be protected from bright sunshine. Straight thermometers are used for 0.5 and 1 m depths, hung in iron tubes in the soil, though heat conducted down a tube may create an error of a fraction of a degree. It is important to prevent water entering the holes in the soil, since water would conduct heat from other depths.

Cheap thermometers are filled with a spirit like pentane, toluol or, usually, ethyl alcohol, instead of mercury. These hydrocarbons are safer than mercury, whose vapour is poisonous, and they expand about eight times as much with heat, which increases sensitivity. Also, they permit measurements below −36°C. However, spirit thermometers are less reliable than mercury thermometers, partly because the calibration may change through the liquid becoming polymerised by light or time, causing the fluid's volume to change.

Thermometer lag

One cause of error is the delay of the thermometer in following abrupt changes of air temperature. The lag is due to the thermal inertia of the instrument and to its poor thermal connection to the air. The delay is greater if the thermometer is massive or ventilation poor.

When the air temperature changes abruptly by ΔT, sluggishness of the thermometer causes an error (e), as follows:

$$e = \Delta T.\ \exp\ (-\ t/L) \qquad C^\circ \qquad (2.2)$$

where L is the 'response-time' or 'lag'. It is the time taken for adjustment by 63 per cent of the abrupt change (i.e. 1 − 1/2.718, the number 2.718 being the 'exponential constant'). The error falls to half the change after 0.69 times the lag has elapsed, and 10 per cent after 2.3 times (see Note 5, p. 315). The lag may be 85 seconds for an alcohol thermometer in still air, or about 50 seconds for a mercury thermometer of the kind commonly used by meteorological services. Lags are less if the air is moving briskly or if the bulb glass is thin.

So far, we have considered the hypothetical case of an instantaneous change of temperature, whereas in reality there is a change over a definite period. Reported extreme cases of rapid change include falls of 19C° in an hour at Kansas City, 7C° in 30 minutes in Singapore, and 10C° in 20 minutes in Melbourne. There have been rises associated with the onset of chinook winds of 22C° in 15 minutes and even 27C° in 2 minutes. Templeman *et al.* (1988) reported changes by 2.5C° in 5 minutes and more than 4C° in 10 minutes during a storm in southern England.

With a constant rate of change, the thermometer reading is what the air temperature was L seconds beforehand. Thus the error would be about 0.4 C°, if the lag is 45 seconds and the rate of change 0.5 C°/minute. With a more common change of 1.5 C°/hour, the error would be only 0.02C°.

A thermometer with too rapid a response to the second-by-second fluctuations of surface air temperature, wavering randomly by a degree or two, would be unsuitable for determining the mean temperature, unless readings are taken repeatedly and averaged. In the case of measuring the temperatures of soil at 0.5 m and 1.0 m depths, the lag is deliberately enhanced by enclosing the thermometer bulb within a wax-filled glass tube so that the reading hardly changes whilst the thermometer is being withdrawn and observed.

A particular problem arises from thermal lag in the case of measurements of the daily maximum temperature on still, sunny days at low latitudes. Measurements with an Assmann psychrometer (described on p. 42), which has practically no lag because of strong ventilation of the thermometers, show the surface air to warm gradually for a few minutes to a temporary maximum, and then fall suddenly by a degree or two as a huge bubble of air heated by the ground detaches and is replaced by unheated air. A sluggish unventilated thermometer shows a daily maximum which is lower, near the average of the swings, on account of the smoothing caused by the lag.

Maximum temperature

A form of mercury thermometer developed by 1852 has a constriction in the mercury column just above the bulb. Any increase of temperature makes the liquid expand through the constriction. On the other hand, cooling makes it contract and then the column of mercury breaks at the constriction, leaving mercury above in the stem, separated from the bulb and registering the highest temperature reached since the thermometer was last reset. The thermometer should be supported almost horizontally, with the bulb only 12 mm lower than the other end, to avoid gravity helping to return mercury down through the constriction.

The maximum thermometer is reset after a reading, to reconnect the mercury column. To do this, the upper end of the thermometer is grasped firmly by the thumb and forefinger and the stem steadied with the fingertips. Then the extended arm is swung back and forwards without jerking. After resetting, the reading should not differ by more than 0.2C° from that of a nearby normal dry-bulb thermometer. If the mercury thread becomes broken above the constriction, the column is reunited by gently heating the bulb.

It is usual to take maximum temperature readings along with the other early morning measurements, although the maximum then normally refers to conditions during the *previous afternoon*. So care must be taken in allotting a date when recording a reading, to make clear whether the date is that of the occurrence or of the observation.

Minimum temperature

Minimum thermometers most commonly contain alcohol, because of its low freezing point, high surface tension and transparency. At any given moment, the meniscus of the alcohol should present a reading close to that of the climate station's dry-bulb thermometer. Within the liquid column is a 20 mm dumb-bell-shaped index of dark glass or metal, whose upper end shows the lowest temperature since the last resetting. The surface tension of the liquid's meniscus pushes the index towards the bulb when temperatures fall, but the liquid passes by the stationary index into the upper part of the thermometer when temperatures rise.

After a reading has been taken, the index may be reset higher up the thermometer, nearer to the meniscus, by very gentle tapping on the thermometer when it is slightly tilted. The index must not be allowed to penetrate the meniscus, and it is important that the minimum thermometer be fixed almost horizontally in the screen, like the maximum thermometer, to prevent the index being moved by gravity.

The faults of minimum thermometers are those common to all spirit thermometers. There is the change of volume caused by polymerisation of alcohol, mentioned earlier, and errors may also be caused by the breaking of the liquid column during transit or adhesion of the spirit to the glass. Drops of alcohol often form in the upper part of the stem by distillation. Careful warming of the upper part of the stem may suffice to remove condensed spirit from the walls of the tube, though the operation may have to be repeated several times.

At any climate station there is one alcohol thermometer within the screen for minimum air temperatures and another outside to measure the lowest ground-surface temperature. The latter is called a 'grass-minimum' or 'terrestrial-minimum' thermometer. It is exposed horizontally, with the bulb just touching the tips of the blades of trimmed grass. The thermometer should not be put out in the daytime, to avoid its being trodden on or the risk of condensation of alcohol on surfaces above the liquid's meniscus.

A thermometer devised by James Six in 1782 conveniently gives both the previous maximum and minimum temperatures, as well as the present temperature. A glass tube in the shape of a 'U' is sealed at the left end where there is alcohol. This expands with the temperature and pushes down a column of mercury at the base of the 'U', so that it rises up the right limb, which has more alcohol and is effectively open at the top. Each alcohol/mercury interface moves an index, as in the usual minimum thermometer, registering the minimum on the left and the maximum on the right. Unfortunately, the separate indications of temperature in left and right limbs may not agree, because alcohol wets the glass and eventually bypasses the mercury. So the instrument is not recognised as a standard.

Pen recorders of temperature at climate stations are not used for routine observations but simply to determine the times of the extreme temperatures and the rates of temperature change. The normal instrument is based on the distortion of a strip of 'bimetal', coiled almost into a tube. The strip is made by rolling together two thin sheets of metal, one of brass and one of Invar (an alloy of iron and nickel). On being heated, the brass expands about 20 times as much as the Invar, and the difference makes the coil straighten since the brass is on the inside. The initial sensitivity to temperature change becomes reduced with age, but eventually the response is stable. One end of the coil is clamped, and the motion of the other end is amplified by levers and moves a pen, recording on a moving chart.

The main cause of error is the friction of the pen on the recording chart. This is minimised by slight lubrication of the bearings and adjustment of the contact of the pen on the chart. Also, the instrument is lightly tapped before a reading is taken. In all, the error of a bimetal thermograph should be no more than 0.4C° if there is care in its adjustment, maintenance and use.

The temperature lag of a bimetal thermograph is uncertain. A value of 20 seconds has been quoted, which is only half the time for a mercury-in-glass thermometer. On the other hand, transferring a thermograph from a cool room to a hot one shows a lag of about 20 minutes, partly due to the friction of the pen on the paper.

MEASUREMENT OF HUMIDITY

The amount of water vapour in the air can be described in at least five ways, in terms of:

1 the water-vapour pressure (the pressure exerted on any surface by the gaseous water molecules);
2 the relative humidity (the ratio of the actual water-vapour pressure to the saturation water-vapour pressure at the air's dry-bulb temperature);
3 the 'absolute humidity' (i.e. the number of grams of water vapour in a cubic metre of air);
4 the 'mixing ratio' (i.e. the ratio of the masses of water vapour and dry air, respectively);
5 the 'dewpoint', the temperature to which the air must be cooled for the existing water vapour to equal the saturation amount. The equivalent of dewpoint at temperatures below the freezing point is the 'frostpoint'.

The various terms are explained in introductory texts such as that of Linacre and Hobbs (1977).

The different ways of expressing atmospheric humidity are interrelated. For example, if air has a water-vapour pressure of 12.3 hectopascals (hPa) and a temperature of 20°C (so that the saturation vapour pressure is 23.4 hPa – see Notes 6 and 7, pp. 315–16), the relative humidity is 53 per cent (i.e. 12.3/23.4). Also, the dewpoint would be 10°C, since that is the temperature at which the saturation water-vapour pressure equals 12.3 hPa.

Many people customarily use relative humidity (RH) as the measure of atmospheric moisture, though this is undesirable because the RH varies far more during the day than does either vapour pressure or dewpoint, for instance. Moreover, it does not specify the moisture content of the air without supplementary information on the air's temperature. Vapour pressure is more useful in formulae for calculating evaporation rates but less directly comprehensible than the dewpoint.

It is usual to measure the air's humidity at 9 a.m. and at 3 p.m., which may give different results because of changes during the day. Typically, there is a rise of dewpoint from dawn to about 9 a.m. on account of the evaporation of dew, followed by a fall of the air's dewpoint until about 3 p.m. as 'convection' (i.e. vertical stirring) brings drier air down to screen height. When this has abated, there is a slow rise to a second maximum at about 10 p.m., and subsequently a decrease till dawn caused by dewfall from the air.

This is a convenient point at which to explain the two kinds of convection. Firstly, there is *thermal* convection, due to lower air being warmed by the ground after the surface has received heat from the sun's irradiance. The warmed air is lighter and therefore rises, being replaced by descending cooler air elsewhere. This kind of convection is dominant where the surface is warm and winds weak. Secondly, there is *forced* convection, due to wind over surface roughnesses. The resulting turbulence leads to air rising and falling in eddies of various sizes. This kind of convection is dominant where winds are strong and irradiance feeble. Usually it is clear from the context what kind is meant, in what follows.

As regards the changing of dewpoint in mid-afternoon, thermal convection is stimulated by the relatively high surface temperature, and also forced convection by the stronger wind often encountered at that time of day.

Cloudiness affects the diurnal variation of dewpoint, both reducing the daytime evaporation which raises Td and decreasing the thermal convection which lowers it. One outcome is that the cloud prevailing at low latitudes and along mid-latitude coasts during wet seasons leads to a decreased diurnal variation of dewpoint.

The variation of dewpoint depends on location. Tropical oceans cause a daytime increase of dewpoint as a result of continuing evaporation and only modest thermal convection from a surface which is cooled by the evaporation. Ashore, near the coast, there is an increase due to sea-breezes, when the dewpoint may rise by 3–4C°. The daytime increase at Marsfield on calm, cloudy days is about 1.3C° in January and 3.9C° in April. There is also an increase on high land in winter. On the other hand, in dry cloudless climates there may be a daytime decrease of vapour pressure by as much as 2 hPa, as a result of convection.

The psychrometer

There are several instruments for measuring atmospheric humidity, but the standard is a psychrometer. This is a pair of identical vertical thermometers, one of which has the bulb kept wet by means of a wick. One end of the wick dips into a bottle of water and the other is wrapped in wet muslin around the thermometer bulb. Evaporation from the wetted bulb lowers its temperature (Tw) below the air temperature (T), shown on the other thermometer. To obtain the air's water-vapour pressure (e), the two temperatures are entered into Regnault's equation, on the assumption that forced convection of heat to the wet bulb is much greater than its irradiance:

$$e = ew - A\ (T - Tw) \qquad \text{hPa} \qquad (2.3)$$

where ew is the saturation vapour pressure (Table 8.1) at the temperature (Tw) of the *wet-bulb* (this is important), and T is the dry-bulb temperature. The bracketed difference is called the 'wet-bulb depression'. The factor A in Eqn (2.3) depends on the local pressure (p: hPa); in high places, the sea-level value must be multiplied by p/1000. The factor depends also on the ventilation speed. It is 0.80 hPa/C° in winds of about 1 m/s at about sea level, but 0.67 hPa/C° if the thermometer bulbs are ventilated with a flow of at least 5 m/s.

A very rough indication of the dewpoint (Td), except for extreme conditions, can be obtained from the following simple expression:

$$3\ Td = 5\ Tw - 2\ T \qquad ^\circ\text{C} \qquad (2.4)$$

Table 2.3 gives the dewpoint for observed dry and wet-bulb temperatures,

Table 2.3 Effect of psychrometer ventilation on the dewpoint temperatures corresponding to given dry-bulb (T) and wet-bulb (Tw) temperatures. The bracketed values are for use with an unventilated psychrometer

Dry-bulb temperature (T)	*Wet-bulb depression: (T – TW): C°* 1	2	4	8	16
5	3(3)	1(0)			
10	8(8)	6(6)	1(0)		
15	13(13)	12(11)	8(7)		
20	19(18)	17(17)	14(13)	5(3)	
25	24(24)	22(22)	19(19)	12(11)	
30	29(29)	27(27)	25(24)	18(17)	
35	34(34)	33(32)	30(30)	24(24)	9(6)
40	39(39)	37(38)	35(35)	30(29)	17(15)

in either still or ventilated conditions. In general, the effect of ventilation on the derived dewpoint is slight, except in very dry conditions, which entail a large wet-bulb depression.

Ventilation may be achieved by whirling the pair of thermometers through the air at about two revolutions per second or by means of a fan. The latter is used in an Assmann psychrometer, shown in Figure 2.4. The fan provides a flow of 3–10 m/s past the bulbs and allows the thermometers to be stationary (and therefore easier to read) and shielded from the sunlight. Another feature of the Assmann psychrometer is that the thermometers are selected as identical in calibration.

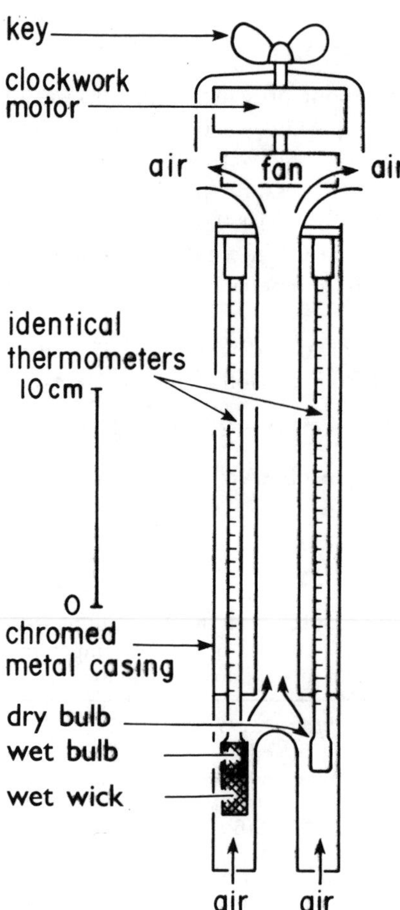

Figure 2.4 An Assmann psychrometer for accurately measuring the dry-bulb temperature (i.e. the temperature of the air) and the wet-bulb temperature, to estimate the air's water-vapour content

Every climate station should be equipped with an Assmann psychrometer for occasional checking of the unventilated standard instrument. This will reduce differences between the measurements in various countries, which, at present, lead to apparent discontinuities of humidity at national boundaries.

Psychrometer errors

Almost all errors in using a psychrometer cause too high a wet-bulb temperature and hence an overestimate of atmospheric humidity. The exception is a temporary cooling if the reservoir is refilled with cold water. A dirty wick, poor ventilation and impure or insufficient water, all reduce the evaporative cooling of the wet bulb. If this leads to an extra 0.5C° in the wet-bulb temperature, the calculated dewpoint is increased by 1C°, assuming psychrometer readings of 21°C and 15°C for instance. This corresponds to a relative-humidity error of 4 per cent.

The wick or muslin must be unbleached and clean. A woven cotton bootlace is satisfactory. Any initial coating of grease should be removed by washing in ammonia or boiling for 15 minutes in soap and water, followed by several rinsings in distilled water. Dust or salt particles from the air, or wax from the fingers, are best dealt with by replacing the wick, not washing it. In principle, it should be replaced about fortnightly. But neither the wick nor the water should be changed within the half-hour before a reading is taken.

The wick should cover 50 mm of the thermometer and be tied both above and below the bulb. The wick must not sag between the bulb and the reservoir, which is a small bottle about three-quarters full of soft or distilled water placed below the wet bulb and about 120 mm to one side. The water could be melted frost from a refrigerator.

Significant errors occur with the kind of unventilated psychrometer which is normal at climate stations. A set of four adjacent, identical instruments near London yielded 1460 daily dewpoint values over a year, of which 29 per cent differed from the average value each day by 0.6C° or more and 1.2 per cent by 1.2C° or more.

The hair hygrometer

It is usual at climate stations to supplement the unventilated psychrometer by a thermo-hygrograph, which continuously records both temperature and relative humidity. The basis of the humidity measurement is the change of length of an array of human hairs at least 200 mm long. The hairs expand about 2.5 per cent in length on changing from complete dryness to saturation. However, the length of the hairs is irregularly related to relative humidity – i.e. 39 units of extension at 20 per cent RH, 64 at 40 per cent, 79 at 60 per cent, 91 at 80 per cent and 100 units at 100 per cent. Changes are amplified by levers before being graphed.

Hair hygrographs give acceptable values only if great pains are taken in their care and use. Errors up to 15 per cent RH can arise from dust on the hairs, so they must be washed with distilled water at least weekly (or daily if subject to salt spray or industrial fumes) and never touched by hand. In arid areas, the hairs should be exposed to a saturated atmosphere each month.

A hair instrument is less accurate than a ventilated psychrometer, especially in very dry conditions. The error can be up to 20 per cent. Even with the best type of hair hygrograph, errors are typically 3 per cent RH, and more when the relative humidity is below 35 per cent or if there is condensation on the hairs.

STEVENSON SCREEN

The psychrometer, the thermo-hygrograph and the maximum and minimum thermometers at a climate station are protected from solar radiation, precipitation and theft by a white louvred box, called a 'screen'. Various designs have been used in the past, with at least five different kinds in use in Australia, for example, prior to 1907. One was the Glaisher stand, which was open on one side. In addition to screens for routine measurements at climate stations there have been many other designs for special purposes. A home-made version was described by McConnell (1988).

Differences between designs can cause discrepancies of 1–3C° between measurements. Daily *maximum* temperatures in London were found to be 0.2C° lower in winter with a Glaisher stand instead of a Stevenson screen, whereas the daily *minima* were 0.3C° lower in winter and 0.1C° higher in summer.

Stevenson's screen was first described in 1864 and developed by about 1880, and now is the standard design (see Figure 2.5). Initially, the box had no base and internal dimensions of only 380 × 350 × 190 mm. Nowadays, the typical size is 700 mm wide, 530 mm from back to front and 600 mm high. Also, the box has a floor to exclude radiation from the ground which could cause heating of the thermometers. The box is oriented exactly north/south, with one wide side facing away from the equator and hinged at the bottom as a door. At latitudes below 23 degrees, the opposite side is also hinged for use when the sun is further polewards than the climate station. In this way, sunlight never enters the box when it is opened, except at high latitudes in summer near dawn and dusk when the effect is unimportant. The roof of the box is double, with a clear space between, and the construction is light in weight to minimise thermal lag.

It was common in Europe in earlier times for the screen to be fixed

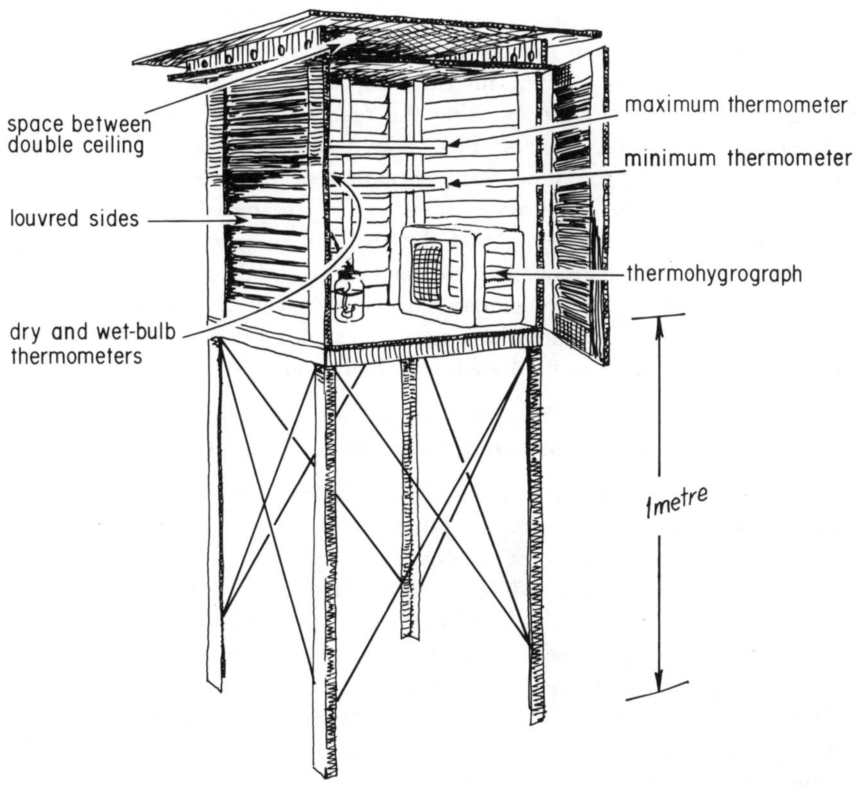

Figure 2.5 A Stevenson screen. The inner ceiling and the floor of the box consist of separated, overlapping slats, to permit ventilation

outside a north window of a building, but now the screen rests on a stand on the ground. The height of the stand depends on national choice, with the base between 1.2 and 2.0 metres above the ground. An extreme example of variety is the case of screens at five different heights within the same city (Milwaukee). In countries like Poland, the thermometers are put higher than any layer of snow that may be expected, so the screen is approached by two or three steps.

The instrument height is significant because of its effect on the daily range of the measured air temperature. The ground is hotter than the air during the day, and at night it is cooler, so the range is less at a greater height above the ground, up to a few hundred metres. It is greatest near the ground, in summer, at high elevation, and with a clear dry atmosphere at low latitudes. On a clear and calm winter's day at Pune in India, the ground surface maximum and minimum temperatures differed by 34C°, whereas at 1.5 m the difference was only 22C°. The effect of screen height on readings is less when the sky is cloudy and the wind is strong.

Whatever the height of the instruments, the screen must be supported firmly to prevent vibration affecting the readings, especially of minimum temperature. Also, the louvres must be kept clear of leaves, spiders' webs, insects' nests, etc., to allow free ventilation of the interior. Otherwise, evaporation from the psychrometer wick raises the measured humidity.

A special difficulty arises in cold climates, where wind-blown snow may be driven into the louvre openings or the box be smothered by wet snow or freezing rain. Some improvement can be achieved by surrounding the screen with fine gauze or treating the surfaces with a non-wettable material like Teflon or a de-icing fluid such as ethylene glycol.

Effect on temperature measurement

Temperatures measured within a screen are affected by the temperature of the screen itself. Observations in Japan have shown temperatures at the centre of a screen to differ by as much as 0.7C° from those by the walls. This explains some of the difference between screen measurements and temperatures obtained with a ventilated psychrometer: differences of 0.6C°, 1.0C° and 1.2C° have been reported. Indeed, thermometers in a well-ventilated verandah or shed may be closer to the psychrometer readings. To reduce the error due to the screen, the thermometers within should be mounted at least 8 cm away from the walls to allow free ventilation, and the screen might be ventilated by a fan.

The screen should be kept white, to reduce the absorption of sunshine. It should be repainted each two years, but more often in a polluted atmosphere. Also, the box should be cleaned at least monthly, and more frequently in a dirty industrial area. It should be brushed or wiped with a wet cloth after any dust storm.

A secondary effect of the screen is to muffle the effect of any sudden change of air temperature. The lag of the temperature of screen surfaces is about 4 minutes if the wind speed is 4 m/s, but 8 minutes if only 1 m/s. This effect, and the reduced wind speed inside the box, dampen the response of the thermometers. If the internal ventilation is only 1 m/s and a thermometer's lag therefore is about 100 seconds, it takes 6 minutes for the error to fall below 0.1C°, in recording an abrupt change of air temperature by 3C°, for instance.

Another problem is the cooling of the screen after rainfall, due to the evaporation of water on the louvres. It lowers the measured temperatures.

Despite this catalogue of the disadvantages of the Stevenson screen, there is no better alternative. In any case, use of this screen has become universal and standard. Any change now would disrupt the comparability of measurements.

MEASUREMENT OF WIND

Systematic observations of wind speed and direction began in 1650 in Italy and in 1667 in England. Nowadays, wind is measured at almost all climate stations, usually by means of a 'cup anemometer', devised by Robinson in 1846. The standard British design has three cups of 12.7 cm diameter travelling round a circle with a 1 metre circumference and a vertical axis. Rotation of the shaft by the force of the wind on the cups either generates a voltage, which is measured remotely, or else causes closures of a switch, the number of closures being counted electrically. The usual mercury switch necessitates a rigid and accurately vertical mast to avoid false registrations. Also, the bearings on the cups' axis need annual overhaul or replacement. Home-made cup anemometers have been described by several authors, e.g. Snow *et al.* (1989: 504).

Propeller anemometers are common in the USA. The wind meter resembles a model airplane without wings, with a single propeller. The tail swings the propeller into the wind, and then the propeller's rotation generates a voltage which is measured. Calibration of such an anemometer in a wind tunnel still permits errors up to 20 per cent.

A smaller instrument is a hand-held 'ventimeter'. The wind pushes a small pith-ball up a vertical clear-plastic tube, wider at the top than at the bottom, and the ball rests at a calibrated height at which the wind's force equals the ball's weight. Fluctuations of the reading make ventimeter measurements very inexact.

The swinging-plate anemometer was first devised in Italy in about 1570. A home-made version (see Figure 2.6) is so cheap that hundreds can be mailed out for simultaneous comparable measurements over a region. The wind speed is approximately proportional to the square root of the angle of deflection of the plate, up to about 55 degrees. As an example, a plate 30 cm long, 15 cm wide and weighing 200 grams is deflected 9° by a wind of 3 m/s, 31° by 6 m/s, and 52° by 9 m/s. The instrument is designed to measure up to 22 m/s, but is not suitable for measuring the quite common winds below 2.5 m/s. Also, it is hard to observe the angle of deflection if the instrument is on top of a 10 m mast and the wind is fluctuating, especially if the plate's oscillations happen to resonate with some of the wind's eddies.

A wind meter which is gradually being superseded by the cup anemometer is the Dines anemograph. This is a high-grade instrument with a tube swinging to face the wind so that the wind's pressure forces water down a U-tube, which raises the level in the other limb, lifting a float carrying a pen. (The pressure of the wind is less on high mountains, since the air's density is less.) The vertical movement of the pen registers on a chart, which is replaced daily. The anemograph is a reliable and sensitive instrument, giving a continuous record of wind fluctuations so that gusts as brief as 5 seconds long can be measured. It is easy to calibrate, operate and

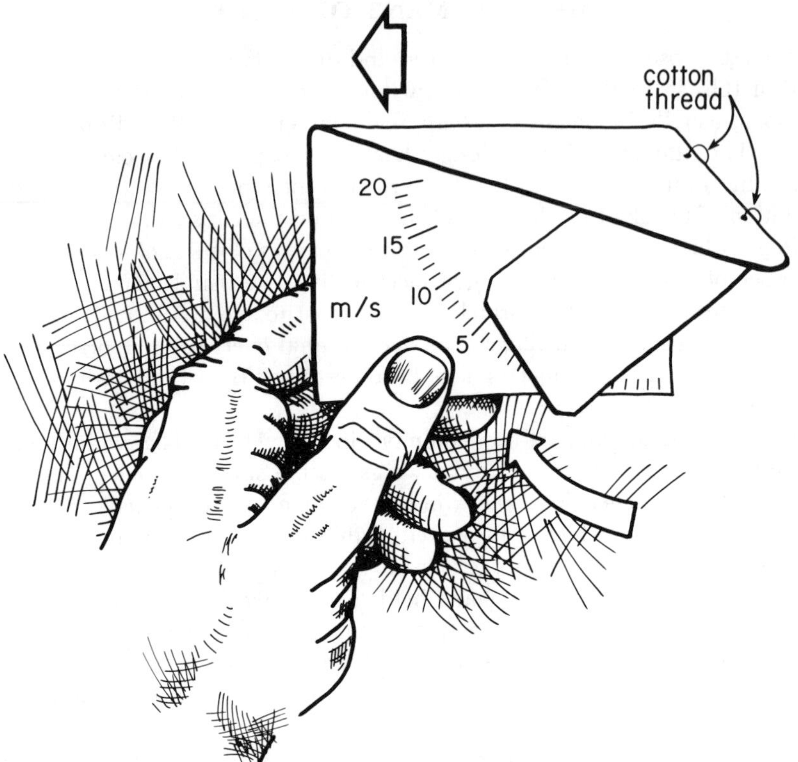

Figure 2.6 Home-made swinging plate anemometer made of aluminium sheet (Linacre and Barrero 1974)

maintain. It reacts more quickly to rapid changes of wind than a heavy cup-recorder does. However, it is expensive to install and less accurate for low speeds because the instrument responds to the square of the velocity.

Apart from the usual instruments just considered, many other anemometers have been described for special purposes. For example, there is a clockwork-driven cup-anemograph made by Woelfle in Germany which records both wind direction and strength for a month without attention. For indoor use, there is the 'kata-thermometer' anemometer which consists of a large-bulb thermometer filled with alcohol or a xylene-toluene mixture, with the bulb silvered to reduce the influence of radiation. One measures the time for the thermometer to cool either from 54.5°C to 51.0°C or from 38°C to 35°C, after taking the bulb from warm water. The rate of cooling shows the wind speed over the bulb for even very small air movements.

Problems

An anemometer for a climate station must satisfy conflicting requirements. It has to be cheap, but at the same time accurate. It should be sensitive to light winds, yet able to withstand the strongest likely to be encountered, such as 88 m/s in Antarctica. The instrument may be subject to corrosion near the ocean, or to snow, or to the freezing of lubricants on moving parts. Icing is another hazard; it can be reduced by electrical heaters, except that these distort light winds and are impracticable in strong winds. In some places there are risks from lightning (especially at low latitudes) and from vandals, notably people with guns.

The price paid for a cup-anemometer's robustness is its insensitivity to winds below a certain speed. The threshold is typically about 1 m/s. As a consequence, winds of less than 1.5 m/s are customarily counted as 'calms'. Actually, friction of the vertical bearing of a cup anemometer may affect measurements even up to 2.5 m/s, and this friction increases in time.

Another problem is the tendency for a cup anemometer to respond only sluggishly to fluctuations of wind speed. This leads to overestimation of the mean speed in gusty conditions, because the instrument accelerates readily when speeds increase but freewheels when the wind dies down, especially if the instrument is heavily built. The lag is quantified in terms of the 'distance constant', which is the distance traversed by the wind after an abrupt change of speed, during the time in which the instrument alters by 63 per cent of the change. The distance may be 0.7–8.0 metres for a three-cup anemometer; 2–5 metres is reckoned acceptable. The wind-speed error resulting from this effect is typically 5–13 per cent of the mean speed.

There are other errors possible in measuring wind. One is the calibration error of the anemometer, which may be up to 7 per cent, for instance. The total of all errors is preferably less than 1 m/s or 5 per cent of the mean. But this requires good maintenance. In practice, the errors associated with wind speed measurements at airfields, for instance, may add up to at least 10 per cent.

One in 20 instruments develop faults each year, so they should all be serviced regularly – every 18 months near the coast or every three years inland. A useful check of the proper operation of an anemometer is a regular comparison of its readings with those from neighbouring instruments.

Taking measurements

The readings are best expressed in Système Internationale units of metres per second, though knots are accepted. The connections between various units are shown in Table 2.4

Most anemometers show only instantaneous speeds. Cup anemometers also show the momentary speed when attached to a small electricity

Table 2.4 Relationships between various wind-speed units

	knot	*m/s*	*mph*	*km/h*	*ft/s*
knot	1.000	0.514[a]	1.152	1.852	1.689
m/s	1.943	1.000	2.237	3.600[b]	3.281
mph	0.868	0.447	1.000	1.609	1.467
km/h	0.540	0.278[c]	0.621	1.000	0.911
ft/s	0.542	0.305	0.682	1.097	1.000

Notes: [a]i.e. 1 knot = 0.514 m/s,
[b]i.e. 1 m/s = 3.6 km/h,
[c]i.e. 1 km/h = 0.278 m/s.

generator. However, anemometers which involve counting the number of revolutions of cups or blades measure the 'wind run' between readings, which is the distance that would be travelled by a balloon in the wind. (The wind run is typically about three times the distance travelled by the centre of each cup in the case of a cup anemometer. This ratio is called the instrument's 'factor'.) Daily 'wind-run' is calculated from the number of cup rotations from 9 a.m. to 9 a.m. A daily wind run of 285 km (or 177 miles), for example, is equivalent to 3.3 m/s, on average.

Nominally 'instantaneous' winds are usually measured at climate stations once or twice a day at the times of measuring temperatures etc. Ideally, the frequency of measurement would depend on the similarity of consecutive readings. If the coefficient of correlation is around 0.8 (see Chapter 4) for pairs of measurements taken a particular number of hours apart, that is the proper interval between observations, for the same reasons as those determining the most suitable spatial separation of observations discussed in the section 'Climate stations', pp. 25–7. As an example, the variation of the coefficient for different times between observations at a site in north-west Tasmania suggests the desirability of measurements there each three hours.

Wind direction

Wind direction is measured by means of a vane, accurately balanced about a truly vertical axis, so that it does not tend to settle in any particular direction. The vane follows wind direction better if it is flat rather than splayed, and a fairly heavy construction helps to smooth out any flutter. The bearing should be lightly damped to prevent oscillations in response to natural fluctuations of the wind. However, it should take no more than 5 seconds to respond at least 63 per cent towards an abruptly imposed new direction. Vanes will not usually turn with a wind of less than 1.5 m/s, and then one uses smoke or a pennant, or faces into the wind, to find its direction.

The vane direction is observed and averaged over a few minutes at the times of taking the routine measurements at climate stations, and then recorded as a compass direction. This is subsequently converted to a code number. If directions are measured in terms of the number of degrees clockwise from north and then converted to compass points, some bias is commonly introduced since conventionally the four cardinal directions each correspond to a range of 30° (e.g. 345°–015° is equivalent to North), whilst the intermediate directions relate only to 20° ranges, e.g. 015°–035° is equivalent to NNE. Also, there is a human tendency to prefer major rather than minor points of the compass (e.g. South rather than south–south-east), revealed by a total frequency for the eight principal directions much larger than the total for the eight intermediate directions. That bias can be removed by respective multiplying factors to make the two totals equal. Alternatively, the preference for cardinal directions can be offset by the human preference for even numbers, if the cardinal points are indicated by odd numbers.

Another bias is due to parallax, because it is uncomfortable for an observer to stand directly under the vane and look vertically upwards. This problem is cured by arranging for electrical detection of the vane direction, with convenient placing of the read-out elsewhere.

Exposure of wind instruments

It is important to have a satisfactory exposure for any anemometer or wind vane. Where an anemometer is mounted at height H above the ground there should preferably be 100 H of uniform land upwind. If there is an object of height h within a distance h from the anemometer, the latter should be at least 2 h from the ground; or at least 1.5 h high if the object is 10 h away, and at least h high if the separation is 30 h.

Where an anemometer is mounted on the side of a tower, its spacing from the tower can significantly affect the measurement. The reading may be reduced by about 2 per cent if the spacing is 1 metre, or 1 per cent if about 2 metres, for instance.

Correct exposure is essential for any indicator of wind direction too. Obstacles and buildings upwind can make considerable difference to the readings, and the direction indicated may vary with elevation above irregular ground. The vane should preferably be in the open at the standard 10 m.

MEASUREMENT OF RAINFALL

Rainfall is measured most simply by noting periodically how much has been collected in an exposed vessel since the time of the last observation. This has been done from the 4th century BC in India, where measurements were taken in order to tax agricultural land accordingly.

With a conventional rain gauge, the depth of rainfall is found by dividing the area of the collector into the volume caught. The collector is usually a funnel, leading the water into a receiving vessel beneath. Both are within an upright tube, the diameter of which is the same as that of the funnel. Neither collector nor vessel must be painted, since paint may eventually absorb moisture. The collected volume (or 'catch') is measured either by pouring it into a graduated cylinder or by the use of a graduated dipstick in a vessel of standard cross-section. The latter procedure tends to be more accurate where daily falls are high or when using a 'storage gauge' which collects the catch over a long period.

Different designs of rain gauge

There are more than 40 designs of rain gauge, which differ as regards the diameter and height of the top opening. A Swedish model has an opening with a diameter of 357 mm at 1.5 m height; in Finland and Czechoslovakia the diameter is 252 mm; in the USA, France and Australia 203 mm; in Germany, Holland, the USSR, Poland and most of South America, Asia and Africa it is 160 mm; the Snowdon gauge in Britain has a diameter of 126 mm and a height of 0.3 m; in Belgium the opening is 113 mm in diameter. The various national gauges collect amounts which are typically between 99.6 per cent and 120.2 per cent of that collected in adjacent gauges of a design chosen as reference.

A wide gauge collects an inconveniently great amount of water in a wet climate, whilst a narrow one collects too little to be easily measured and can become blocked by debris. Also, gauges with a small diameter tend to collect less depth of rain because of the increased turbulence across the opening, deflecting rain away. One of 28 mm diameter collects 7 per cent less than one of 200 mm diameter.

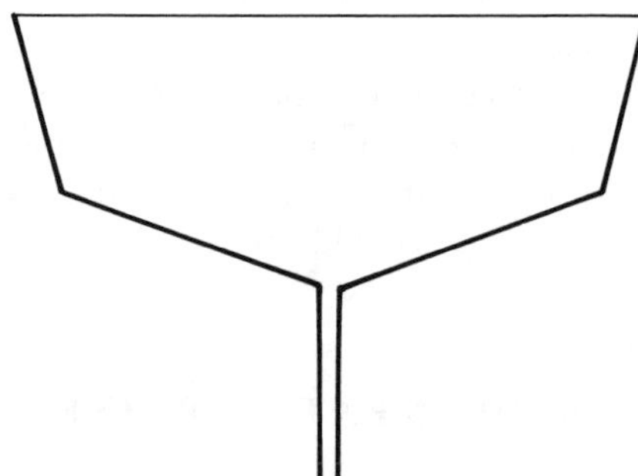

Figure 2.7 A 'champagne-glass' rain collector (Folland 1988: 1507). The diameter of the rim is 437 mm, the sides are 15° from the vertical and 110 mm deep, and a line from the outlet at the bottom to the rim is 40° from the horizontal. (By permission of Dr C. Folland)

The gauge adopted as standard by the World Meteorological Organization is based on the British design, with a cylindrical opening 127 mm in diameter and 130 mm high, above a funnel which takes the collected water into the vessel below. The rim is 1 m above ground to prevent water splashing in from the surroundings.

Simpler designs are sometimes used for economy, involving a plastic funnel into a vessel such as a wine bottle, for example. Cheap gauges allow the use of many of them for assessing the overall rainfall onto a broad catchment area, and the amount caught proves to be within a few per cent of that in a standard gauge. In fact, gauge design may not be important in comparison with other factors affecting rainfall collection. Five different kinds tested in Nebraska showed no appreciable or consistent differences between the amounts of rain collected. So it is probably valid to use a quite different kind that is popular with farmers, made of clear plastic. It has a tapering rectangular section, so that the increase in depth of the collected water is more easily discerned when rainfalls are small. Unfortunately, the plastic can become misshapen through frost or heat, or the instrument may be blown away.

A new design has been proposed by Folland (1988) which omits the usual tube around the collecting funnel and has a slightly conical opening, as shown in Figure 2.7. The amount collected by this gauge is singularly unaffected by wind speeds up to 5 m/s.

Procedure

Rainfall measurements are usually taken each morning, at the same time as other climate-station readings. The time differs in various countries. For instance, it is 9 a.m. in Australia, but 5 a.m. and 7 p.m. in the USSR. Any morning reading has to be associated with the date of the measurement, but one must bear in mind that much of the collected rain may have fallen during the *previous* day.

Where it is not practical to attend to a gauge every day, the collected rain may be accumulated for weekly or monthly measurement. In that case, the receiving vessel has to be larger and stronger and attention given to reducing losses due to evaporation between measurements. The evaporation loss may be equivalent to 0.1–0.5 mm/d of rainfall, if the temperature is 20°C and the relative humidity 60 per cent, for example. The loss is reduced by adding a known amount of oil to the receiver, to float in a layer at least 5 mm thick on the collected water. Paraffin is often used, but it could be any non-resinous, low-viscosity oil, mixed with an equal quantity of water to ensure its saturation. It needs renewing annually to avoid the layer dispersing as an emulsion into the collected rainfall. The evaporation loss from a rain gauge is also reduced by using a narrow-necked receiver, preferably painted white to reject solar radiation, or installing below ground.

Freezing of the collected water may be a problem in a cold climate. The problem is overcome by adding anti-freeze to the gauge after each measurement. Solutions of either calcium chloride or ethylene glycol are suitable. The anti-freeze must have a density slightly less than that of water, to facilitate its mixing with the collected rain. This can be achieved by a mixture of one part anti-freeze to 1.12 parts methylated spirits.

Rainfall recorders

An alternative to a storage gauge is a pluviograph, which is more expensive but gives more information. Whereas a storage gauge which is observed each day or each week tells only the total rainfall during that time, a pluviograph shows exactly how much fell, at precisely what times of each day, and whether there were sudden showers or a prolonged drizzle.

The most common pluviograph involves a small reservoir beneath the collecting funnel, with a float on the water in the reservoir. The float carries a pen which records on a chart fixed to a drum rotating about a vertical axis. When the reservoir becomes full it automatically empties by siphoning within 15 seconds. There must be no tendency for the water to dribble at the start or end of siphoning.

The chart record resembles irregular saw-teeth, each tooth reaching to the top of the chart and representing either 10 mm or 25 mm of rain, with the ascending part of each tooth consisting of steps upwards at the times of rainfall. The time is easily measured within 5 minutes or so, as the chart travels past the pen at about 12 mm/h. The thickness of the ink line on the chart leads to an uncertainty equivalent to 0.3 mm of rain. Pluviograph registrations of a rainfall may differ by 1 mm from measurements in adjacent normal gauges.

A different kind of rainfall recorder is used in the USA and Canada, which involves a pen recorder that registers the increasing weight of water in a collecting bucket. The instrument is suitable for measuring snow and hail, as well as rain, but it usually gives readings with a possible error of 2.5 mm, which is relatively inaccurate. An error of 0.5 mm in measuring hourly rainfall can arise merely from expansion of the chart paper in wet weather.

Another sort of rainfall recorder, the 'tipping-bucket rain gauge', was invented by Christopher Wren (see Table 1.1). It consists of a small vessel which tips empty as soon as the collected rain fills the vessel (Figure 2.8), and a record of the times of tipping tells the rate of rainfall. A modern version can operate for two months without attention, recording every occurrence of rain.

One design has a collecting area of 750 cm^2, i.e. a diameter of 309 mm. The rain is led into a horizontal section of tube, which is partitioned and pivoted. The upper half receives the rain, tips over when full and then drains, remaining down until the other half has filled. Each filling

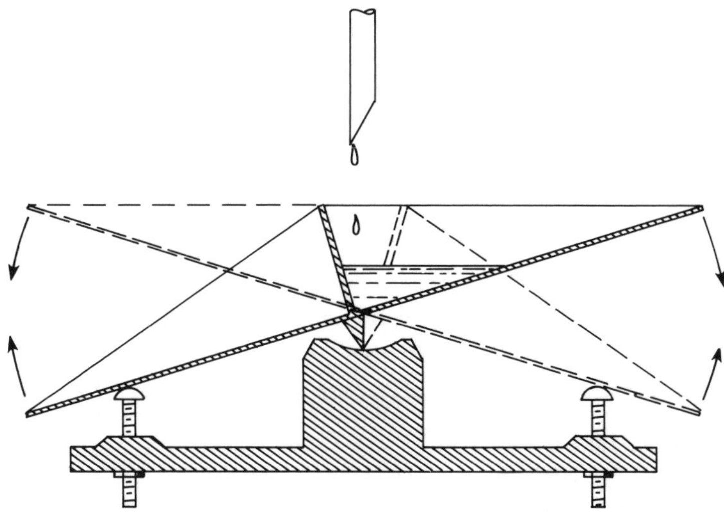

Figure 2.8 A tipping-bucket rainfall gauge

corresponds to a rainfall of either 0.2 mm or 1 mm, and the tipping closes an electrical switch so that a mark is made on a chart, a magnetic tape or an event recorder. The device is suited to remote recording or incorporation into an automatic weather station. The data can easily be analysed by computer, to find, for instance, the frequencies of various intensity of rainfall, the diurnal variation, the tendency for a wet day to be followed by another, and so on. A home-made version has been described by Hookey (1965: 193).

A disadvantage of a tipping-bucket device is the evaporation from the bucket between tippings. Also, the bearing may stick after a long dry period. Furthermore, there is no measurement during the time of tipping, which leads to underestimates when rainfall is intense (see Note 8, p. 317). A tipping-bucket gauge may record 1 per cent too little when there is 50 mm/h, which is described as 'very heavy rainfall' (Meteorological Office 1956: 285).

Such an intensity is uncommon but possible, especially in low latitudes, as discussed in Heavy Rainfalls, pp. 296–300. Global-record downpours averaged 1872 mm/h over one minute, and 120 mm/h over a whole storm. However, the average over 6 minutes in Melbourne, for instance, exceeds 50 mm/h merely once each year. In the wettest parts of New South Wales the rate exceeds 86 mm/h during 10 minutes only once a year, and 310 mm/h once in 200 years.

Ideally, at least 10 per cent of the rain gauges in a network of climate stations should be recorders, preferably in populated areas and places where storm drainage or river control might be needed. In Norway there are 50 tipping-bucket recorders out of 800 gauges. Unfortunately, pen-recording gauges generate an embarrassing number of charts to store, and many automatic rain gauges do not work well in the tropics because of interference by insects, fungi or large animals, and because of vandalism, poor maintenance and lack of spare parts.

Effect of wind

Wind has been a common cause of underestimating rainfall, as pointed out by Luke Howard in 1811. The air movement tends to scoop away the raindrops as they are settling into the collecting funnel. As a result, there is significant underestimation when winds exceed about 5 m/s. Table 2.5 shows that conventional gauges are associated with a reduction of catch by about 2 per cent for each increase of wind speed by 1 m/s. A fairly well protected gauge in the mountains of Utah, for instance, collected 30–50 per cent more rain than one in a windy place nearby.

The effect of wind on the accuracy of measuring precipitation is greater when a gauge is used to measure snowfall instead of rain, as the flakes fall so slowly. On the other hand, the larger raindrops that fall at low latitudes, and the weaker winds there, tend to increase the rate of descent and reduce the sideways deflection so that rainfall measurement there is more accurate. Figure 2.9 shows how the rainfall intensity and wind speed affect the underestimation. Wind has most influence where the rainfall is light.

To lessen the effect of wind, it is common in the USA to fit a shield around the rain gauge, making the air flow horizontally across the opening, instead of scooping raindrops out. Alternative designs of shield are illustrated in Figure 2.10. It was found in a comparison in Nebraska that a Nipher shield increased the catch by 3.5 per cent and an Alter shield by 1.5

Table 2.5 Effect of wind speed on rainfall measurement by a conventional gauge

Reference	*Place*	*Wind* *m/s*	*Reduction of measured rainfall* *(%)*	*per m/s (%)*
Stanhill 1958	England	4	20	5.1
Harbeck and Coffey 1959	Oklahoma	3	4	1.7
Showalter 1962	USA	–	–	1.0
Mueller and Kidder 1972	wind tunnel	15	89	5.9
Dahlstrom 1973: 14	Sweden	11	24	2.2
Chaine and Skeates 1974: 13	Canada	18	33	1.8
Smith 1975a: 118	UK sheltered sites	–	5	2.5
	exposed sites	–	20	2.5
Sevruk 1982: 81	Australia	8	12	1.5
			Median	2.2

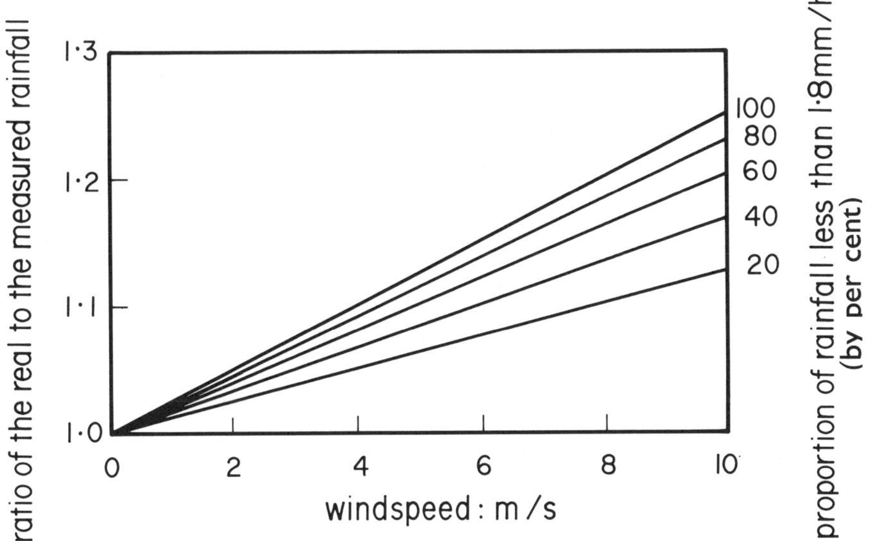

Figure 2.9 Effect of rainfall intensity and wind speed on the underestimation of rainfall (Sevruk 1982: 7, by permission of the World Meteorological Organization)

per cent, over the course of a year. So the Nipher shield is generally reckoned superior, except when there is snow. However, the effectiveness of any shield is uncertain, and they are not fitted in Britain, for instance.

An alternative to the shields shown in Figure 2.10 is a turf wall around the gauge, the wall to be 0.3 m high and 3 m in diameter. A disadvantage is that the wall collects snow in winter and there are problems of drainage. Also, the wall attracts sheep and deer, which trample it down. A better way of reducing the effect of wind is to mount the gauge in a pit, with the gauge-rim level with the ground's surface and the rest of the pit covered by a grille which accepts rain without splashing. A measurement taken with this elaborate arrangement can be regarded as the true rainfall.

Winds can be weakened by sheltering the gauge, though this would also protect it from being open to rain from all directions. In general, it is recommended that no obstruction of height h be within a distance of 4 h of the gauge. In no circumstances should any obstruction be within a distance h. Also, the gauge in a climate station should be over 3 m from the screen. Whatever the situation, the location of the gauge should always be described completely in the records.

Gauge elevation

Elevated gauges catch less rain, as observed by Benjamin Franklin in 1771, because the wind is stronger away from the ground. Gauges at ground level

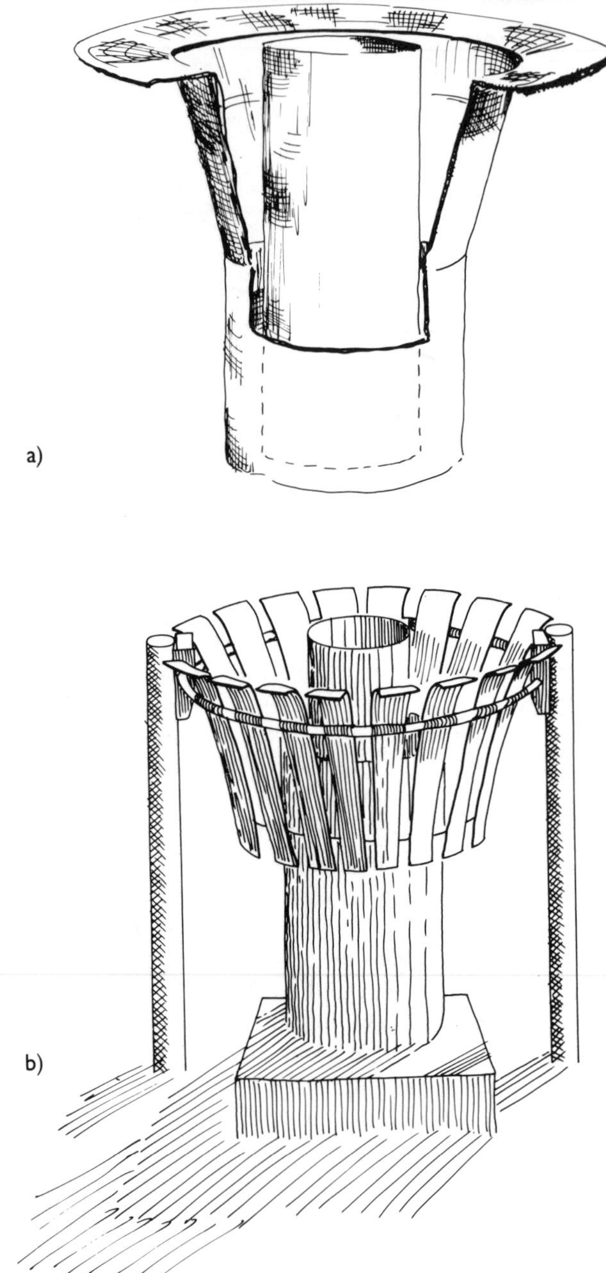

Figure 2.10 (a) A Nipher shield (1878) and (b) an Alter shield (1937) for fitting around a rain gauge. The Alter shield has leaves which hinge to release any snow collected around the gauge. (By permission of McGraw-Hill Book Co.)

and at 1.5 m collect the same rainfall in still air, but in a wind of 10 m/s there is a reduction of 30 per cent in the higher gauge. A gauge at 1.5 m at Uppsala collected about 4 per cent less than one at ground level. Two adjacent gauges in Holland, respectively at 0.4 m and 1.5 m elevation, differed in their catch by 12 per cent. Overall it has been found that there is a reduction of 2–3 per cent for each extra metre elevation of a normal gauge. So a gauge at 2 m catches only 95 per cent of that into a gauge whose rim is at 0.3 m, and one at 8 m catches only 75 per cent.

Standard heights of the rims of rain gauges differ in various countries, being 0.8 m in the USA but 0.3 m in Australia, for example. A greater elevation is desirable in cold countries to avoid collecting blown snow. A common practice is to mount gauges on a convenient roof, where they are safe from interference, although obviously vulnerable to stronger winds.

It has been queried whether the gauge rim should be parallel to the sloping ground of a hillside or accurately horizontal. Upright gauges collect 30 per cent less than gauges perpendicular to a windward slope on a Japanese mountain, but 10 per cent more than gauges perpendicular to the leeward slope. However, the upright gauge is standard, and it should be on a level site.

The surrounding surface should be soft enough to reduce splashing, and grass is preferred. The gauge must be firmly anchored to prevent it being overturned by strong winds.

Other errors in rainfall measurement

There are several possible causes of error, apart from the wind and improper height or exposure and the effects of evaporation mentioned already. Other causes will be considered in turn:

1 Hail may block the collecting funnel, so that it overflows.
2 In light rains, a significant amount of moisture may be held on the funnel surface, unmeasured. Likewise, not all the water collected within the gauge is transferred to the measuring cylinder, and this can cause an error of up to 0.2 mm. Wetting errors can total 2–10 per cent of the rainfall.
3 There can be an error in reading the water in the measuring cylinder equivalent to 0.2 mm of rain. Also, in heavy rain it is easy to miscount how many times the measuring cylinder has been filled.
4 If the reading is not taken for a few days, all the collected rain may be attributed to the last storm.
5 There may be splash of rain either into or out of the collecting funnel, amounting to several per cent of the rainfall. This is most likely with heavy rainfalls which involve large raindrops, and gauges with a low rim.
6 The sampling error may also be appreciable, resulting from the inherent

> irregularity of rainfall over an area, even a small area. Oliver (1959: 290) quoted 8.3 per cent differences between daily readings of six gauges and their mean, all the gauges being within 17 m of each other. Likewise, Reynolds (1965: 114) measured differences of 7 per cent for four gauges within 20 m, recording individual storms. Court (1960) reported the case of identical gauges only 3 m apart on a windy ridge, which collected hourly amounts differing by 50 per cent of the smaller catch.

When adjacent gauges give different values, the highest is likely to be the most accurate, since errors of measurement tend to cause underestimates, except for splashing in with a gauge too near the ground.

Errors can also arise, subsequently, in handling the measured data, and they need to be checked. Rainfall values below 1 mm are sometimes not included or rainfalls may be rounded off to the nearest 5 'points', i.e. within 0.05 inch or 1.3 mm. Rainfalls measured to the nearest tenth of a millimetre may be recorded without the decimal point, giving an error of an order of magnitude. Other problems are the recording of readings on the wrong day and evaporation loss during dry spells. Examination of 20,000 rainfall values printed out from computer storage showed that about 4 per cent were suspect in one way or another. About half of these could be amended manually, but 40 had to be discarded.

One way to detect errors due to an abrupt change of circumstances at a rainfall station involves what is called 'double-mass analysis'. This consists of comparing the accumulated measurements at the station with those from another nearby, which is reckoned as reliable. The accumulations are compared graphically, as in Figure 2.11. In this example, there was clearly some disturbance of the relationship between the stations A and B in 1973. A similar comparison for stations A and C can be used to determine whether circumstances at A changed or whether those at B did.

Gauges should be inspected regularly, preferably each two years, to check the site exposure, uniformity in the methods of observation, leaks and other faults in the equipment, and to maintain observer interest. Half the gauges in England were found faulty at the time of a first inspection after installation, and a quarter of them were still faulty three years later.

If the correlation between measurements from adjacent gauges is high, then one of them merely duplicates the other and is not needed. For instance, analysis of data from 333 gauges in an English catchment showed that the total rainfall in the area could be determined equally accurately by a more rationally located network of only 220. Similarly, the number in a certain area of Wales could be reduced by 30 per cent. Fewer gauges allows better gauges, and less but superior equipment reduces measurement errors and observation costs, which more than offsets the increased sampling error.

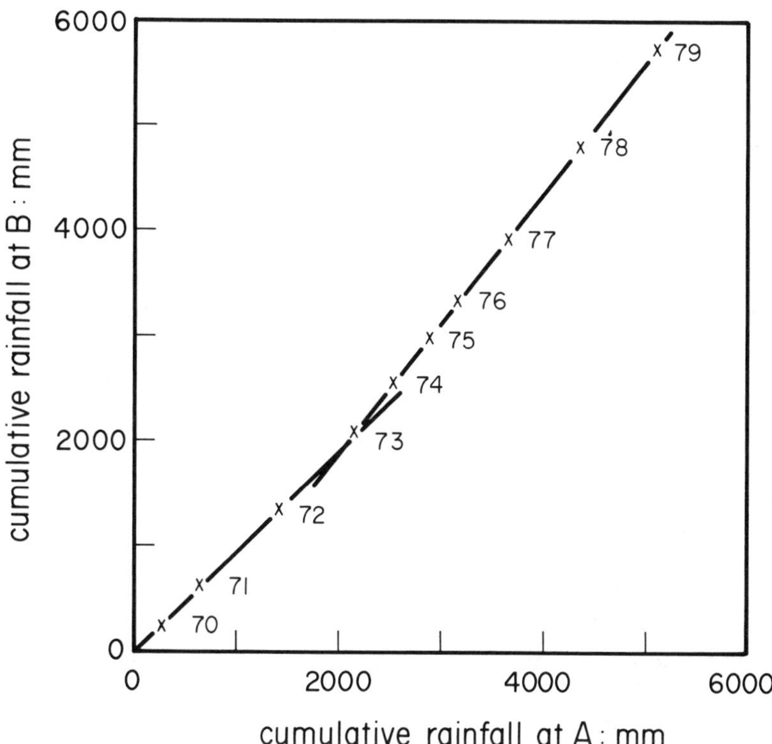

Figure 2.11 Double-mass analysis of rainfall measurements at stations A and B, between 1969 and 1979

Total error

The total error in measuring rainfall can be appreciable, chiefly due to faulty exposure of the gauge. Apart from that, the error may well be up to 10 per cent, or even more in strong winds. So an error of 2 mm has to be regarded as normal for rainfalls of below 40 mm, or 5 per cent for larger amounts. These are about twice the errors specified in Table 2.2 as the maximum desirable. One should ignore small differences between rainfall measurements.

ARCHIVING

It is wasteful to measure what is not then adequately stored for later analysis and use. Therefore it is important to archive the endless flow of climatic observations. Unfortunately, the task is complicated by the variety of the forms of data, e.g. hand-written records, recorder charts of various shapes and sizes, punched or magnetic tape, sunshine recorder charts, rolls of printout paper, etc.

Minor manual checking of data should be carried out soon after measurement, at least monthly, and before archiving. Filipov (1968: 9) reckoned that 1–2 errors per hundred observations was about average. If there is no time to check doubtful figures immediately, they should be flagged for later attention to reduce the archiving of rubbish and to help monitor the equipment and procedures.

The process of archiving may be as simple as chronological shelving of an observer's measurement books, or it may be highly elaborate in order to permit easy retrieval and use by others later. It seems that the total cost of (i) entering data into an archive, plus (ii) extracting the data from it, is roughly constant. In other words, cheap entry of the data leads to expensive access subsequently. The better the quality of the measurements, and the more urgently or frequently the data may be needed, the more care is warranted in storing them.

The archiving done by national meteorological offices involves transforming the data into a common form, suitable for computer manipulation and storage. However, the process of encoding the data incurs mistakes due to human error. In the era of punched cards, up to 9 per cent of cards were reckoned to contain at least one error; there were typically two errors for each 1000 digits encoded. They can be found by comparing the encoding done by one operator with that done independently by another operator from the same data.

All stored data must be labelled as regards place, time, instrument, accuracy, units and coincidental events. Pen-recorder charts should be annotated before filing, with the date of filing (including the year), the times of any marks on the chart, and the dates and times of the start and end of the record.

Data editing

Quality control is possible once the data are entered into a computer. This identifies missing figures, brings gaps to the attention of a human scrutineer for checking, and checks the internal consistency of the data (see Note 9). Simultaneous maximum and minimum temperature values from nearby places can be compared for similarity. One day's minimum cannot exceed the previous day's maximum. Data may be compared also with long-term means, to ensure that the values differ only to an extent which is reasonable. Also, checks can be made that dry-bulb temperatures exceed wet-bulb temperatures, that minima are above the dewpoint temperatures, and that unusually low temperatures occur only when winds come from a polar direction. The difference of temperature from the previous value must lie within the known maximum change.

Checking is important, and shows that improvement of present measurements (along with better observer morale and training, and better data management) are generally more necessary (and cheaper) than any increase of the number of climate stations. Several per cent of data collected by 'co-operative' (unpaid) observers in the USA are inaccurate or missing. Robinson (1990: 830) quoted the case of a farmer almost deprived of $70,000 drought insurance because a rainfall of 0.07 inches was keyed into the computer as 0.17 inches.

Testing the procedure for validating data in the USA involves deliberately replacing some raw data by known errors and then counting how many survive the checking system.

Once the data are checked, they may be supplemented by calculations of averages, the amount of scatter and the extreme values (see Chapters 4 and 6).

The location of the archive must be publicised. Generally a single central archive is much better than many local stores of data. Most national meteorological agencies keep archives of climatological data, such as is the US National Climatic Data Center at Asheville (North Carolina 28801). In addition, there are occasional published compilations, chiefly of monthly mean values, including the World Survey of Climatology published by Elsevier in several volumes.

RELEVANCE TO CLIMATE CHANGE

Consideration of the errors associated with measuring various elements of climate shows that discerning a real change of global climate is not easy. Even an apparent change might instead be due to alterations in the exposure of the instruments, or to changes of the instruments used or to the procedures for taking observations. The growth of cities causes warmer microclimates around climate stations, without necessarily altering the global climate. Such artificial variations are added to inherent fluctuations of climate (see Figure 1.3), combining as 'noise', which obscures any underlying change. The consequence is that time has to be allowed for the change to match the noise in size, for it to be seen clearly. We are currently at this stage. Unfortunately, it will be even harder to find a remedy by the time the warming becomes more obvious, if we do nothing now.

As regards the increases of dewpoint to be expected from global warming, it is clear that the customary measurements with unventilated psychrometers or hair hygrometers are too crude to detect small changes.

Another consideration concerns the network of measurements needed to establish any global change. What has been said earlier in this chapter deals with land-based stations, but 70 per cent of the globe is covered by oceans whose surface temperatures are not evenly sampled. Satellite measurements will provide total coverage of surface temperatures, but not of air temperatures at screen height.

3

ESTIMATING CLIMATE DATA

There are at least four situations in which it may be necessary to estimate climate data instead of measuring them. Firstly, data may be needed for times in the past when measurements were not taken, or for times in the future. Secondly, values may be required for places having no measuring equipment. Thirdly, it may be useful to check the plausibility of suspect measurements by means of an estimate based on reliable data. Lastly, the ability to derive an estimate which accurately and consistently agrees with independent measurements demonstrates an understanding of the physics of the process being studied.

In this chapter we will consider the nature of estimates in general and then discuss ways of estimating temperature, dewpoint, net irradiance and evaporation rate as particular examples. Methods of estimating solar irradiance, wind and rain are dealt with in Chapters 5, 6 and 7 respectively.

ESTIMATING IN GENERAL

To estimate the evaporation from a water surface on a particular day, for instance, one could use some suitable equation relating the evaporation to another climatic element which is more easily measured. The evaporation is the wanted (dependent) element on the left side of the equation, and values of the known, measured (independent) variables are inserted on the right. If there is a single independent variable it is referred to as a 'proxy' or 'surrogate', standing in for the desired element. The equation may be derived from simultaneous measurements previously, like those of evaporation Ep and temperature T in Australia, plotted in Figure 3.1. A straight line through these points is represented by the following equation:

$$Ep = 0.4\ T - 3 \qquad \text{millimetres/day} \tag{3.1}$$

However, the line links the measurements only crudely. Perhaps a curved line would give a better fit to the measurements? The equation for the curved line in Figure 3.1 is as follows:

$$Ep = 0.013\ T^2 \qquad \text{mm/d} \tag{3.2}$$

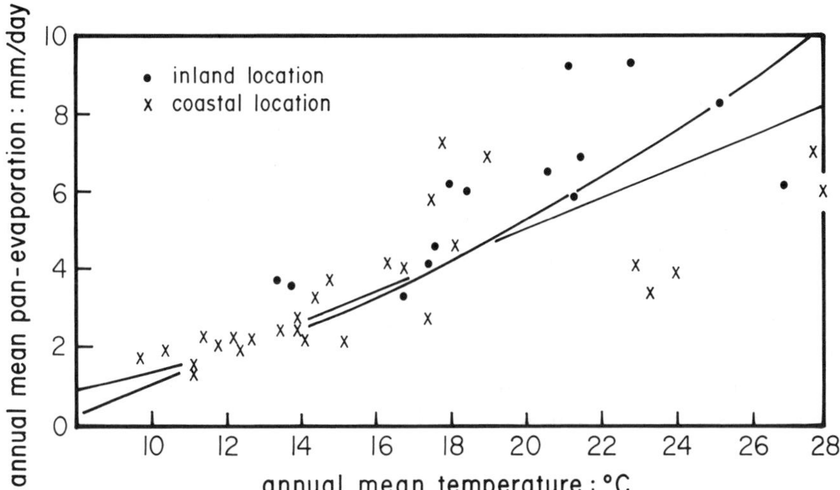

Figure 3.1 Relationship between the annual mean evaporation rates, Ep, from pans at various places in Australia and the mean temperatures there. The data comes from Gentilli, J. 1971

Which is better? One simple way of comparing Eqns (3.1) and (3.2) is to total the vertical deviations of the points in Figure 3.1 from the straight line and from the curved line, respectively. The lower the sum, the better the fit. In this case, Eqn (3.1) is more accurate than Eqn (3.2).

Equations (3.1) and (3.2) are both 'empirical' because they are derived from a set of simultaneous measurements of the independent and the dependent variables, not from theory. This has both advantages and disadvantages. On the one hand, an empirical equation is generally easy to derive, remember and use. Moreover, it is based on reality. On the other hand, it may represent reality only crudely if a simple equation is used to represent a complicated process. For instance, there would be considerable errors in using either Eqn (3.1) or (3.2), since both ignore factors other than temperature which can affect evaporation. Figure 3.1 shows that a location inland, where the climate is dry, tends to enhance water loss, quite apart from temperature. In addition, it is arbitrary to select any particular proxy as representative. Instead of Eqns (3.1) or (3.2), we might have chosen to relate evaporation rates to either the daily maximum temperature, the wind speed, or the atmospheric humidity, for example. In practice, one tends to choose the most readily available data, with no guarantee that they are the most appropriate. A third disadvantage of empirical estimation is that it relies on a *description* of the link between input and output data rather than an explanation.

The alternative to an empirical expression is a theoretical one such as Penman's well-known equation for the rate of evaporation from a water surface, discussed in 'Estimating evaporation', (pp. 105–6). The derivation of the equation does not rest on previous measurements of the dependent variable but arises from laws of physics that are universal. So the equation applies always, everywhere. This distinguishes a theoretical formula from an empirical one and is clearly a great advantage. Unfortunately, the theoretical formula may require data which are not usually available. For example, Penman's formula requires data on four climatic elements, not all of which are commonly measured. Also, theoretical expressions demand an understanding of the underlying processes linking the unknown element to known measurements, and tend to be relatively complex. Fortunately, the complexity may be reduced by incorporating a modicum of empiricism into the theory, as in the section on Estimating evaporation, pp. 105–6. In fact, an alternating emphasis on theory and empiricism is an essential characteristic of the scientific method.

Filling a gap in a series of data

One purpose in estimating climate data is to fill gaps in a series of measurements taken at a particular place. Maybe values are missing on account of instrument malfunction, which can be inconvenient if you need to know conditions during the interruption or if a long continuous series is required to establish the frequency of high temperatures or of rain, say. The difficulty can be overcome by any of four methods:

1 *Climatological estimation.* In this case, one assumes that a missing daily value equals the average of measurements taken at the same time in other years.

2 *Interpolation.* A missing March value, say, may be deduced from a smooth curve on a graph of measurements in January, February, April and May. A superior procedure is described in Note 1, p. 317. Another is 'kriging'; for a missing monthly value, kriging consists of combining interpolation within a year, as just described, with interpolation in a series of values for that month over several years.

3 *Parallel estimation.* Available data from one climate station (B) are compared with those from another station (A), and the hypothesis is made that simultaneous measurements at A and B bear a constant relationship. The station B should be as near to A as possible and similar in circumstance. In the case of rainfall, evaporation or wind speed, one would expect a constant *ratio,* and for temperature, dewpoint and wind direction, a constant *difference.* The ratio or difference would be determined as the average of the values at times when data are available from both places. Subsequently, missing figures for A could be obtained by applying the ratio or difference to the simultaneous readings at B. Parallel estimation is more accurate than interpolation for filling gaps in records of daily extreme temperatures, for example.

A refinement of this method of filling a gap is to estimate the value at A (i.e. Va) as a weighted average of simultaneous values from several nearby climate stations (B, C, D, etc.), instead of from B alone:

$$Va = a + b.Vb + c.Vc + d.Vd + \text{etc.} \tag{3.3}$$

where a, b, c and d are constants which are selected by statistical analysis of measurements when all the climate stations give values. An example of this procedure is given in Chapter 6 (p. 243), concerning the estimation of wind strength at a place which is between anemometers various distances apart.

Another modification of parallel estimation of an element involves replacing it by a related variable. Instead of directly estimating the monthly mean temperature T, one might estimate the ratio (T − Tmin)/(Tmax − Tmin), where Tmin is the mean of the coldest months and Tmax that of the hottest. This is likely to be more accurate than considering T directly, because the ratio varies less over space, i.e. it is more 'conservative'. The ratio is a 'normalised' version of T, which means that it has been adjusted to allow for long-term differences between the ranges at various sites. An estimate of the ratio leads to the required value of T.

4 *Proxy estimation.* An alternative to relating the missing value to simultaneous measurements of the same element at other climate stations is to link it to other elements at the same station. This involves some empirical equation like Eqn (3.1), which can be used subsequently, when only one element is measured, to estimate the other.

It is best to use more than one method to fill the gaps. Then the scatter of the respective estimates indicates the precision of their mean value, assuming all methods are equally valid.

ERRORS OF ESTIMATION

The error of an estimate is indicated by the difference from measurements. However, not all the difference is due to estimation error – there is also the contribution due to the errors of measurement, discussed in Chapter 2.

Estimation errors can arise in three ways. Firstly, there may be untrue assumptions implicit in the adopted method of estimation, e.g. a wrong choice of proxy or an invalid assumption of homogeneity. Secondly, the independent variables (on the right-hand side of the estimation equation) contain measurement errors (see Chapter 2). The effect of these on the estimate, is discussed below. Thirdly, the equation may involve averaged measurements of factors which are not independent. This creates a special kind of error in long-term estimates which also will be considered.

The effect of measurement errors

Errors can be expressed either (i) in absolute units, e.g. a temperature may be measured with an error of up to 0.2C°, or (ii) as fractions or percentages, e.g. the evaporation error may be quoted as typically 5 per cent of the measured rate. Obviously, one cannot speak of the percentage of a temperature since the value would depend on what scale was used – a change of 6C° is different as a percentage if there is cooling from +3°C to −3°C, from what it is when the cooling is from 36°C to 30°C. Absolute errors should be considered in dealing with either the addition or the subtraction of independent terms in a formula, whilst fractional errors are relevant to the multiplication or division of terms as shown in Note 2, p. 317.

The error in estimating the element on the left of an equation is a combination of the separate errors of the data inserted on the right. The combination may be the algebraic sum (treating all errors as positive), as in the earlier comparison of Eqns (3.1) and (3.2). Or, preferably, it is expressed in terms of the 'root-mean-square-error' (RMSE), which is slightly larger. To calculate the RMSE, each individual error is squared (which makes all errors positive), the various squares are summed and averaged, and then the square root calculated. The RMSE is like the 'standard deviation' discussed in the next chapter, and therefore compatible with standard statistical theory. The combining of errors in calculating the RMSE resembles the mathematician's combination of mutually perpendicular vectors.

The fractional combined error of a difference between inaccurate measurements is especially large. For example, consider the wet-bulb depression (T–Tw), where T is the dry-bulb temperature and Tw the wet-bulb temperature. If values are 20°C and 15°C respectively, and each is measurable to within 0.5C°, the mean absolute error of (T–Tw) is $(0.5^2 + 0.5^2)^{0.5}$, i.e. 0.7C°, and therefore the fractional error is 0.7/(20 − 15), i.e. 14 per cent. On the other hand, the fractional error for the sum of the temperatures would be only 0.7/(20 + 15), i.e. 2 per cent, which is far less. This shows the relative inaccuracy of estimation formulae (or methods of calculating climate change, for instance) which involve difference terms.

An alternative way of finding the effect of a measurement error on the accuracy of an estimate is to insert slightly different values of the measurement in the equation to see how much the resulting estimates change. This is called 'sensitivity analysis'.

Errors due to averaging

Another kind of estimation error arises from the customary use of measurements which have been averaged over a certain period, like the annual mean temperatures in Eqn (3.1). Averaging over a long time is

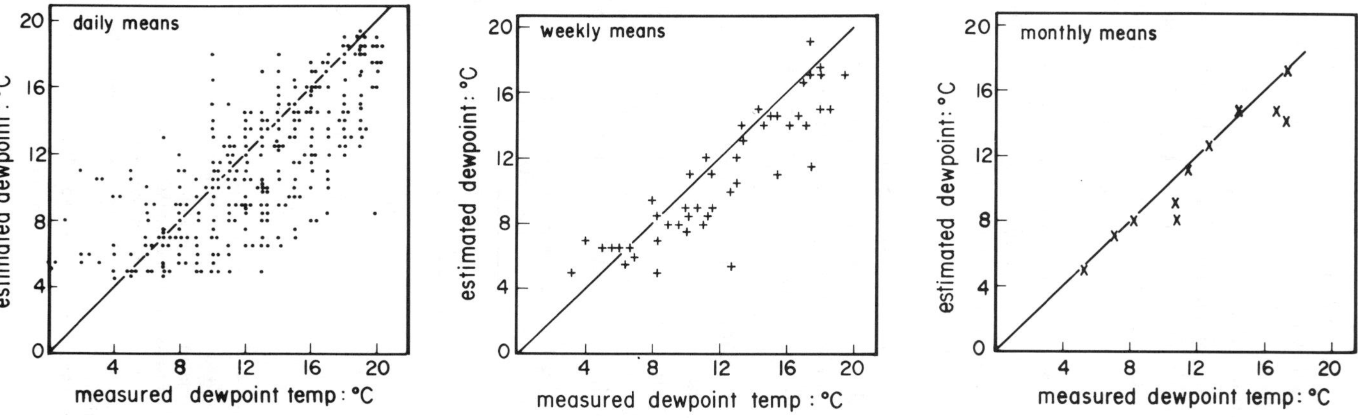

Figure 3.2 Effect of the period of averaging on the errors of estimating the surface dewpoint at Marsfield, using measurements there of the daily maximum and minimum temperatures inserted into Figure 3.10. The data are 9 a.m. values in 1978. The error is indicated in each case by the vertical distance from the diagonal. The average error is 2.7C° for daily mean values, 1.8C° for weekly means and 1.0C° for monthly means

advantageous in cancelling out any short-term variations, including many random fluctuations. This is shown by the reduced scatter in Figure 3.2 for monthly means, compared with values for shorter periods. Also, the amount of calculating involved is much less than in finding the sum or average of numerous estimated values for a series of short periods.

However, a particular disadvantage of a long period of averaging arises when there are two or more elements on the right-hand side of a formula and they are not independent of each other. Take the case of the Dalton equation for estimating the evaporation rate, mentioned in Note 2, p. 317. The wind speeds and humidity values near the ground are often related to each other, as the usual reduction of speed at night is accompanied by a reduction of the 'saturation deficit'. (The latter is the difference between the water-vapour pressure of the atmosphere and the pressure at saturation at the prevailing air temperature.) The association of wind and deficit creates an error in using 24-hour mean values to calculate the daily evaporation. It can be seen in Table 3.1 that the evaporation calculated from daily mean values of the wind speed and saturation deficit is 4.2 mm, which differs substantially from the more accurate sum of estimates for a 12-hour day and a 12-hour night, i.e. 5.8 mm. The reason for this is that 'the product of averages is not the same as the average of products', as explained in Note 3, p. 318. Using short-term data and subsequent averaging gives more accuracy, at the cost of more work.

The effect of altering the period of averaging the primary data was studied experimentally by Jobson (1972). He compared measured evaporation from Lake Hefner in the USA with estimates based on averaged

Table 3.1 Effect of the averaging time on the estimation of the daily evaporation (Eo) by means of a simple version of the Dalton equation[a]

Period	*Given conditions*			*Derived values*		*Evaporation rate*	
	u *m/s*	*T* *°C*	*e* *hPa*	*es* *hPa*	*(es − e)* *hPa*	*0.2u(us − e)* *mm/d*	*mean*
Day[b]	4	20	10	23.4[d]	13.4	10.7	5.8
Night[b]	2	10	10	12.3[e]	2.3	0.9	
24 h[c]	3	15	10	17[f]	7	4.2[g]	4.2

Notes: [a] Eo = 0.2 u (es − e) mm/d, where es is the air's saturation water-vapour pressure, e the ambient water-vapour pressure, and u the wind speed (Linacre and Hobbs 1977: 37).

[b] In these cases the evaporation rate is calculated separately for uniform day and night conditions, and then averaged to obtain the 24-hour mean.

[c] In this case, the 24-hour evaporation rate is calculated from 24-hour averages of wind, temperature and humidity.

[d] i.e. the saturation water-vapour pressure (svp) at 20°C, from Notes 6 or 7 in Chapter 2.

[e] The svp at 10°C.

[f] The svp at 15°C,

[g] i.e. 0.2 × 3 × 7.

climate data. The estimates erred by more than five per cent on 3 per cent of occasions when three-hour averages were used, but on 20 per cent when using daily averages. Errors become unacceptably large if the averaging period is a month.

Other work confirms the greater accuracy of short-term averaging. Errors in calculating daily evaporation in June in Maryland were found to be less than 3 per cent when the input data were averaged over six-hour periods. On the other hand, use of monthly averages of conditions at Edmonton in Canada in winter creates an error up to 25 per cent in calculating the monthly evaporation.

Another error arises if an independent variable affects the estimate other than proportionally. For example, Eqn (3.2) gives an overall mean of 3.3 mm/d for a daytime at 20°C (when the rate would be 5.2 mm/d) and a night at 10°C (1.3 mm/d), whereas the equation would give 2.9 mm/d for a 24-hour period with a mean temperature of 15°C. The latter differs by over 10 per cent from the more accurate 3.3 mm/d.

Overall accuracy of estimation

Any empirical formula should be tested against measurements which have not themselves been used in deriving the formula. Otherwise the test is simply one of fit between the equation and the basic measurements, not a test of the formula's accuracy. The need to obtain a fresh set of measurements may be avoided by the following procedure. One takes a complete set, removes one value and then uses the estimation procedure to fill the gap that was created. The difference between the removed and estimated values is an indication of accuracy. This is repeated for each value in the set in turn to derive the average difference. The procedure can be used to compare the relative merit of alternative estimation methods in terms of the average difference for each method.

It is not possible to determine an error of estimation less than the error of the check measurements. Furthermore, it is not worthwhile trying to estimate a climatic element much more accurately than other elements with which it is to be associated. For example, consider estimating the actual evaporation rate (Ea) from a rainfall catchment in order to calculate the catchment's mean rate of run-off (R). The latter is the difference between Ea and the mean precipitation rate (P). If the possible error in determining P is p millimetres/day and the error in estimating Ea is e mm/d, the error (r) in calculating run-off is given by $(p^2 + e^2)^{0.5}$ mm/d. Let us assume that p and e are both 1 mm/day. In that case, r is 1.4 mm/d. But if the error in estimating Ea were somehow reduced to 0.5 mm/d, r would be 1.1 mm/d, i.e. halving the evaporation error leads to a reduction of run-off error by only 21 per cent. A further halving of the error (e) would cause only another 7 per cent reduction of the run-off error. In other words, an accuracy for estimating Ea much better than that for measuring P (which Chapter 2 showed is rarely measured well) hardly improves the determination of run-off.

Another point in considering what is reasonable for the estimation accuracy is the degree of variation of the estimated element. For example, if the daily mean temperature fluctuates by as much as 10C° during a month it would be absurd to try to estimate the monthly average within 0.01C°.

ESTIMATING SCREEN TEMPERATURE

The screen temperature is a basic climatic element and sometimes has to be estimated. We will consider the following aspects:

1 The long-term mean temperature.
2 The monthly mean temperature, especially for January and July, the difference between which is called the 'annual range', assuming that they are the hottest and coldest months.
3 The daily range.
4 The hourly temperature.

These are variously affected by five geographic factors – latitude, elevation, distance downwind from the sea, the offshore sea-surface temperature, and the landform. In the case of short-term conditions there is also considerable carry-over from the previous state, due to thermal inertia, i.e. the storage of heat in the ground and lakes.

Long-term mean temperature

Effect of latitude

Tobias Mayer pointed out in the 18th century that the long-term mean sea-level temperature Tc at latitude A differs from the equatorial temperature by an amount approximately proportional to the square of the sine of the latitude angle. A later empirical formula is the following:

$$Tc = 44.9\,[\cos(A - 6.5)]^2 - 17.8 \qquad {}^\circ C \qquad (3.4)$$

The right-hand side of Eqn (3.4) is approximated by $[27 - 0.008\ A^2]$. The dependence on a *squared* term can be explained on the assumption that the temperature depends primarily on the solar irradiance of the ground, rather than on advection. The irradiance of the ground is the product of *two* factors, each depending on the angle of latitude: (i) the irradiance of the top of the atmosphere, and (ii) the attenuation of the solar beam as it traverses the air.

Eqn (3.4) implies symmetry about the equator, which is not exactly true, because of more land in the northern hemisphere, bringing the 'thermal equator' to 5–10 degrees north. However, the equation is in accord with the gradient across Australia of about 0.6C° per degree of latitude. Errors of Eqn (3.4) at some places are due to peculiarities of the ocean-surface temperature upwind, the distance inland and the landform locally.

Effect of elevation

The decrease of temperature with increased elevation is known as the 'lapse rate'. Hann (1897: 245) reported cooling by 4.5–7.5 C°/km at 29 different places, i.e. about 6 C°/km. Similarly, 27 more recent studies of temperatures on mountains throughout the world show a median value of 5.7 C°/km. All but three values are between 5.1–6.5 C°/km. So we have the following approximate relationship between the measured screen temperature (T) at height h (metres) and the equivalent at sea level, Tc:

$$T = Tc - 0.006\ h \qquad ^{\circ}C \qquad (3.5)$$

The value of 6 C°/km in Eqn (3.5) is only approximate, for the following reasons:

1 The lapse rate increases with *elevation* in some places but decreases in others. It is 4.2 C°/km below 1400 m in the European Alps and 6.1 C°/km at higher elevations, and it is 4.2 C°/km below 600 m in New Zealand, and 7.3 C°/km higher up. However, there is a reduction of the lapse rate at the tops of high mountains, as in the free air. The rate is 5.5–7.5 C°/km in the concave terrain of a valley bottom in the Polish Carpathians, but 4.3–4.8 C°/km on the convex landform of the mountain tops. The rate is 6.0 C°/km below 1500 m in Java, but 5.5 C°/km above that, and in Norway the mean lapse rate is 7.5 C°/km at sea level but 5.5 C°/km at 2 km.

2 Lapse rates vary with *latitude*, being about 4.9 C°/km over much of the Earth (up to 3600 m) but 6.6 C°/km at latitudes between the Tropics. Likewise, rates in the south of England are steeper than in the north of Scotland, though this may be due to more sheltered climate stations in the north.

3 Slope *orientation* affects the lapse rate. The rate has been found to be most on mountainsides facing north, away from the sun, in Poland and Mongolia. In India and Sri Lanka the rate is 5.5 C°/km on the wet west sides of hills, but 7.6 C°/km on the dry east where the skies are less cloudy.

4 There is a *seasonal effect*, as the lapse rate is usually less in winter than in summer. For example, in Colorado the rate is 1.8 C°/km in January but 6.3 C°/km in June. Similarly, the gradient in Hainan Island in China is 6.7 C°/km in summer but 2.1 C°/km in winter.

5 The lapse rate varies *diurnally* when the sky is clear, especially in low-latitude highlands, where the steepest lapse rates occur at night (Note 4, pp. 318–19). The lapse rate of long-term average *maximum* temperatures at different elevations in Colorado is 6.7 C°/km, like the lapse rates mentioned above. On the other hand, the lapse rate of daily *minimum* temperatures in Ecuador, for instance, is about 4.6C°/km, which is less than the rate for mean temperatures.

Lapse rates up mountains are comparable with rates in the free air above flat ground nearby. The value taken as standard for the free-air lapse rate by the International Civil Aviation Organization is 6.7 C°/km in the first 2 km above sea level and 6.3 C°/km between 4–6 km. These figures roughly match Eqn (3.5) for mountainside temperatures. Also, the levelness of cloud-base across the sky, even near mountains, implies approximate similarity of free-air and mountainside lapse rates. Temperatures at 9 a.m. at 870 m on a mountain in Japan are about 1.0C° colder in winter than at the same height over a town 20 km away, and 0.5C° warmer in summer.

More precisely, the comparison is affected by the time of day. For instance, mountain temperatures in the European Alps are commonly 2–3C° colder than the free air at the same level, except on sunny afternoons. Likewise, radiosonde measurements above a valley in Austria are 1.5C° warmer than at the same level on the adjacent mountainside at 7 a.m., but only 0.8C° warmer at 7 p.m. Finally, summertime data from eight weather stations in California show values 2.0–4.8C° cooler than at the same height in the free air nearby at 4 a.m., 1.9–5.1C° warmer at 10 a.m. and 0.5–2.9C° warmer at 4 p.m. As a result, the mountainside lapse rate was about the same as in the free air, for the day as a whole.

Effect of the ocean upwind

Long-term mean air temperatures are usually affected by the temperature of the ocean surface upwind. The notable exception is central Asia in summer and northern Asia in winter (when the polar seas are covered by ice), because of the distance from the ocean. Generally, the effect depends on the coastal sea-surface temperatures, the pattern of the global winds and the consequent distance that the incident wind has travelled overland.

Global surface temperatures in January and July have been depicted by several authors (e.g. Lamb 1972: 149). The patterns depend on the ocean's surface currents, which are similar in the two seasons but displaced by the sun's latitudinal movement. Temperatures along the west coasts of the Americas and southern Africa are several degrees below the latitudinal norms as a result of ocean currents from the poles and upwelling near the coast. For instance, the temperature off Peru is about 8C° cooler than the latitude average, so that chilled west winds blow on to the land. In

a)

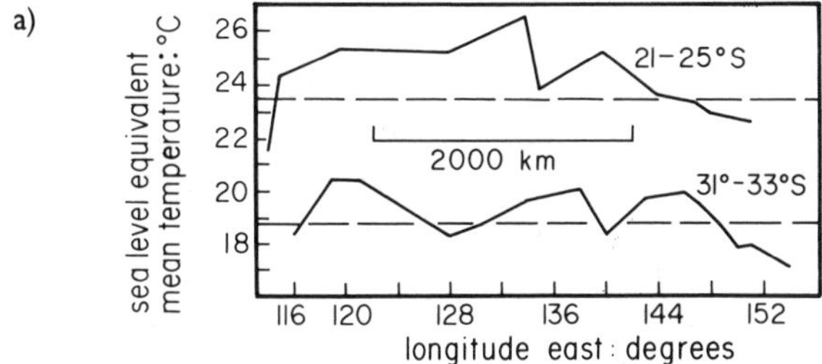

Figure 3.3a The variation of the sea-level-equivalent annual mean temperature, calculated from Eqn (3.5), at about 23°S and about 32°S across Australia. The horizontal lines show the latitudinal mean calculated from Eqn (3.4)

b)

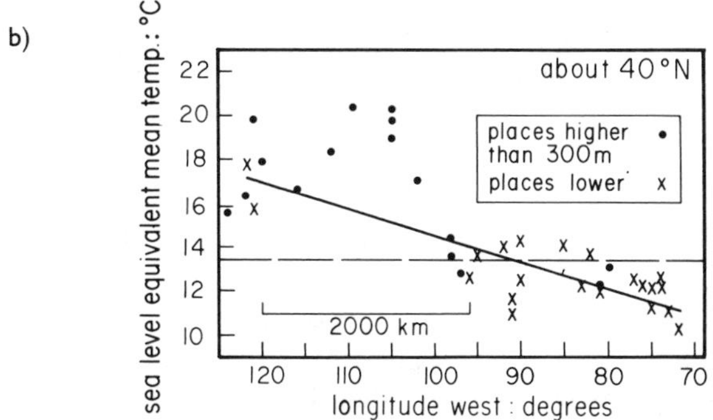

Figure 3.3b As for (a), but at about 40°N across the USA. The circles relate to high places, and their relatively excessive Tc values suggest that the lapse rate for those places differs from the 6 C°/km assumed in Eqn (3.5)

c)

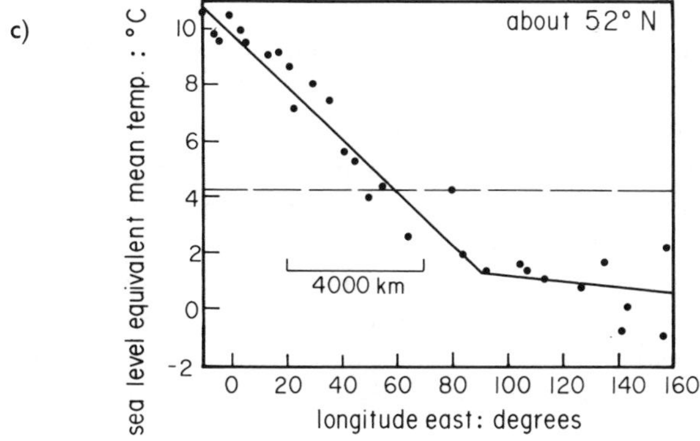

Figure 3.3c As for (a), but at about 52°N across Eurasia

summer, the deviation of sea-surface temperatures from the latitudinal average is about −8C° off the coasts of California and Japan and −4C° off south-west Africa. But in winter it is over +20C° west of Norway, because of the greater stability of sea-surface temperatures compared with land temperatures and the warmth in the ocean current from the Mexican gulf.

The day-to-day variations of gradient winds average out during a month, resulting in patterns (Linacre and Hobbs 1977: 112) from which it is usually possible to estimate the overland distance upwind to the coast, the 'fetch'. On-shore winds from an extensive ocean are initially at about the coastal sea-surface temperature, and then the air's temperature approaches and overshoots the latitudinal norm during subsequent travel overland. Thus, the annual mean temperature rises with distance inland, if the initial coastal temperature is below the latitudinal average, as shown in the upper curve in Figure 3.3. However, the case of westerly winds from the north Atlantic is different, because that ocean is warmed by the Gulf Stream and so the winds *cool* as they blow across Europe and then Asia. There may be cooling in the USA also. More details are given in Note 5, p. 319.

Effect of cloudiness on the long-term mean temperature

Measurements in Ecuador show that places with a rainfall of less than 600 mm/a are about 1°C warmer than wetter places. Presumably the extra warmth is due to more solar irradiance, caused by less cloud. Variations of cloudiness may also be partly responsible for the apparent effects of latitude, elevation, season, hour of the day, and the ocean upwind, discussed already.

Annual range of temperature

The annual range is the difference between the mean temperatures of the hottest and coldest months, usually January and July. It increases with latitude, as shown in Figure 3.4, but seems largely unaffected by elevation. For example, annual ranges in a part of New Zealand are 13.3C° at 800 m, 11.7C° at 1380 m and 12.8C° at 1830 m.

The annual range is considerably influenced by the fetch of the wind over the land. That is why the range at Irkutsk (inland in the USSR at 52°N) is 40C°, which is much more than the 8C° at Valentia (Ireland) at about the same latitude but on the Atlantic coast. The thermal stability of an ocean moderates temperature fluctuations nearby, as a result of the great specific heat of water, cooling due to more evaporation with increased irradiance, the spread of heat to greater depth, and the mixing induced by ocean currents.

Data from 25 places at about 40°N in various continents show annual ranges of 19C° at the coast, 16C° at 20 km inland, 28C° at 60 km, 26C° at

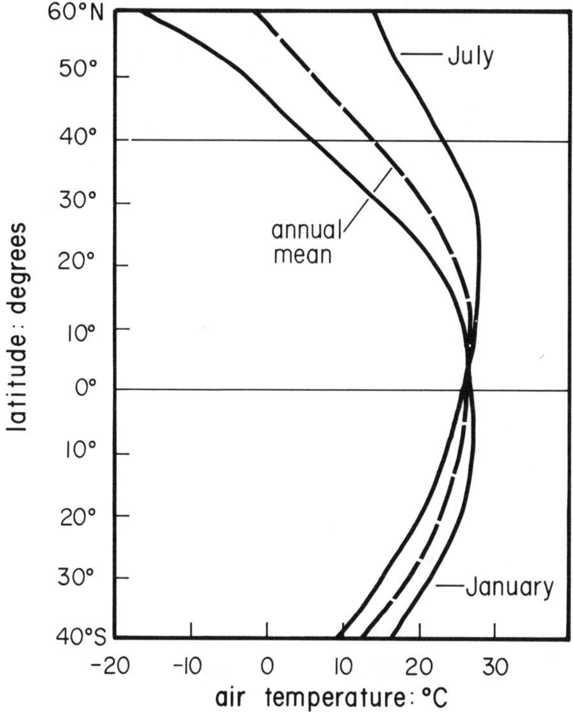

Figure 3.4 Monthly mean temperatures in January and July, showing the effect of latitude on the annual range of temperatures (Mihara and Ando 1974: 3). The smaller ranges in the southern hemisphere are due to there being more ocean than in the north. (By permission of the University of Tokyo Press)

150 km, 24C° at 230 km, 29C° at 950 km, and 26C° at 1500 km. So the sea's moderating influence is felt mostly within 50 km of the water, which is comparable with the extent of the sea-breeze (see Chapter 6).

An empirical expression for the range, derived from data from 67 places around the world, is as follows (Linacre 1969a: 16):

$$Ra = 0.4 + 0.35A + 0.0033d \qquad C^\circ \tag{3.6}$$

where A is the latitude, and d is the distance inland. However, this is only an approximation, since Ra varies non-linearly with d. This is shown in Figure 3.5, which indicates a relationship of the following form:

$$Ra = k.d^n \qquad C^\circ \tag{3.7}$$

where the exponent n in that case is about 0.16. Data from northern Victoria and Queensland in Australia, and from the northern USA and Eurasia give values of 0.15–0.44 for n, with a median of 0.25. Such a consideration of the effect of distance inland on the annual range omits the usual but undesirable concept of 'continentality', dealt with in Note 6, p. 319.

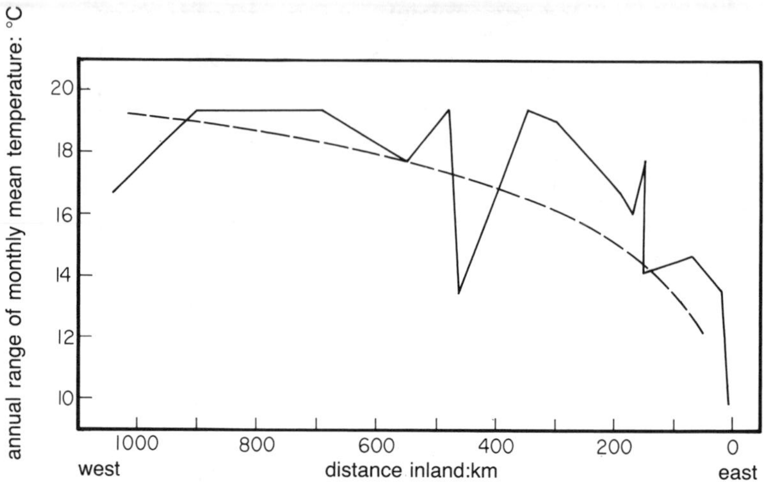

Figure 3.5 Variation of the annual range of temperature in New South Wales, westwards from the east coast

Monthly mean temperatures

A knowledge of the long-term mean temperature and the annual range Ra makes it possible to estimate approximately the long-term mean temperature for any month of the year in temperate climates. The procedure involves drawing the kind of irregular irradiance-temperature loop shown in Figure 3.6, where the parallelogram is an estimated equivalent of the loop of measurements. The parallelogram is derived from the annual mean temperature and Ra, and from the observation that the vertical side equals 0.4 times Ra as a consequence of the temperature's lag of about a month on changes of irradiance (see Note 7, p. 320). It is also necessary to use empirical expressions for the extremes of global irradiance in the hottest and coldest months, i.e. $[207 + 6.45A - 0.108A^2 + 0.014h]$ and $[165 + 0.013A - 0.059A^2 + 0.014h]$ W/m^2, respectively, where A is the latitude (degrees) and h the elevation (metres). The highest and lowest points of the parallelogram represent conditions in January and July, according to which hemisphere the place is in. Mean values for other months can be derived by marking off equally spaced points, as in Figure 3.6 In this particular case, the average error of the estimates of monthly mean temperature is 2.3C°.

The procedure assumes that temperature depends primarily on irradiance, i.e. the effect of advection is ignored. So it is inapplicable near the equator, where there is a seasonal reversal of the prevailing wind. Also, the sun makes two transits overhead each year at low latitudes, which complicates the loop. In addition, the method is inapplicable above 60 degrees latitude,

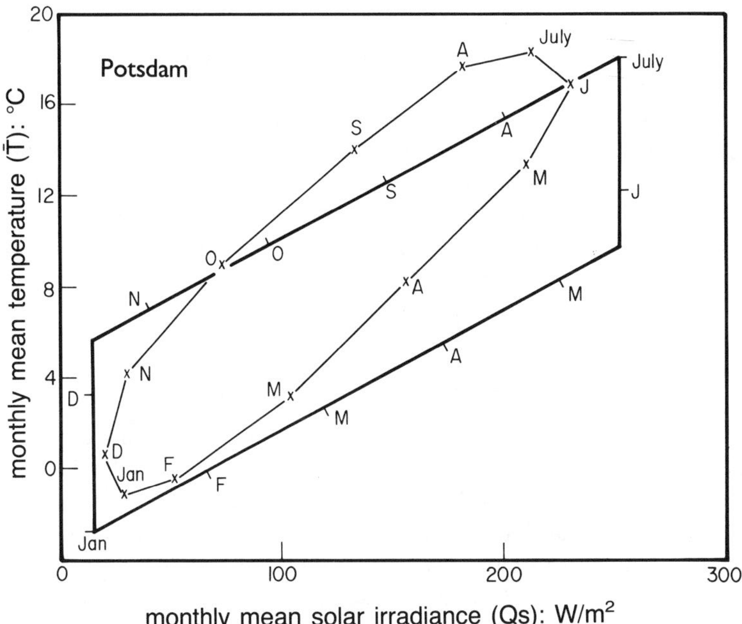

Figure 3.6 Climagram for Potsdam in Berlin (at 52°N, 81 m elevation and about 700 km inland). The loop shows actual measurements from (Wallen 1977, 63). Estimation of the parallelogram is described in Note 7, p. 320

where the high reflectivity of snow alters the effect of irradiance on temperature.

Empirical expressions for the monthly mean temperature can be derived statistically for any region where there are long-term records from numerous climate stations, in terms of latitude, longitude, altitude and perhaps distance from the coast. The variation of monthly mean temperature with latitude changes during the year, the dependence being greater in winter than in summer. For that reason, a different statistical relationship applies each month. Examples of such a procedure involve data from 206 places in Canada, 232 stations in the north-eastern USA, 62 stations in Greece, 22 places in south-east New South Wales, 21 in West Virginia, and 12 stations in Wyoming, respectively. The derived expressions can be used only in the nearby regions.

Another method for estimating monthly mean temperature is occasionally possible in arid or semi-arid areas, based on a correlation with monthly rainfall. Temperatures are lowered by increased cloud, since cloud reduces solar irradiance of the ground, and leads to rain and hence evaporative cooling. An example is the connection between January temperatures and rainfalls at Alice Springs in Australia, where temperatures in January are about 39.5°C when there is 2.5 mm of rain, 35°C when 25 mm, and 30°C when 250 mm. The correlation is useful because there are many more rain gauges than places where screen temperature is measured.

Sometimes it is necessary to infer monthly mean temperature when observations are missing for some days of the month. A study in New Zealand showed that if one has only 15 days' values of the daily maximum gathered at random during a month, their average is within 1.2C° of the true monthly mean daily maximum in 90 per cent of months, or within 2.5C° in 99 per cent. For 25 days' readings each month, these figures become 0.6C° and 1.1C°, respectively.

Daily range of temperature

The difference between daily maximum and minimum temperatures depends on the elevation (h metres), the amount of cloud (expressed as either *C* oktas (eighths) or C tenths of the sky's area), the dewpoint temperature (Td) and the screen temperature (T). An analysis of 357 monthly mean values from the World Survey of Climatology (published by Elsevier) yields the following empirical equation for the annual mean daily range (Rd) over a wide range of places:

$$\mathrm{Rd} = 11.2 + 0.00106\,\mathrm{h} - 0.634\,C + 0.315\,\mathrm{T} - 0.355\,\mathrm{Td} \qquad \mathrm{C}^\circ \quad (3.8)$$

Eqn (3.8) shows that elevation and temperature increase the range, whilst cloud and humidity decrease it. However, such regression analysis assumes linear relationships, which is a gross simplification. Nigerian data, for example, imply a proportionality between Rd and the natural logarithm of (T − Td). Further details follow.

Effect of elevation

Eqn (3.8) implies a general increase of daily range with elevation, as measured in many places. However, a smaller range occurs at the tops of mountains in Tennessee, Appalachia and elsewhere, and whilst the range in New South Wales (in Australia) is most at about 400 m elevation. In fact, the daily range varies with elevation in a complex way, as shown in Figure 3.7.

Various processes dominate at different elevations. Sea-breezes reduce the daily maximum at low levels near the coast, and the greater cloudiness at moderate height on mountains also reduces the range. In addition, the range tends to be compressed by strong winds at the tops of mountains. However, it tends overall to increase with distance from the sea (i.e. with elevation), and there is less atmospheric attenuation of radiation fluxes at the highest levels so that maximum and minimum temperatures are both more extreme on high mountains.

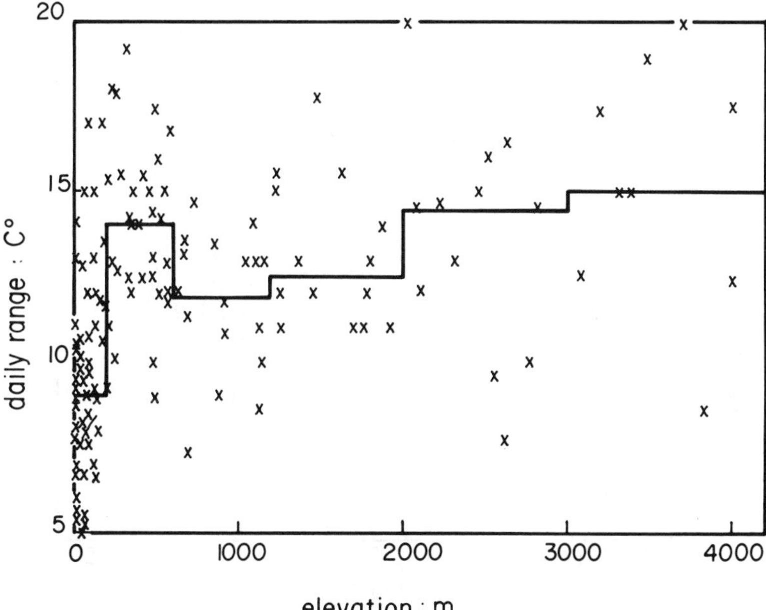

Figure 3.7 The effect of elevation on the daily range (Linacre 1982: 379). The horizontal lines show the median values in ranges of height chosen on the basis of the processes involved. (By permission of the Royal Meteorological Society)

There are inversions in valleys on clear nights, i.e. low daily minima, which again increase the daily range. A valley at 3658 m in China has a range of 13.7C° and one at 778 m has 13.0C°, while a summit between, at 3047 m, has a range of only 6.3C°.

Other effects

1 The *ocean* hardly affects the daily range at places well inland. Figure 3.8 implies that the effect reaches 400 km at most. Across the USA, the July mean daily minimum temperature varies only within 100 km of the ocean. The weaker sea-breezes induced by the smaller areas of the Black Sea and the Baltic Sea reduce their effect on the daily range to within 10 km of the coast.

A transect across Africa at 30°S shows a bigger oceanic effect on the annual range than on the daily range. The annual range is about 5C° at the west coast, 10C° at 100 km inland, and about 15C° beyond 400 km, i.e. an increase by 10C°. On the other hand, the daily range varies from about 10C° at the coast to 13C° at 100 km and 16C° beyond 200 km, which is only 6C° more than at the coast.

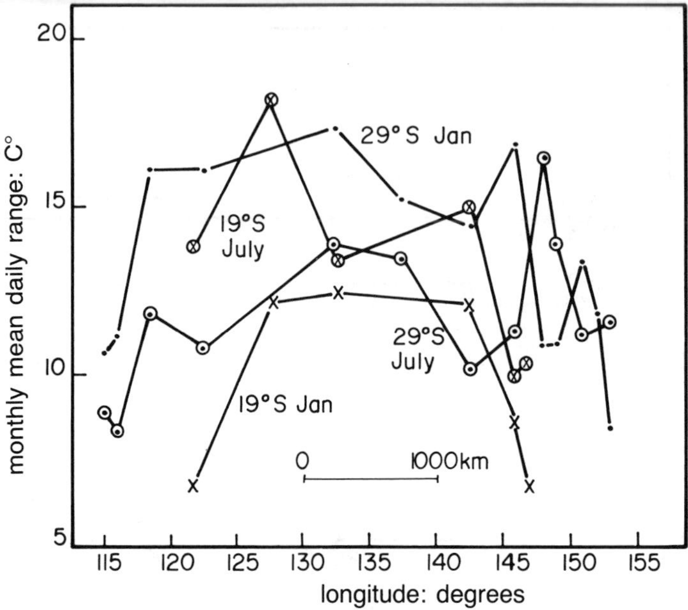

Figure 3.8 Daily ranges of temperature at places across Australia at about 19°S–29°S in January and July

2 At the lowest *latitudes*, where cloud prevails, the daily range is generally low, though greater than the annual range. It tends to be less than the annual range at latitudes above 25 degrees. In dry places near the Tropics, the range can exceed 18C°.

3 Increased cloudiness makes the daily range less in some *seasons* than in others, which explains the reduced range in summer at 19°S in Australia, and in winter at 29°S (see Figure 3.8). The range in New South Wales is about 14C° in January (summer) and 12C° in July.

4 The *weather* affects the daily range, because of alterations of cloudiness and windiness. The greatest ranges occur during the calm, clear days accompanying the passage of anticyclones.

5 The occurrence of *rain* (or of the cloud required for rain) alters the daily extremes. Nocturnal rain raises the minimum temperature, whereas daytime rain reduces the maximum. Thirty years of data from 70 places in the USA show that 1 mm or more of rain in summer reduces the maximum by about 3C° and increases the minimum by 1–2C°, so the daily mean is slightly reduced in summer. However, rain in winter lowers the maximum by less than 2C°, whilst increasing the minimum by about 5C°, so that the mean is slightly raised. Also note that rain wets the ground and therefore increases its heat capacity, which helps stabilise

surface air temperatures.

6 The *terrain* affects the daily range by altering the surface wind speed. A place with stronger winds usually has a reduced range. Also, a valley collects cold air at night, lowering the daily minimum and thereby increasing the daily range. The range in a frost hollow in south-east England can be as much as 27C°.

7 A transect at 40°N across the USA shows there is an effect of *mountains* on the daily range, but not on the annual range. Both are about 5C° on the west coast and increase to more than 13C° within 300 km. Then the daily range decreases on the east side of the mountains, being about 10C° from halfway across the continent as far as the east coast. On the other hand, the annual range continues to rise from 18C° at 300 km inland to 26C° at 800 km, remaining so high right across to the east coast, 3000 km further downwind.

Hourly temperature

Figures for the monthly mean temperature and the daily range can be used to estimate the temperature at any time of the day, assuming that one knows the date and latitude, which determine the daylength, given in Note 8, p. 320. The diurnal variation is typically a sinusoidal increase of temperature from dawn to about 2–3 p.m., followed by an approximately linear fall until the next sunrise (Figure 3.9). The accuracy of estimating hourly temperatures, assuming any of five other patterns of daily variation instead, proves much the same whichever pattern is chosen, so long as the sky is clear. None is accurate when there is irregular cloud.

The time of maximum temperature depends on location. It occurs at about 1 p.m. at the top of mountains in China, but an hour or two later in the valleys.

The daily pattern is altered also by the sea-breeze near the coast in temperate latitudes. This is discussed in Chapter 6. The time of arrival of the breeze depends on the distance inland and is followed by an abrupt cooling, so that the normal daily maximum is not reached. The daily range is consequently reduced if the sea-breeze arrives before 2 p.m. or thereabouts, i.e. if the place is near enough to the coast.

The daily maximum temperature

Spatial variations of the daily maximum temperature make it unsafe to assume that the same conditions obtain beyond a few kilometres away. Observations at Sydney show a correlation coefficient of only 0.75 between daily readings at places 12 km apart, and 0.64 across 25 km. However, measurements near Canberra show that maximum temperatures on any particular day are closely related to the elevation, the lapse rate being

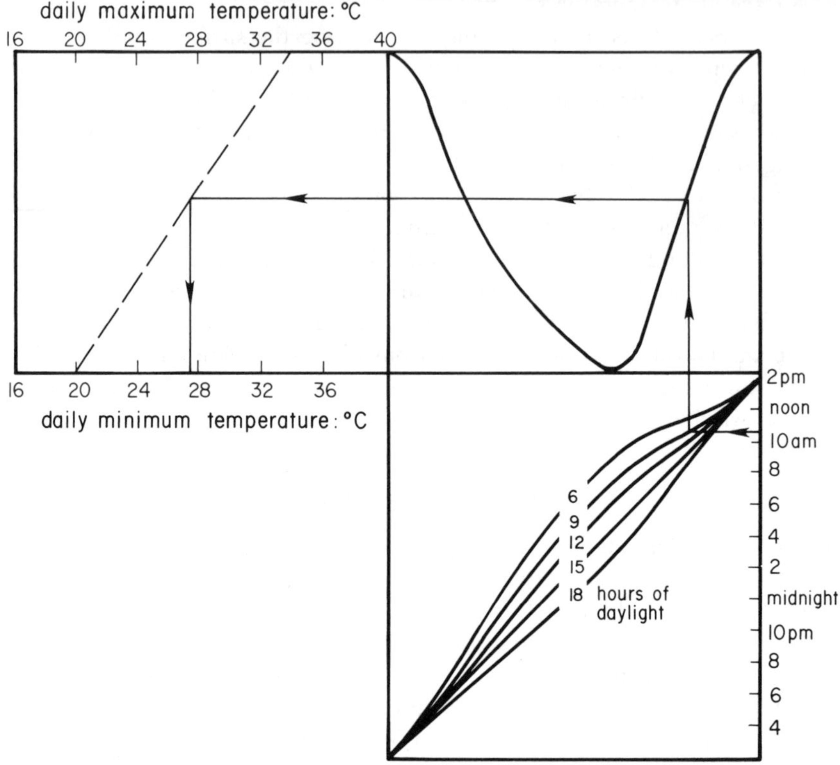

Figure 3.9 Nomogram for estimating the temperature at any time of day from the daily extreme values, assuming no change of cloudiness or wind direction. It was developed from the work of Walter (1967: 141) and Evans (1980: 34). The arrowed lines show the estimation of the 10.30 a.m. temperature, given daily extremes of 20°C and 34°C and a daylength of 9 hours: the answer is 27.4°C. The daily minimum temperature is taken as occurring at sunrise and the daily maximum at 2.30 p.m. (Linacre and Hobbs 1977: 28). Note that the 'time of day' is solar time (see Chapter 5), not summertime daylight-saving time, for instance

about 9.8C°/km. This equals the 'dry adiabatic lapse rate', the cooling due to air being lifted without cloud formation, because the atmosphere at the time of maximum temperature is highly convective and has a low relative humidity.

Another relationship may be useful in estimating the monthly mean daily maximum. It has been observed at Marsfield that the temperature can be linearly related to the number of raindays in the month.

The daily maximum temperature of the top few millimetres of exposed soil or of the leaves of short grass (Tgx) is not what is measured in a Stevenson screen (Tmax). The surface maximum Tgx has been found equal to [2 Tmax − 12]°C, provided it is above 11°C, in places ranging from Finland to Morocco. Actually, Tgx depends also on the amount of irradiance (i.e. cloud) and wind, rather than Tmax alone, but the empirical expression serves to indicate the considerable difference possible between the two maxima, e.g. 13C° if Tmax is 25°C.

The daytime wind's stirring and thermal convection reduce the difference between ground and screen maxima. The difference at night-time is opposite, and is greater, because of neither stirring nor convection. As a result, the daily *mean* temperature at the surface tends to be slightly lower than the average at screen height.

ESTIMATING DEWPOINT TEMPERATURE

Dewpoint temperature (Td) is an index of the atmosphere's water content (see 'Measurement of humidity', pp. 39–44). A first approximation can be obtained by equating it to the daily minimum temperature (Tmin), since nocturnal cooling is often arrested at the dewpoint owing to the release of latent heat when condensation occurs. However, this applies only in humid climates. There is no dew in a dry climate, so that Td there may be substantially less than Tmin, e.g. 8C° less in arid parts of India.

The dryness of the climate is indicated by the daily range, which is the difference between Tmin and the daily maximum temperature (Tmax); the range is larger in a climate with a low dewpoint. The relationship has been derived empirically by examining monthly mean data from 37 places in Australasia and 90 places spread over the rest of the world. The outcome is Figure 3.10. A test of the diagram using January and July values from 46 other places around the world gave a mean error of only 0.2C°, though the error's 'standard deviation' (discussed in Chapter 4) was 3.8C°, so Figure 3.10 can be inaccurate by a few degrees, on occasion. It is instructive to consider the causes of such inaccuracies.

Figure 3.2 shows that some errors in using Figure 3.10 depend on averaging the various temperatures over a month, so that processes lasting only a few days are averaged out. Relevant processes include: (a) the daily fluctuation of the dewpoint (see 'Measurement of humidity', pp. 39–44), (b) changing winds, which alter the difference between screen and ground temperatures, (c) varying cloudiness, and (d) wetness of the ground. As regards the last, the ground is a better conductor of heat when it is wet after rain, so then there is less daily variation of surface temperature.

The difference between screen and ground minimum temperatures also affects the accuracy of Figure 3.10. The diagram involves the screen minimum, but the arrest of nocturnal cooling by the formation of dew occurs at the surface of the ground. Average differences at Marsfield are shown in Table 3.2. The ground minimum in summer is about 2.7C° lower than the screen minimum on days when dew forms, and in winter the difference is about 4.2C°. (On days without dew, the respective figures are 1.6C° and 5.5C°). Also see Note 9, p. 320.

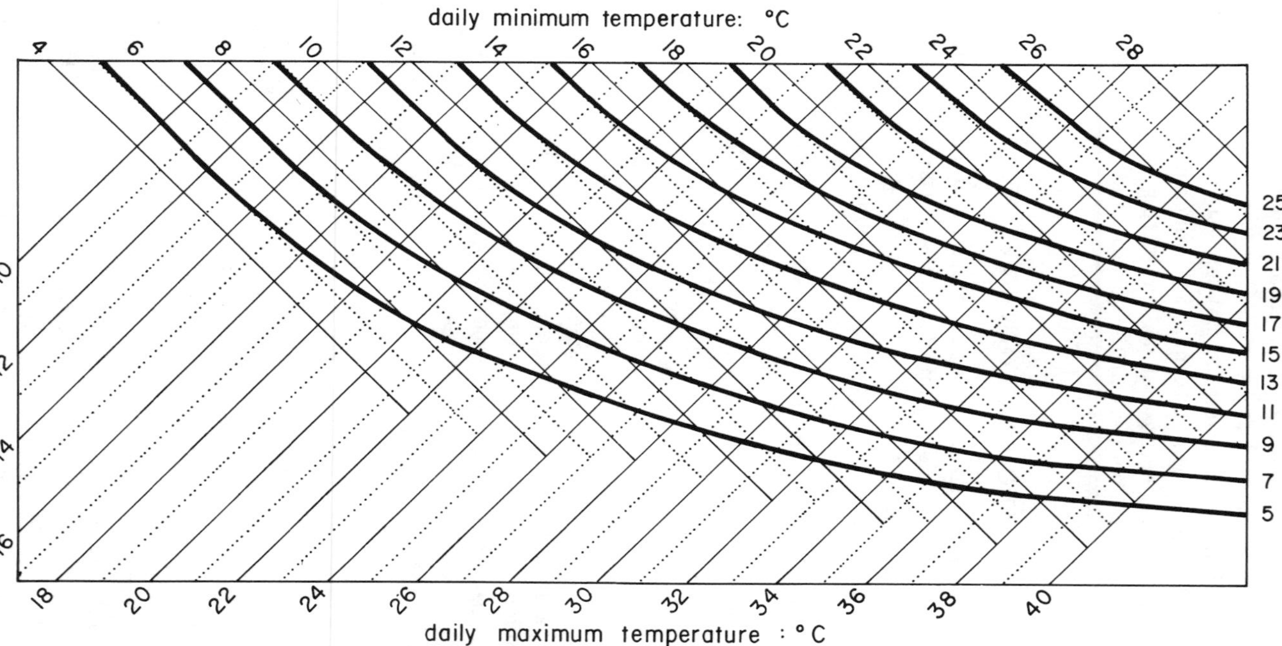

Figure 3.10 Empirical diagram for estimating the dewpoint temperature from daily extreme temperatures

Table 3.2 A comparison of average screen dewpoint temperature at 9 a.m. (Td), screen minimum temperature (Tmin) and grass minimum (Tgn), on rainless days at Marsfield.

Month	*Year*	*No. days*	*Td*	*Tmin*	*Tgn*	*Td − Tmin*	*Tmin − Tgn*
(a) Days *with dew*							
January	1978	8	18.1	16.4	14.0	1.7	2.4
(summer)	1981	6	18.2	15.2	12.3	3.0	2.9
		Means	18.1	15.8	13.1	2.3	2.7
July	1978	15	3.3	1.3	−1.7	2.0	3.0
(winter)	1981	9	5.7	2.8	−2.4	2.9	5.2
		Means	4.5	2.1	−2.1	2.4	4.2
(b) Days *without dew*							
January	1978	5	15.0	18.2	17.5	−3.2	0.7
	1981	5	18.4	20.0	17.5	−1.6	2.5
		Means	16.7	19.1	17.5	−2.4	1.6
July	1978	9	2.8	5.6	1.0	−2.8	4.6
	1981	13	2.4	5.0	−1.4	−2.6	6.4
		Means	2.6	5.3	−0.2	−2.7	5.5

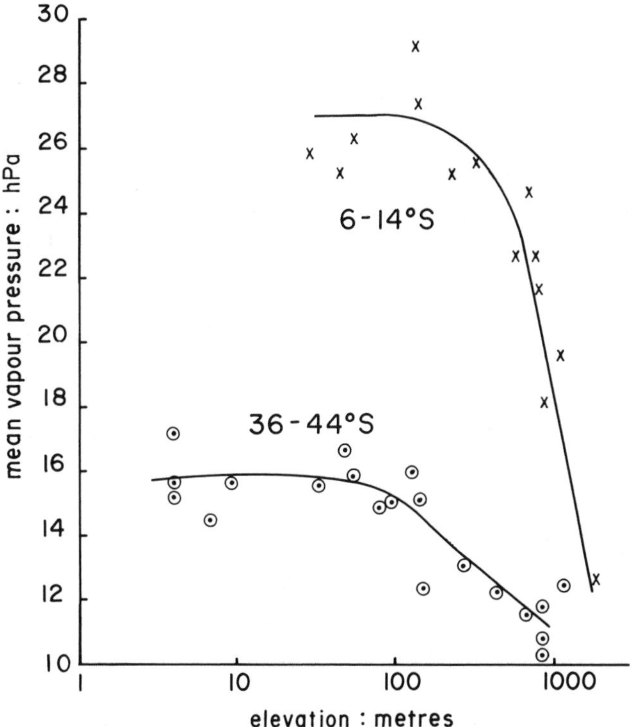

Figure 3.11 Effect of elevation in Australia on the annual mean water-vapour pressure in two bands of latitude

Winds reduce the difference and inhibit dew formation. In addition, winds increase the ventilation of the screen psychrometer and hence the measurement of dewpoint (see 'Measurement of humidity', pp. 39–44). So estimates from Figure 3.10 of the dewpoint at Marsfield are found to be too high when there is a strong wind. The effect of cloud at Marsfield is to make estimates of Td from Figure 3.10 too low.

The dewpoint, and hence the water-vapour pressure (e), are less at high elevations. Figure 3.11 shows the connection between e and elevation in Australia. There is good mixing in the lowest 100 m of the atmosphere, but an exponential decrease of Td at higher elevations, so that e varies as the power 3.7 of the atmospheric pressure p. Other reports show that e varies according to a power of p between 1.5–3.5. Likewise, a relationship between water-vapour density and elevation in central Asian mountains implies a power of about 4. From these values, we may take a power of 3 as typical. This would mean, for example, that a decrease of air pressure by 10 per cent due to a rise from sea level to 1000 metres is accompanied by a reduction of water-vapour pressure by about 27 per cent. That is confirmed by data from the western USA, showing that e falls exponentially with an increase of elevation, e.g. it tends to be 31 per cent less at a place 1000 metres higher.

Other methods of estimating dewpoint temperature

An improvement on Figure 3.10 might be achieved by allowing for the effects of altitude (h metres), and distance inland (or its proxy, the annual range of temperature Ra). So an alternative method of estimating the monthly-mean dewpoint involves the following equation, derived statistically:

$$Td = 10.9 + 0.63\,T - 0.53\,Rd - 0.35\,Ra - 0.0023\,h \qquad °C \qquad (3.9)$$

where T is the monthly mean temperature and Rd the average daily range. However, Eqn (3.9) assumes linear relationships, whereas Figure 3.10 has curved lines. So the equation can be in error by 2–3C°.

A similar equation was obtained by Baier and Russelo from Canadian data (Williams 1985: 14):

$$Td = 0.52\,Tmin + 0.60\,Tmax - 0.009\,(Tmax)^2 - 2.0 \qquad °C \quad (3.10)$$

where Tmin is the daily minimum temperature and Tmax the maximum. The quadratic term makes some allowance for the non-linearity of the dependence of dewpoint on daily maximum temperature. A quadratic term also proves necessary for data from Peshawar in Pakistan, but not in southern England where temperatures vary less widely. This illustrates again that such expressions are peculiar to particular places.

Differences between estimates indicate the inaccuracies involved. Eqn (3.10) gives a dewpoint value of 18.3°C for daily extremes of 30°C and

20°C, for instance, whereas Figure 3.10 gives 18.7°C and Eqn (3.9) gives 17.1°C, if there is an annual range of 12C° near sea level. The span of estimates is 1.6C°.

The sort of statistical analysis which yielded Eqns (3.9) and (3.10) can also be used to represent monthly mean dewpoint temperatures at places over a region, in terms of the month, latitude, longitude, elevation, distance to the sea, and rainfall, as for temperature (see 'Estimating screen temperature', p. 79). A separate equation may be required for each belt of latitudes. However, this requires numerous data from many places and it is tedious.

An empirical and crude equation is derived later, pp. 141–2 from the correlation of afternoon cloudiness (C oktas) with the difference between Td and the daily maximum temperature at Marsfield:

$$Td = Tmax + 0.8\,C - 14 \qquad °C \tag{3.11}$$

This gives a dewpoint of 16°C if the daily maximum temperature is 30°C and there is no cloud. The respective accuracies of Eqns (3.9)–(3.11) are likely to be in descending order, on account of the decreasing amount of information used as input.

Once the dewpoint and dry-bulb temperatures are known, it is a simple matter to calculate the relative humidity (RH), the ratio of saturation water-vapour pressures at the respective temperatures. Eqn (3.12) is an approximation:

$$RH = 100 - 4\,(T - Td) \qquad \text{per cent} \tag{3.12}$$

Thus, dry-bulb and dewpoint temperatures of 30°C and 16°C, respectively, correspond to about 44 per cent. It is actually 42.9 per cent.

ESTIMATING NET IRRADIANCE

Our third case of estimating climate data in the absence of measurements concerns the net irradiance (Qn). This is the overall amount of radiant energy received by the ground from the sun and sky, after subtracting what is lost upwards from the ground. It governs the air temperature and the rate of evaporation from water surfaces, so it is important in agriculture, for instance. However, it is rarely measured on account of the expense and inaccuracy of suitable instruments and the difficulty of defining a standard surface receiving the radiation (see Note 10, p. 322). There is also the problem of maintaining the instrument. So it customary to estimate net irradiance.

Definition

It is defined in terms of the radiant-energy fluxes in the atmosphere shown in Figure 3.12, the nomenclature being explained in Note 11, p. 322.

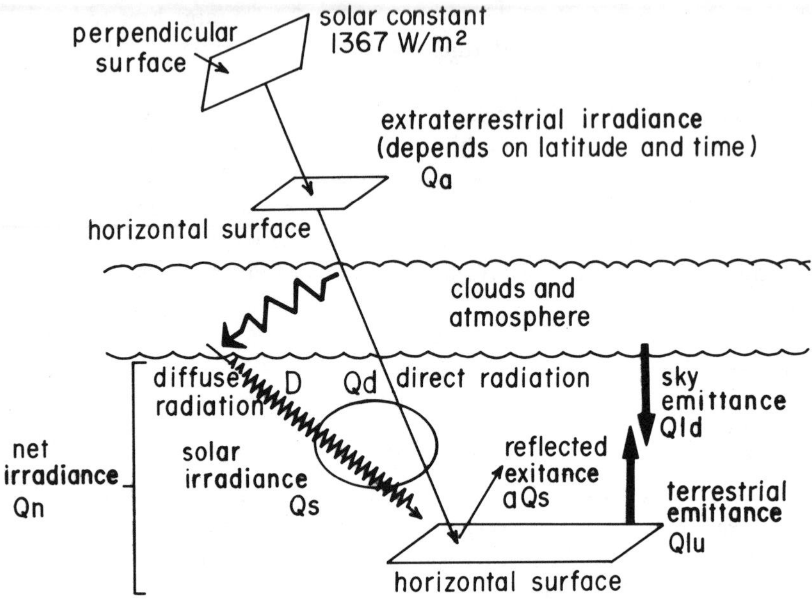

Figure 3.12 Various radiant energy fluxes and their relationships. The 'reflected exitance' is the product of the ground's albedo α, and its incoming irradiance Qs

Basically, there are two sorts of radiation, 'short-wave' (with a wavelength of about 0.2–2.5 micrometres or 'microns') and 'long-wave', with a wavelength of 2.5–25 microns. Short-wave radiation comes initially from the sun, with a mean power per unit area called the 'solar constant' (Io), see 'Nature of solar radiation', pp.151–2. A fraction of this, the 'extra-terrestrial irradiance (Qa), is the power on to unit area of the ground in the absence of an atmosphere. Monthly-mean values of Qa are given in Note 12, p. 323; the highest is 508 W/m^2 in December at 40°S. The variation with latitude is much more in winter than in summer.

As the solar beam traverses the atmosphere, some is scattered by cloud, aerosols and the molecules of the air, creating diffuse radiation. Together, the direct and diffuse short-wave irradiance of the ground form the 'global solar irradiance' of the ground (Qs). Part of this (i.e. α.Qs) is reflected, the fraction 'α' being the 'albedo', from the Latin word for whiteness. The reflected radiation comes upwards from the ground and therefore is called an 'exitance'. It may be appreciable if the surface has a high albedo, as in the case of snow, and some may later be reflected down again by a layer of cloud, augmenting the primary global radiation (see 'Measuring solar irradiance', pp. 174–5).

There are also fluxes of long-wave radiation, up from the ground and down from the lower atmosphere, both ground and atmosphere being at

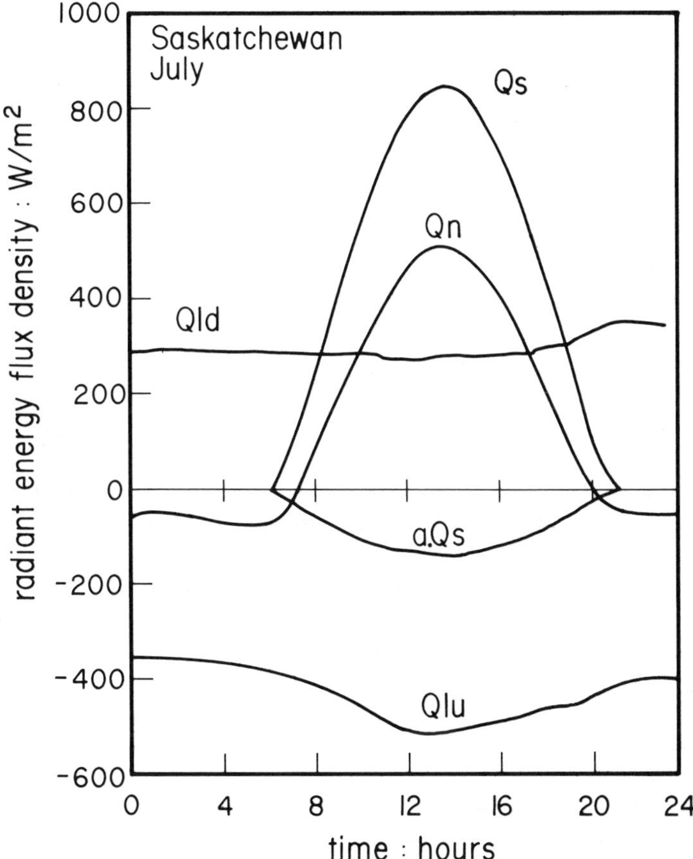

Figure 3.13 Radiation fluxes over a stand of native grass 0.2 m high at Matador (Saskatchewan) on 30 July (Oke 1978: 21, by permission of Methuen & Co.)

Kelvin temperatures in the range 250–320K. (The temperature in Kelvin is given by adding 273.15 to the Celsius temperature.) The downwards 'atmospheric irradiance' (or 'counter radiation') is symbolized by Qld in Figure 3.12, and is somewhat less than the upwards 'terrestrial emittance' (Qlu) from the surface, because the sky is colder than the ground. The daily variations of the two long-wave fluxes and the two short-wave fluxes are shown in Figure 3.13 for a typical case, where the difference between the long-wave fluxes (i.e. the 'net long-wave irradiance') is seen to be fairly constant.

The 'net irradiance' (Qn) is the collective name for the four fluxes discussed so far:

$$Qn = (1 - \alpha)\ Qs + Qld - Qlu \qquad W/m^2 \qquad (3.13)$$

It is called the 'radiation balance' in some publications, but that is to be deplored (see Note 11, p. 322).

We will now examine each of the components on the right-hand side of Eqn 3.13, in order to contrive formulae for estimating net irradiance.

Albedo

Eqn (3.13) shows that one influence on Qn is the albedo α, the reflectivity for short-wave radiation. It is chiefly a characteristic of the irradiated surface and, to a small extent, of the solar elevation and the cloudiness. It is the *ratio* of measurements taken with a radiometer facing down and up, respectively, over the given surface. (A radiometer is an instrument for measuring radiation – see Chapter 5.) However, errors may be incurred by not allowing the radiometer to reach equilibrium when facing in each direction – this inflates the deduced albedo. Secondly, there may be an error due to convective cooling from the instrument surfaces which become heated by the radiation energy, because the cooling may differ in the two directions. Thirdly, the instrument may not be level, so that some solar radiation reaches the underside of the sensor. Fourthly, the lower glass cover of the radiometer may refract light on to the lower sensor surface. With such errors, the total can be appreciable, and albedo values may range 23–31 per cent for the same crop surface, for instance.

Another variation of values can arise from the manner of averaging. Reddy (1987) compared daily average albedo values obtained in each of three ways from measurements on the same patch of ground. The first involves taking the albedo as the slope of a regression line linking simultaneous brief measurements of the solar irradiance (Qs) and the reflected radiation (α.Qs). This gave a value of 0.19. In the second method, the albedo was taken as the ratio of the daily totals of α.Qs and Qs, respectively, which gave 0.24. The third value was the average of the ratios of measurements taken at each daytime hour, which was 0.27. The range of these values indicates the need for caution in using published figures for the albedo.

There have been many reports of albedo measurements. Typical values are given in Table 3.3, and further information is available about particular surfaces. For instance, the albedo of snow decreases from about 90 per cent just after it has fallen, to about 40 per cent when it has become dirty, as discussed in Chapter 7. The albedo of a crop canopy is appreciably less than that of a single leaf, because of reflections and absorption within the foliage. A crop's albedo also is not constant – the value for a wheat crop is initially that of bare soil, later changes from 11 per cent at the start of ear shooting to 17 per cent at completion, and is 22 per cent when the crop is mature. The albedo of grass lessens as it grows taller, e.g. it is 25 per cent when 30 cm high and 21 per cent when 3 m. The albedo of soil depends on

Table 3.3 Values of the albedo (α), representative of published measurements for various ground surfaces

Surface	*Typical albedo values (%)* Range[a]	Median	Number of references
Snow	42–95[b]	70	14
Desert	28–35[c]	32	9
Grass	16–28	22	33
Field crops	15–24	20	23
Forests	10–18	15	18
Soil	8–19	14	23
Towns	13–15	14	9
Water	4–13	7	15

Notes: [a] Excluding the highest two and lowest two published values, i.e. this is a trimmed range.
[b] The albedo exceeds 80 per cent for most of the time in Antarctica (Schwerdtfeger 1984: 20), but is commonly 60 per cent in New Zealand (B. Fitzharris 1987, private communication). The value is 59 per cent in Alaska (Wendler and Ishikawa 1973).
[c] Excluding the single highest and lowest values.

its wetness, being about 11 per cent when wet and 18 per cent when dry. A wind of 10 m/s lowers the albedo of a lake to 9 per cent, from 13 per cent when there is a calm, the sun being 30° above the horizon.

The albedo of a city is either reduced by the canyon structure of the central business district, or increased by light-coloured roofing. Central Tokyo has an albedo of about 10 per cent, whilst a garden suburb in Adelaide has an albedo of 17 per cent, and the inner city there an albedo of 27 per cent. For St Louis the figure is about 13 per cent, and the residential, business and industrial areas in Sydney have albedos of about 14 per cent, alongside 19 per cent for natural bushland and farmland. A few measurements over Sydney immediately after rain suggest that the albedo is lower then. Removal of snow or the presence of air pollution reduces a city's albedo.

The albedo of a pan used for measuring the evaporation from an open water surface is only 9 per cent, if the inside of the pan is coated with black paint. However, a normal pan is galvanized, giving a higher albedo. Measurements at Griffith (New South Wales) showed a variation from 21 per cent at 9 a.m. in July to 12 per cent around 1 p.m. and 18 per cent at 4 p.m., with an average of 16 per cent. Measurements at Marsfield in Sydney over 8 days in spring gave about 13 per cent (R. Nurse 1986, priv. comm.). So 14 per cent is representative.

Several authors have reported a slight reduction of albedo as a result of cloud. Figure 3.14 shows that whether there is a decrease or an increase depends on the time of day, i.e. the angle of the sun.

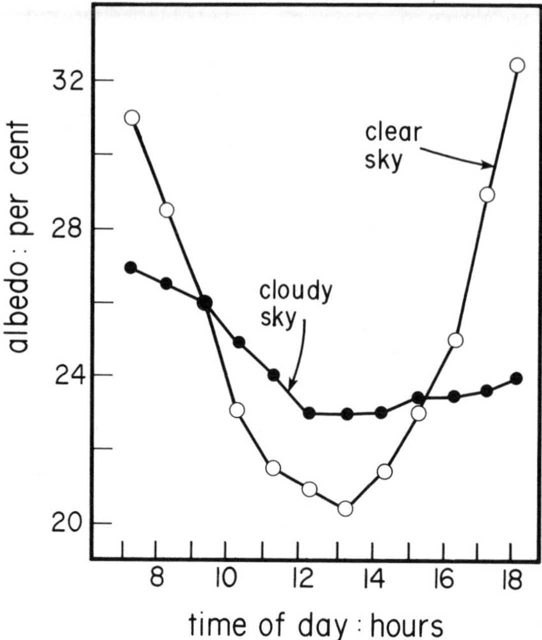

Figure 3.14 Effects of cloud and time of day on the albedo of crops of maize, oats and beans (Impens and Lemeur 1969: 266, by permission of Springer-Verlag Inc.)

Effect of the sun's angle on the albedo

The albedo is greater when the sun is low in the sky, which occurs either at high latitude, at dusk or dawn, or in wintertime. This is shown in Figure 3.14 and Table 3.4. Six different crops in Canada were found to have an albedo of 33 per cent when the sun was 10° above the horizon, but 24 per cent when at 60°. The figure for wheat is 28 per cent when the solar elevation is 20°, but 16 per cent when 65°. A banana field in Kenya has an albedo of about 26 per cent at 9 a.m. and 4.30 p.m., but one of only 12 per cent at noon. Similarly, the albedo of snow at the South Pole is 95 per cent if the sun is less than 20° from the horizon, but below 82 per cent with higher elevations.

The diurnal change of albedo is especially great with a water surface, such as that of the oceans which cover 70 per cent of the Earth. For either an ocean or a shallow pond, the albedo is 48 per cent when the sun is about 10° above the horizon, 9 per cent when at 40°, 6 per cent at 60°, and 5 per cent at 72°, though the effect of solar elevation is less when the sky is cloudy. Overall, the albedo of a smooth sea is 5 per cent when the sun is unobscured and about 10 per cent if the sky is overcast. The average for

Table 3.4 The effect of the sun's angle above the horizon (i.e. time of day or year) on the albedo of natural surfaces

Place	Surface	Albedo at various times (%)					Source
		Dawn or dusk	10 a.m.	Noon			
				Winter	Mean	Summer	
India	dry soil	35	21		19		(a)
Sudan	cotton crop			23		14	(b)
Minnesota	grass	43			22		(c)
Alberta	grass	27			19		(d)
Australia	grass	40			22		(e)
Australia	forest	30	15		14		(f)
England	grass		29		23		(g)
Thailand	forest	18–21				11	(h)
N. Australia	grass	30			20		(i)
	Median	30	21		19	13	

Sources: (a) Padmanabhamurty and Subrahmanyam 1964: 294; (b) Rijks 1967; (c) Idso *et al.* 1969: 246, (d) Nkemdirim 1972; (e) Kalma and Badham 1972; (f) Paltridge 1975: 37; (g) Schwerdtfeger *et al.* 1975: 24, (h) Pinker *et al.* 1980; (i) Rosenberg *et al.* 1983: 29.

open water is taken as about 12 per cent at 60° latitude, 8 per cent at 30° latitude and 7 per cent at the equator (Cogley 1979: 779).

The practical importance of the daily fluctuation of albedo is less than might be expected, because solar irradiance is weak when the albedo is high, at dawn or dusk. Consequently, the daily total reflected exitance, as a fraction of the daily global irradiance, is close to the noon albedo. For example, the albedo of a crop in Sydney varies from 26 per cent at 8–9 a.m. to 21 per cent at noon and 44 per cent at 7–8 p.m., whereas the daily reflected exitance divided by the daily solar irradiance is 23 per cent, which is close to the midday value.

Long-wave radiation

It is not easy to measure the long-wave radiation components of net irradiance, so they are usually calculated by considering separately the downwards atmospheric irradiance (Qld) and the upwards terrestrial exitance (Qlu). The nature of Qld is considered in Note 13, p. 323. There are many formulae for calculating it, but the simplest for the case of clear skies is that of Monteith (1973: 35):

$$\text{Qld} = 208 + 6\,\text{T} \qquad \text{W/m}^2 \tag{3.14}$$

where T is the daily mean screen temperature. In effect, this is the irradiance from a surface around 20C° cooler than the screen. However, the screen temperature is poorly related to the effective temperature of the radiating layers of the atmosphere when daytime sea-breezes or nocturnal ground inversions affect the screen temperature. The inversions can lead to underestimates at night by about 20 W/m^2, in using screen temperatures in Eqn (3.14). If the downwards long-wave irradiance Qld is related to the surface temperature Ts, rather than the screen temperature, the right-hand side of Eqn (3.14) should be replaced by [258 + 3 Ts], according to Greene and Nelson (1983: 275).

The long-wave irradiance of the ground is increased by the sky when there is cloud, because cloud base is warmer than the effective radiation temperature of a clear sky (see Note 13, p. 323). As a result, if the sky has *C* oktas of cloud (i.e. *C* eighths of the sky is covered by cloud), the atmospheric emission is given approximately by the following:

$$\text{Qldc} = \text{Qld}\ (1 + 0.0034\ C^2) \qquad \text{W/m}^2 \qquad (3.15)$$

Combination of Eqns (3.14) and (3.15) for the case of a temperature of 15°C shows that an increase of cloudiness from zero to 8 oktas increases the downwards long-wave irradiance by about 57 W/m^2, i.e. by 7 W/m^2 per okta, increasing the net irradiance (Qn) equally. This value approximates the 6 W/m^2 per tenth of cloud reported by Paltridge and Platt (1976: 140), i.e. 7.5 W/m^2 per okta. However, the net irradiance derived from formulae developed in a review by Linacre (1968: 53), for the cases of clear skies and overcast respectively, differ by about 90 W/m^2, i.e. 11 W/m^2 per okta of cloud, and data of Nielsen *et al.* (1981: 260) imply 10 W/m^2 per okta. Presumably, the different figures are due to various kinds of cloud. In what follows, we will use a central value of 9 W/m^2 per okta.

The terrestrial-exitance component of the net irradiance is given by the Stefan-Boltzmann equation:

$$\text{Qlu} = \mathbf{e}\ (\text{Tk})^4 \times 5.67 \times 10^{-8} \qquad \text{W/m}^2 \qquad (3.16)$$

where **e** is the surface emittance and Tk is the screen temperature in Kelvins. The emittance may be taken as about 0.97 for most natural surfaces (see Note 14, p. 323). (To remember Eqn (3.16), note the adventitious sequence of digits from 4 to 8.)

The use of screen temperatures in Eqn (3.16) assumes that screen and surface temperatures are the same, whereas the surface is cooler at dawn and relatively warm in the daytime, especially if the surface is dry. The result is an underestimate of the terrestrial emittance in the daytime, and an overestimate at night. The error is about 5 W/m^2 for each degree by which screen and surface temperatures differ; this is derived by comparing emittances calculated from Eqn (3.16) with two surfaces at about room temperature, but differing by one degree.

Net long-wave exitance

The difference between Qlu and Qld is the upwards net long-wave exitance Qnl. It tends to be greater at higher elevations. For a clear sky, it may be approximated thus:

$$Qnl = Qlu - Qld = 107 - T \qquad W/m^2 \qquad (3.17)$$

Eqn (3.17) and the discussion following Eqn (3.15) imply the following, when there are *C* oktas of cloud in the sky:

$$Qnlc = 107 - T - 9\,C \qquad W/m^2 \qquad (3.18)$$

This means that the net long-wave exitance varies between 20 W/m^2 when the sky is overcast, to 92 W/m^2 when the sky is clear, assuming a screen temperature of 15°C. So cooling is almost five times faster on a clear night than when the sky is overcast.

Another expression for Qnlc refers to the cloudiness in terms of the hours of bright sunshine (n) in a daylength of N hours. (The connection between cloudiness and the ratio n/N is discussed in 'Estimating the solar irradiance of level ground', p. 180, and values of N are in Note 8, for this chapter.) The cloudy-day long-wave exitance (Qnlc) is as follows:

$$Qnlc = Qnl\,(a + b.n/N) \qquad W/m^2 \qquad (3.19)$$

where Qnl is the clear-sky flux and a and b are constants which are derived empirically. Various authors have proposed either 0.1 and 0.9, or 0.2 and 0.8, for a and b, respectively. The latter values are preferable because they imply that, when the sky is overcast Qnlc is only a fifth of Qnl, which is in accord with Eqn (3.18).

At this stage, we have considered the estimation of net irradiance (Qn) by deducing most of its parts – the albedo and the two long-wave energy fluxes. The remaining component is the global irradiance (Qs), whose measurement and estimation are considered in Chapter 5. Here it will be assumed that Qs is known, so that we can now proceed to the estimation of net irradiance.

Estimating net irradiance from global solar irradiance

There are at least four methods of estimating net irradiance (Qn):

1 Estimation of the various component parts (which have been discussed already) and then adding them.
2 Using an empirical correlation between Qn and the measured global solar irradiance (Qs).
3 Determining empirically a relationship between Qn and the ratio of Qs to the extra-terrestrial irradiance (Qa).
4 Using a relationship between Qn and some other proxy.

The second is the simplest, and there have been numerous papers published on the constants c and d in the following equation:

$$Qn = c\,Qs - d \qquad W/m^2 \tag{3.20}$$

A close connection between Qn and Qs is shown by the remarkably high values of the correlation coefficient. Published values range between 0.96–0.99.

Twenty-three values of the factors c and d in Eqn (3.20) are given in Table 3.5, but the list could have been much longer. The median values of c and d in the bottom row of the table yield the following:

$$Qn = 0.66\,Qs - 28 \qquad W/m^2 \tag{3.21}$$

Confirmation comes from measurements published by Berliand (1970), taken world-wide during the International Geophysical Year in 1957/1958. Those data give Figure 3.15, and imply the following relationship, like Eqn (3.21):

$$Qn = 0.63\,Qs - 40 \qquad W/m^2 \tag{3.22}$$

It might be thought that the constants in Eqn (3.20) would be different for averaging periods of less than 24 hours, because of the difference between daytime and nocturnal radiation conditions. (There are no short-wave fluxes at night.) However, reduction of Qs, by including the zero value after sunset to obtain a 24-hour average, offsets the negative net irradiance during the night. So the constants in Table 3.5, which are for daily or longer periods, resemble values in Table 3.6, for conditions in the daytime alone. Other work shows that the constants for periods of only 10 minutes or an hour are also similar.

Table 3.5 Values of the constants c and d in Eqn (3.20) for green crops like grass.

				Calculated Qn: W/m²	
Source	*Place*	*c*	*d*	*Qs = 100*	*300 W/m²*
Shaw 1956	Iowa	0.75	10	65	215
Monteith and Szeicz 1961: 167	England	0.60	65	−5	115
and likewise for 19 more sets of figures, and then:					
Clothier *et al.* 1982: 304	New Zealand	0.62	7	55	179
Hu and Lim 1983	Malaysia	0.62	9	53	177
	Medians	0.66	28	37	167

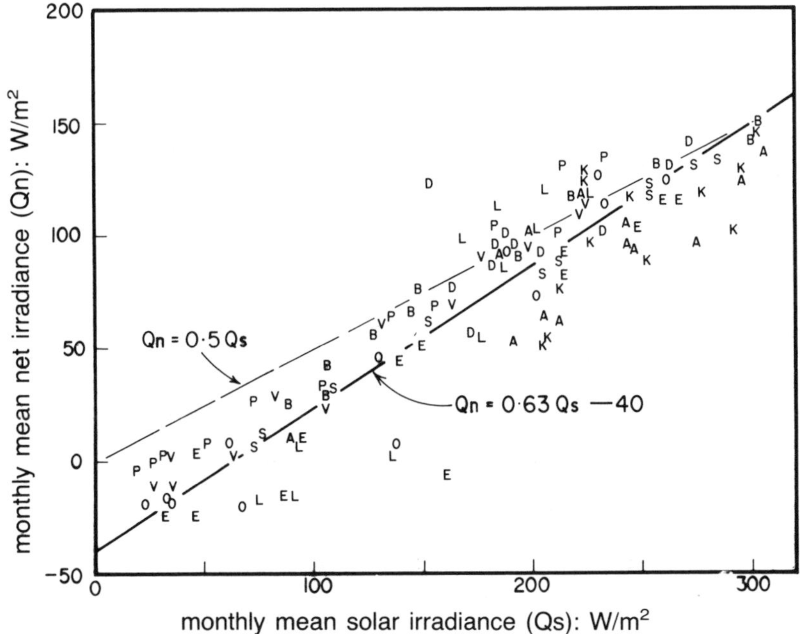

Figure 3.15 Relationship between monthly mean values of solar and net irradiance given by data of Berliand (1970) for the following places:
A – Pune, B – Buenos Aires, D – Dum Dum (India), E – Edmonton (Canada), K – Dakar, L – Vladivostok, O – Omsk, P – Potsdam, S – Aspendale (Australia), V – Vienna.

It follows that the period of averaging hardly affects the ratio Qn/Qs. Mean values over 10 days from measurements over grass at Marsfield were 0.58 for the daytime, and 0.53 for 24-hour periods. Likewise, for millet in India the daytime ratio is 0.67 and the 24-hour ratio is 0.61, and equivalent ratios for maize in Nigeria are 0.65 and 0.57. In other words, net irradiance is reduced only slightly more than global irradiance by including the night-time within the period of averaging.

The ratio Qn/Qs is called the 'radiation efficiency' and is generally about half, e.g. typically 0.53 in the Philippines. Eqn (3.20) shows that the ratio decreases when irradiance is low. There is a reduction also when the ground is covered by snow, the high albedo of which significantly lowers Qn. There is a linear decrease of the ratio at great elevation in the Caucasus mountains, from 0.8 at 2000 m, to 0.3 at 4000 m.

Eqn (3.13) shows that a low albedo increases the constant 'c' in Eqn (3.20). Accordingly, the constant for water surfaces has been found to be around 0.85, instead of the 0.63 for a green surface.

Eqn (3.13) also implies that the constant d in Eqn (3.20) is connected

Table 3.6 Values of the constants c and d in Eqn (3.20), using *daytime* measurements of net and global irradiance of crops, instead of 24-hour means.

Source	*Place*	*Surface*	*c*	*d: W/m²*
Shaw 1956	Iowa	grass	0.75[a]	14[a]
Monteith 1965: 250	England	grass	0.55	0
Davies 1967	W. Africa	grass	0.61	14
Fritschen 1967: 58	Arizona	various	0.73	83
Asuncion 1971: 70	Philippines	–	0.52	6
Kalma 1972: 272	Australia	pasture	0.61	40
Nkemdirim 1972: 39	Canada	prairie	0.69	32
Penney 1981: 174	S. Australia	vineyard	0.66	68
Clothier *et al.* 1982: 304	New Zealand	oats	0.62	7
D. Williams 1982[b]	Marsfield	grass	0.63	49
	Medians		0.63	23

Notes: [a] Overcast sky
[b] private communication

with the net long-wave exitance, which is linearly related to the cloudiness by Eqn (3.18). It is shown by the Prescott formula discussed in 'Estimating the solar irradiance of level ground', pp. 177–9 that cloudiness is in turn linearly related to the ratio Qs/Qa, which is the amount of solar power at ground level divided by the amount above the atmosphere. So Eqn (3.20) becomes the following:

$$Qn = 0.63.Qs + e\,Qs/Qa + f \qquad W/m^2 \qquad (3.23)$$

where e and f are constants to be determined.

A similar equation can be derived from the following, due to Linacre (1969: 61):

$$Qn = (1-\alpha)\,Qs - 1.12\,(100 - T)(0.2 + 0.8\,n/N) \qquad W/m^2 \qquad (3.24)$$

This may be combined with the Prescott formula just mentioned to replace n/N. Then we assume that the albedo is 0.22 and temperatures are about 15°C, to obtain the following:

$$Qn = 0.78\,Qs - 152\,Qs/Qa + 19 \qquad W/m^2 \qquad (3.25)$$

Similar equations were derived empirically by Fitzpatrick and Stern (1973: 406) and Murtagh (1976: 111).

A much more elaborate formula for the net irradiance was developed by Penman (1948) for use when considerable meteorological information is available. It requires not only data on the global irradiance (Qs), the albedo (α) and Kelvin temperature (Tk), but also on the sunshine-duration/daylength ratio (n/N), and on the ambient vapour pressure (e hPa):

$$Qn = (1 - \alpha)\, Qs - 5.5 \times 10^{-8}\, Tk^4\, (0.56 - 0.08\, e^{0.5})\, (0.1 + 0.9\, n/N) \quad W/m^2 \qquad (3.26)$$

Accuracy of estimating net irradiance from global irradiance

In view of the number of formulae available for estimating net irradiance, which should be used? The answer depends on what input data are to hand. Eqn (3.26) needs information on four climatic elements, whereas Eqn (3.22) depends on only one – the global irradiance – and that itself can be estimated (see Chapter 5). So Eqn (3.22) is more versatile and convenient than the others, but likely to be less accurate since it ignores variations of temperature, vapour pressure or sunshine duration.

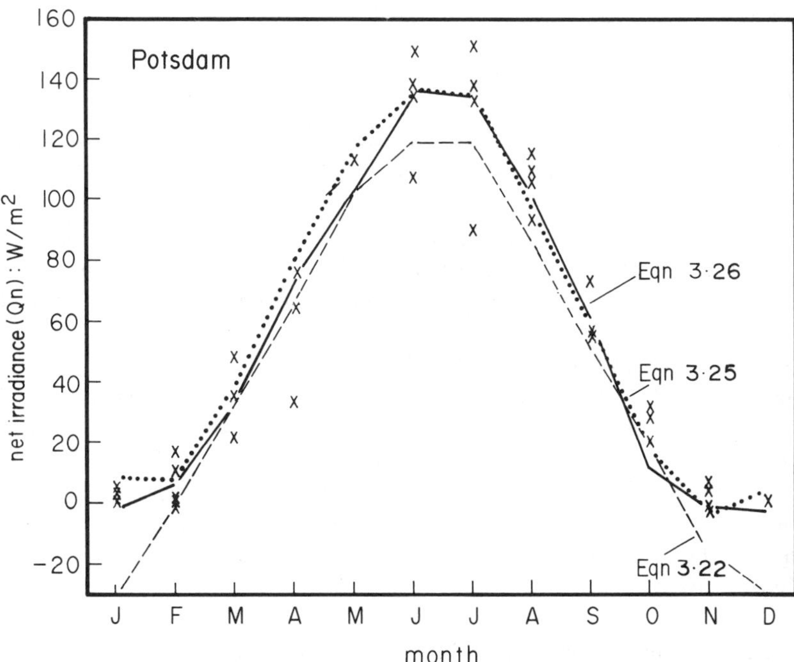

Figure 3.16 Comparison of estimated with measured monthly mean net irradiance at Potsdam (i.e. Berlin). The lines give the estimates from the respective equations. The global irradiance and temperature values for the equations were taken from the estimated parallelogram in Figure 3.6. The vapour pressure values required for Eqn (3.26) were long-term means published in the World Survey of Climatology (1977, *6*: 63). The required values of the ratio n/N (i.e. the fraction of daytime with a clear sky) were taken as [2 Qs/Qa − 0.5] from the Prescott formula discussed in Chapter 5, and values of the extra-terrestrial irradiance (Qa) came from Note 12. The crosses in the graph show *measured* monthly mean values during 1964–68 (Berliand 1970: 82)

Table 3.7 Errors (shown in brackets) involved in estimating the daily net irradiance (Qn) by means of alternative equations, and measurements of daily global irradiance (Qs) at Marsfield. Extra-terrestrial irradiance values Qa are taken from Note 12

	Measured values: W/m²			*Estimates of Qn: W/m²*	
Date	*Qa*	*Qs*	*Qn*	*Eqn (3.22)*	*Eqn (3.25)*
27 Jan 1981	496	311	175	156 (19)	169 (6)[a]
27 Feb	447	266	139	128 (11)	138 (1)
27 Mar	376	209	96	92 (4)	99 (3)
27 Apl	291	148	59	53 (6)	58 (1)
27 May	226	116	44	33 (11)	33 (11)
27 June	191	124	14	38 (24)	20 (6)
27 July	204	147	28	53 (25)	27 (1)
27 Aug	255	93	25	19 (6)	37 (12)
27 Sept	334	257	118	122 (4)	105 (13)
27 Oct	416	307	152	153 (1)	149 (3)
27 Nov	473	177	89	72 (17)	101 (12)
27 Dec	508	143	73	50 (23)	89 (16)
			Mean error	13	7

Note: [a] 6 W/m² is the difference between the measured 175 W/m² and the estimated 169 W/m²

As explained in Chapter 5, the loop method described on p. 78 for estimating monthly mean temperatures can also be used to obtain values of the irradiance (Qs), again from geographical data alone. The figures for Qs can then be inserted into Eqns (3.22), (3.25) or (3.26) to obtain the net irradiance. In this way, figures for Qs from Figure 3.6 provide the graphs in Figure 3.16. It may be seen that Eqn (3.25) is nearly as accurate as the much more elaborate Eqn (3.26) – the mean error of the estimate (compared with the measured value for each month) is 7 W/m^2 for Eqn (3.25) and 6 W/m^2 for Eqn (3.26). Eqn (3.22) is less accurate, especially in winter, the mean error being 14 W/m^2. But in each case the graphs lie within the year-by-year scatter of the measured values shown in Figure 3.16.

Estimation accuracy is generally worsened by averaging over a day instead of a month (see Figure 3.2). Nevertheless, the mean errors shown in Table 3.7 are similar to those indicated by Figure 3.16. Eqn (3.25) appears preferable to Eqn 3.22, presumably because it allows for variations of cloudiness.

Another comparison of the errors arising from various equations is given in Table 3.8. In this case, the data refer to averages over about 20 days from Sweden, instead of the daily values from Australia in Table 3.7. Surprisingly, the simplest Eqn (3.22) proves the most accurate on the whole, perhaps because of a fairly constant degree of cloudiness. As before, the mean error of estimation was 6–12 W/m^2, which is less than the

Table 3.8 Comparison of estimates with summer-time measurements of net irradiance of short grass at Uppsala, which is at 60°N and 24 metres elevation (Rodskjer 1978). Estimation errors are shown in brackets, i.e. the differences between estimates and measurements

	Measurements					*Estimates of Qn: W/m²*			
						various equations			
Date	*Qa W/m²*	*T °C*	*n/N (%)*	*Qs W/m²*	*Qn W/m²*	*(3.22)*	*(3.22)[a]*	*(3.24)*	*(3.25)*
1975									
11–31 May	430	11	48	211	110	93(17)	108(2)	108(2)	111(1)
1–20 June	460	14	70	278	135	135(0)	142(7)	146(6)	145(10)
21 June–10 July	465	16	79	289	127	142(15)	147 (20)	153(26)	152(25)
11–31 July	440	18	57	233	109	107(2)	119(10)	124(15)	122(13)
1–20 Aug	385	19	58	200	88	86(2)	99(11)	100(12)	98(10)
1976									
11–31 May	430	12	58	233	113	107(6)	118(5)	118(5)	120(7)
1–20 Jun	460	13	58	249	123	117(6)	128(5)	132(9)	132(9)
21 June–10 July	465	17	61	268	123	129(6)	138(15)	144(21)	142(19)
11–31 July	440	17	50	213	100	97(3)	112(12)	114(14)	113(13)
1–20 Aug	385	16	71	213	99	94(5)	104(5)	104(5)	103(4)
			Mean errors: W/m²			6	9	12	11

Note: [a] These values were calculated by means of Murtagh's version of Eqn (3.22)

possible 10 per cent error of measured values of Qn. By contrast, the relatively elaborate Eqn (3.26) was found in error by about 13 per cent, on an occasion in Austria.

A further comparison of some of the various equations involves either daily or weekly mean data from South Africa. This shows differences of about 15 W/m^2 between estimates and measured daily values for Eqn (3.24), and 13 W/m^2 for Eqn (3.26). Over several periods each of a month's duration, data from Australia, California, Canada, England, Norway and the USSR in Eqn (3.24) give net irradiance values with an average error of about 6 W/m^2. Tests of Eqn (3.24) in India show an average error of only 4 W/m^2 for 14 monthly mean values of daytime net irradiance, and less than 2 W/m^2 for 24 particular days. Such mean errors are even less than those in Tables 3.7 and 3.8.

Other estimates of net irradiance

A different method of estimating Qn is used in the USSR, based on the data shown in Figure 3.17. This indicates an equation of the following form:

$$Qn = 1.1\,[\Sigma(T - 10)]^{0.5} \quad W/m^2 \qquad (3.27)$$

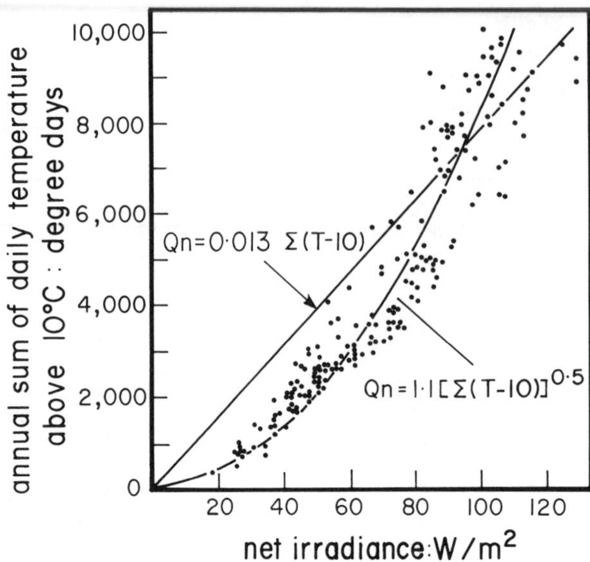

Figure 3.17 Relationship between the annual mean net irradiance and the annual sum of the daily excess of the mean temperature above 10°C (Budyko 1974: 342). The data refer to hundreds of places between 71°N–46°S, but not island, coastal or mountain sites

where $\Sigma(T - 10)$ is the annual sum of the excess of the daily mean temperature above 10°C. It differs from the usual expression, indicated by the straight line, and appears an improvement on it. However, the scatter is obviously large, even for annual mean values.

Alternatively, Webb (1975: 212) suggested that net irradiance could be estimated from the rate of evaporation from an open surface of water, because the net radiant energy received by a water surface (Qnw) is often almost equalled by the amount of latent energy (L.Eo) consumed in the evaporation process. (The term Eo is the rate of evaporation expressed as the lowering of water level in millimetres per day, and L is the latent heat required, which is about 28 W/m^2 per mm/d.) However, Qnw is greater than the net irradiance of a land surface, because of (a) water's lower albedo, and (b) its lower temperature, due to evaporative cooling. Also, Qnw may not equal L.Eo for a small volume of water where evaporation is appreciably augmented by advection and by heat into the sides and base of the water volume, nor in autumn or spring when the change of evaporation rate from a large body of water is much slower than the change of irradiance.

ESTIMATING EVAPORATION

We can draw together various aspects of this chapter by further consideration of the estimation of evaporation. It involves several of the climate elements we have been discussing.

There are plenty of empirical formulae for estimating Eo (see 'Estimating in general', pp. 64–6), but they are suited only to particular environments. On the other hand, the formula of Penman (1948) comes from a consideration of the physics of the process and therefore is universally applicable. In terms of the energy required to evaporate water, the rate is given as follows:

$$L.Eo = (\Delta.Qnw + d.c.S/r)/(\Delta + K_s) \qquad W/m^2 \qquad (3.28)$$

where Δ is the change of saturation water-vapour pressure per 1C° (obtainable from Table 8.1 in units of hPa/C°), Qnw is the net irradiance of the water surface, d is the density of air at the ambient pressure and temperature T, c is the air's specific heat, S is the saturation deficit of the air (hPa), r is the diffusion resistance between the water's surface and the atmosphere (s/m), and K_s is the psychrometric 'constant', i.e. 0.67 hPa/C° at sea level. (It is 0.59 hPa/C° at 1000 m, 0.53 at 2000 m, 0.46 at 3000 m, and 0.38 hPa/C° at 4000 m.)

Eqn (3.28) can be simplified in five ways. First, replace S by its equivalent [Δ (T − Td)], where Td is the dewpoint temperature. Second, replace $\Delta/(\Delta + Ks)$ by an expression with an equal numerical value, which is [$0.42 + 0.011\ T + 3.10^{-5}\ z$], where z is the altitude in metres; the expression resembles one of Rouse *et al.* (1977: 909). Third, replace Qnw by a version of Eqn (3.21) adjusted for the difference between the albedos of water and a vegetated surface, i.e. by [0.85 Qs – 40]. Fourth, replace Qs, the solar irradiance of the evaporating surface, with an expression from Chapter 5, i.e. by [600 (T + 0.006 z)/ (84 − A)], where A is the latitude. Finally, replace the resistance r by d.c/h (which is equivalent by definition), where h is the 'transfer coefficient', the forced convection of sensible heat between surface and air when they differ in temperature by 1C°.

Forced convection is proportional to the wind speed (u m/s, at about 2 m height), as shown by Dalton's equation, discussed on pp. 317–8. The convective heat-transfer coefficient can be derived either from the rate of evaporation (which also is a convective process), or from the wind profile above the surface, or by direct measurements of heat transfer. Data from 36 publications show that for a large water surface h is equivalent to 2.3 u $W/m^2.C°$, with a standard deviation of 0.8 u $W/m^2.C°$.

The five simplifications of Penman's formula yield the following:

$$Eo = [0.015 + 4 \times 10^{-4}\ T + 10^{-6}\ z] \times [480\ (T + 0.006\ z)/(84 - A) - 40 + 2.3\ u\ (T - Td)] \qquad mm/d \qquad (3.29)$$

This requires only geographical data on elevation and latitude, and climate values of daily mean temperature, dewpoint and wind speed. If necessary, the climate values can be estimated, as discussed in 'Estimating screen temperature' and 'Estimating dewpoint temperature' (both in this chapter), and Chapter 6.

The constants in Eqn (3.29) would have to be changed to estimate the rate of evaporation Et from a wet vegetated surface instead of from water, because of the different albedo and roughness. Qn becomes [0.63 Qs − 40], according to Eqn (3.22), and h is 4 u $W/m^2.C°$, according to 12 various publications. After these adjustments to Eqn (3.29), it becomes the following:

$$Et = [0.015 + 4 \times 10^{-4}\,T + 10^{-6}\,z] \times [380\,(T + 0.006\,z)/(84 - A) - 40 + 4\,u\,(T - Td)] \quad \text{mm/d} \qquad (3.30)$$

This takes a simpler form for any particular place after inserting values of height and latitude. As an example, it becomes this for a place at sea level at 38°:

$$Et = [0.015 + 4 \times 10^{-4}\,T] \times [8.2\,T - 40 + 4\,u\,(T - Td)] \quad \text{mm/d} \qquad (3.31)$$

It is difficult to find reliable measured values of evaporation rate, with which to compare estimates from these equations. One cannot test them against measurements with a small pan of water, where the water loss is augmented by heat into the sides of the vessel and the albedo is different. Nor should comparison be made with crops having an appreciable diffusion resistance within the canopy. Amongst the few suitable sets of data is that of McIlroy and Angus (1964: 204), which allows a test of Eqn (3.31) against measurements from frequently watered grass near the coast at Melbourne. The average error of 36 monthly mean values is 0.62 mm/d. The average measured rate over three years was 3.47 mm/d, compared with an average estimate of 3.37 mm/d. Figure 3.18 shows that the formula estimates low in summer and high in winter, since the relationship between temperature and irradiance depends on the distance inland, perhaps on account of sea-breezes (Linacre 1969a: 11).

ESTIMATING AND CLIMATE CHANGE

Consideration of future global climates necessarily involves estimation, based on earlier measurements, which is vulnerable to the inaccuracies that have been discussed in this chapter, e.g. errors due to differences and averaging, the choice of proxies, and the uncertain realm within which empirical relationships remain valid. The reliance that can be put on calculations of future warming depends on the congruence of values obtained in different ways, retrospective agreement of estimates with recent climates, successful prediction of the magnitude of alterations associated

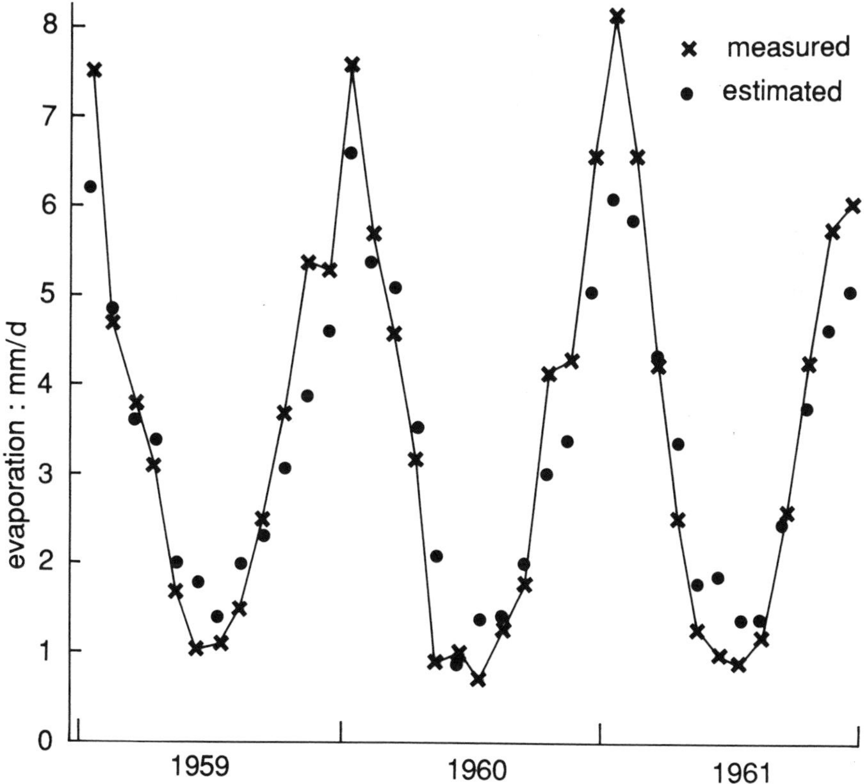

Figure 3.18 Comparison of estimates of the evaporation from wet grass, using Eqn 3.31, with measurements by McIlroy and Angus (1964: 204)

with surface-climate warming (e.g. cooling in the stratosphere, and increased cloudiness), and conclusive explanation of anomalies, like the cooling that occurred between 1940–65 (Figure 1.2). The accumulating evidence gradually strengthens our confidence in the estimates.

The amount of global warming is calculated from the 4 W/m^2 increase of Qld (and thus of net irradiance Qn), reckoned to result from doubling the carbon dioxide in the air. Eqn (3.14) implies that a change of Qld by 4 W/m^2 is associated with a change of surface temperature by 0.7C°, but that is for clear skies. The right-hand side of Figure 3.17 (corresponding to conditions where temperatures are always above the threshold of 10°C) suggests that such a change of Qn would lead to warming by about 500 degree-days annually, i.e. an average of 1.4C°. However, such estimates are very approximate and may be amplified by positive feedback.

One feedback mechanism arises at the poles from the large difference between the albedos of ice and water (see Table 3.3). Melting of ice (e.g. snow) changes radiation absorption from less than 30 per cent to more than 90 per cent, i.e. over threefold. This leads to yet more melting. And so on.

A serious consequence of warming by 3C° is the faster drying of soil beneath crops of grain, i.e. a greater vulnerability to drought in the inland (Mitchell 1989). Eqn (3.31), for instance, shows that raising the temperature by 3C° from 15°C and assuming a wind constant at 3 m/s and a dewpoint rising from 10°C to 12°C, results in an increase of the potential evaporation rate from 4.3 mm/d to 5.4 mm/d, i.e. by 27 per cent.

4

ANALYSING CLIMATE DATA

Measurements and estimates of climate elements are of little use until they are 'analysed'; this includes being organised into meaningful order, reduced to manageable quantity and described succinctly, so as to reveal anomalies in the data for subsequent examination and to permit inferences of consequences. Analysis is the step between obtaining the data and applying it to solve practical problems.

The procedures of analysis are chiefly statistical, and this chapter includes comments on elementary statistics applied to the problems of climatology. However, this is not a statistics textbook.

CLIMATE DATA

Climate data are used to answer problems like the following: What are the normal and extreme conditions to be withstood by equipment at a particular place? Which place has the driest climate for certain industrial activities? How should one best timetable the building of a bridge or growing of a crop? What premiums should be levied when insuring against rain? What is the largest rainfall to be expected once each 50 years? Is the amount of sunshine enough to justify solar-energy collection?

Answers to such questions necessitate concise descriptions of a mass of numbers, estimation of the chances of particular extreme events, testing the similarity of circumstances in different times and places, estimation of one phenomenon from observation of another, or the ascertaining of regularities in the occurrences of events.

Climatological series

A sequence of measurements or estimates may form a 'climatological series'. For example, the January averages of temperature taken at the same place over a period of 40 years would form a climatological series of 40 items. Alternatively, the mean of the 40 January averages, along with a similar mean for the other months of the year, would form an 'annual climatological series' consisting of twelve items. However, 40 January averages along with some February averages, or data from another place, could not be combined to form a climatological series because of the lack of 'homogeneity', a concept which is discussed later.

A climatological series is usually in chronological order, but the data may consist of figures from different places rather than of figures taken at different times. In this circumstance it is not obvious how to arrange the figures in a logical sequence, because location on the ground is defined by two factors (longitude and latitude), whereas time has a single dimension. In some cases the problem may be solved by identifying a single distance as the most important (e.g. the latitude or distance from the coast) and sequencing the data accordingly.

The values in a climatological series are usually ranked in *ascending* order. They can then be divided into 100 equal-sized groups. The highest value in each group is called a 'percentile', so the 25th percentile, for example, is larger than a quarter of the values in the whole series. The percentile number for a particular value in position m (the 'rank order') in a set of N values, arranged from smallest to largest, is commonly given by the following expression:

$$C(V) = 100\,m/(N + 1) \qquad \text{per cent} \tag{4.1}$$

The percentage C(V) is called the 'cumulative relative frequency' and represents the probability of a value *less* than the particular one in position m (see Note 1).

The 'recurrence period', 'recurrence interval' or 'return period' is the average time (T) within which a given value will be exceeded just once. For example, if N annual total rainfalls are measured at a certain place, the average period between years when the annual rainfall equals or exceeds P mm/a just once is given thus (see Note 1, p. 324):

$$T = (N + 1)/(N + 1 - m) \qquad \text{years} \tag{4.2}$$

where m is the rank order of the value P.

PRESENTATION OF DATA

Information on climate elements can be displayed either in a list (such as a climatological series), or a table, a diagram, a graph or by means of an equation. Each has certain advantages.

Tables

One virtue of tables is that an enormous amount of information can be given compactly and precisely in a way that makes obvious any gaps or inconsistencies. An example is Table 4.1. By the way, this illustrates that

Table 4.1 Effects of an increase of various factors on the annual mean, annual range and daily range of surface temperatures (see 'Estimating screen temperature, pp. 72–85).

Factor which is increased	*Effects on the*		
	Annual mean	*Annual range*	*Daily range*
Latitude	reduces[a]	increases	decreases
Distance inland	reduces	various	increases
Altitude	reduces	minor	various
Cloudiness	zero	reduces	reduces
Rainfall	reduces	reduces	reduces
Surface albedo	reduces	reduces	reduces

Note: [a] Increasing the latitude *reduces* the annual mean temperature

the data need not be numerical. The proximity of various data in a table facilitates cross-referencing horizontally and vertically. Also, the frequency of the coincidence of values in various rows and columns can be shown to indicate the degree of association between row and column variables. For example, Table 4.2 shows the extent to which high wet-bulb temperatures tend to occur simultaneously with high dry-bulb temperatures. A table like this is called a 'contingency table'. If it has only two rows and two columns of figures (e.g. Table 4.3) it is described as 'dichotomous' (Oliver 1981:153).

The values in Table 4.2 have been grouped, since temperature is a continuous variable. So the question arises – how many groups are appropriate? Too few would destroy information in the homogenisation of data within each, whilst too many groups would each contain too few items to be a reliable sample and would lessen the simplification which is one purpose of a table. It seems reasonable that there should be fewer rows than columns in Table 4.2, because the dry-bulb temperatures span more than 25C° whereas the wet-bulb temperatures span only about half that. Then there is a rule that the number of equal-sized classes should be less than five times the common logarithm of the number of observations (e.g. a set of 3650 values of dry-bulb temperature should be divided into less than 18 classes) to leave sufficient numbers in each group for meaningful statistics. Table 4.2 has only 11 classes horizontally, which meets this test. The number 11 is also less than the 13 specified by another rule, that there should be no more than $(1 + 3.3 \log N)$ groups for N values.

Tables should be designed with care to fit the page, to show some logical sequence of columns and rows, and to bring close together any data which are to be compared. Vertical lines are unpopular with printers. The units of all values must be specified. Each table needs a number and an adequate

Table 4.2 Number of days of particular dry-bulb and wet-bulb temperatures at Perth at 3 p.m. during 1950–9

Dry-bulb temperature	*Below 12.8°C*	*12.8–15.5*	*15.6–18.2*	*18.3–21.0*	*21.1–23.7*	*23.8–26.5*	*26.6–29.3*	*29.4–32.0*	*32.1–34.8*	*34.9–37.7*	*Above 37.7°C*	*Total of days*
Wet-bulb temp.												
below 10°C	30	63	1									94
10–12.7	31	273	302	47	2							655
12.8–15.5		107	387	400	158	26	4					1085
15.6–18.2			64	201	281	192	77	27	4	6	1	853
18.3–21.0				24	71	205	143	112	49	21	8	633
21.1–23.8					5	28	70	72	62	35	18	290
above 23.8°C						1	4	12	8	10	8	43
Totals	61	443	754	672	517	452	298	223	123	72	35	3650

Source: Muffatti 1966

Table 4.3 Effect of daily mean temperature on student attendance at certain evening courses at Liverpool Technical College (New South Wales) in 1981

Daily mean temperature	*Number of student-days*	
	Attendances	*Absences*
Over 20°C	1395	216
Below 20°C	2296	551

Source: G. Bankhead, private communication

caption, and should be entirely comprehensible without reference to the text. As with diagrams, each table must stand on its own and support the text, not vice versa.

Diagrams

An alternative to a table is some sort of diagram. Diagrams clearly reveal outliers within a set, for checking. Compared with a table, a diagram is more vivid but less accurate, as can be seen by comparing Table 4.4 with Figure 4.1, which displays the same information. Figure 4.1 is an 'isogram', with lines called 'isopleths' linking points representing similar values. Another isogram is the customary synoptic chart for weather forecasts, with 'isobars', which are isopleths of pressure.

There are various kinds of diagram. Figure 4.2 illustrates a 'frequency histogram' showing the relative frequency of particular values. This requires the same consideration of the appropriate number of categories, as for tables. A histogram demonstrates whether the frequency distribution has a single hump, or is 'bimodal' with two humps (Figure 4.3). It is

Table 4.4 The variation of the typical[a] monthly rainfall (mm), at places along the east coast of Australia[b]

Location	*Latitude*	*Jan*	*Feb*	*Mar*	*Apr*	*May*	*Jun*	*July*	*Aug*	*Sep*	*Oct*	*Nov*	*Dec*
Thursday Island	11°S	399	393	321	168	20	11	8	4	2	1	12	163
Cooktown	15	290	334	340	169	55	40	22	18	11	15	33	114
Townsville	19	249	245	169	42	15	14	5	3	5	11	21	82
Rockhampton	23	121	109	83	39	28	32	21	13	17	34	58	92
Brisbane	27	127	123	115	58	43	41	36	28	41	61	74	107
Port Macquarie	31	115	150	154	135	112	98	80	63	65	74	79	100
Nowra	35	61	56	63	66	59	53	61	36	41	50	42	67
Flinders Island	40	38	47	49	62	79	56	80	73	59	49	54	48
Cape Bruny	43	54	49	60	74	80	79	91	77	70	76	72	67

Notes: [a] These are long-term median values (see pp. 117–9)
[b] Data from *Rainfall Statistics* (Australian Bureau of Meteorology 1977)

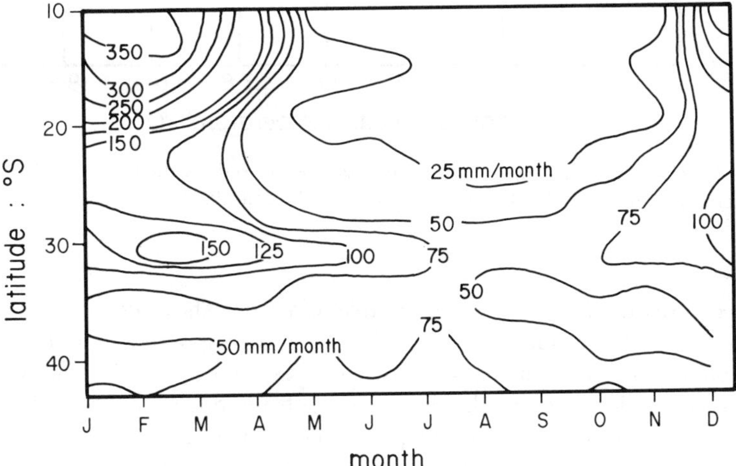

Figure 4.1 Isogram derived from Table 4.4

especially suited for sets of more than 300 values.

A particular kind of diagram illustrated in Figure 3.6 is much used in climatology. It is called a 'climagram' (not 'climogram') and relates long-term average monthly mean values of two climate variables in a loop, to show the association of one with the other. Another popular diagram is exemplified by Figure 4.4, showing the variation of a quantity at various times of the day and year.

The histogram labelled D in Figure 4.5 shows the number of years with a rainfall at Sydney equal to that shown on the horizontal axis. But the graph labelled C avoids homogenising data within classes, and every year's value

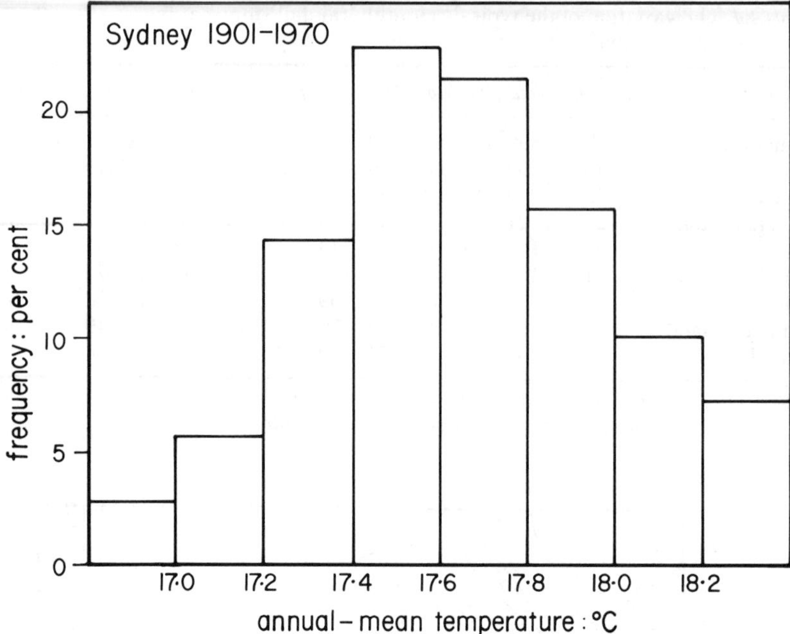

Figure 4.2 Histogram of annual mean temperatures at Sydney during 1901–70 (Linacre and Hobbs 1977: 322, by permission of Jacaranda Wiley Ltd)

is used individually. It is a 'cumulative' diagram, showing the number of years when the rainfall was *less* than the corresponding value on the horizontal axis. In terms of calculus, curve C is the integral of curve D. A smoothed version of C can be drawn through the middle or the right-hand end of each step, according to the precise labelling of the vertical axis. That axis can be labelled for C either in terms of the *number* of years (N in Figure 4.5) or the *percentage* of years, shown at the right of the diagram.

Different is a 'nomogram' or 'alignment chart'. This consists of three scaled lines, roughly parallel, arranged so that a ruler connecting known values on two of the lines intersects the third at a point which gives the desired unknown value.

Graphs

Graphs are more expensive to prepare than tables and less precise, but they reveal trends more clearly. They may reveal up to five dimensions of a variable by means of the two axes, isopleth lines, choice of symbol design and colour.

The picture should lie within a complete four-sided frame, preferably

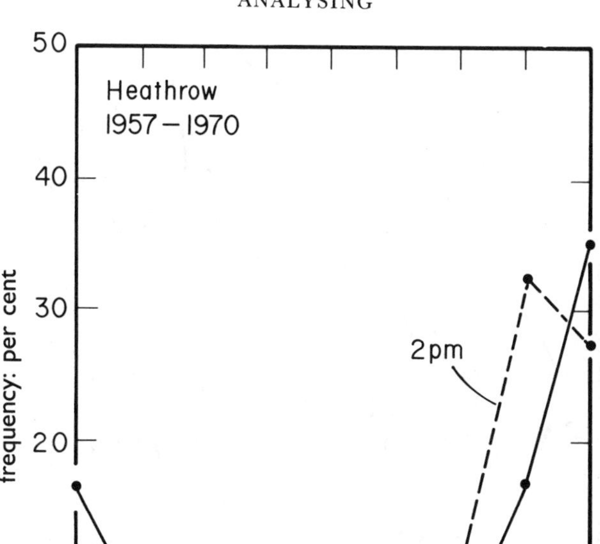

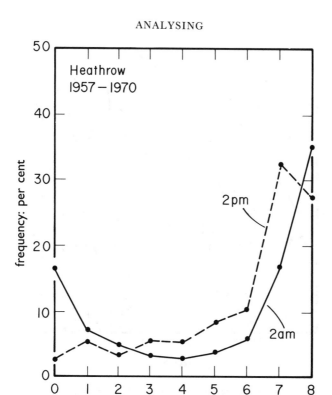

Figure 4.3 The frequency of various degrees of cloudiness at Heathrow, near London (Barrett 1976: 205, by permission of Longman Ltd). Strictly speaking, this is a 'frequency polygon', derived by joining the midpoints of the tops of the columns of a histogram (Lowry 1972: 104)

with calibration gradations on the top and right, as well as on the bottom and left. The latter should each be labelled with the name of the variable, its symbol and units. These axes should intersect only if at least the vertical scale starts from zero. Scale labels should be large and well spaced.

The lower horizontal axis indicates the 'abscissa' (preferably the independent variable, such as time) and the left vertical axis shows the 'ordinate', which should be the dependent variable, such as rainfall. An exception to this rule arises when graphing air temperature etc. against height; it seems more natural to graph height on the vertical axis, as in Figure 4.6. A graph of conditions varying with height is called a 'profile'.

In plotting points for a graph, small crosses or hollow circles are better than dots, because a large dot is imprecise, whilst a small one is hard to see and may be an accidental speck. Lines should not be drawn between symbols unless interpolation is meaningful. Points should be joined by straight lines, except when there are many points and a smooth curve can

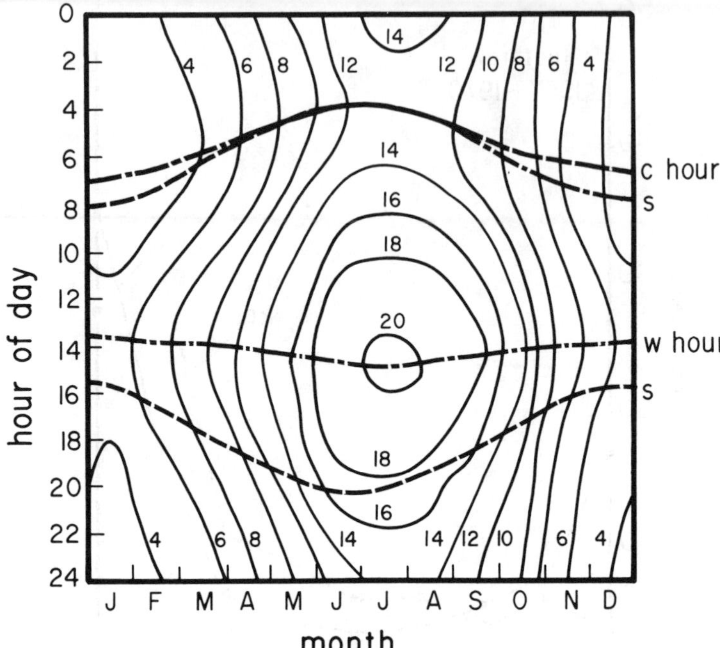

Figure 4.4 The variation of screen temperature at Heathrow near London, during 1957–66 (Lacy 1977: 4). 'c' on the right edge shows the coldest hour, 'w' the time of highest temperature and 's' that of sunrise or sunset. (By permission of the Building Research Establishment, UK Crown copyright.) This kind of diagram is called a 'chrono-isopleth', and preferably the hours should read upwards, so that zero is at the origin in the bottom left, as customary (Court 1989: 3)

be justified logically. For clarity, a graph should normally contain no more than four curves. Grid lines are unnecessary. It is important to restrict the number of words on a graph, but better to label each curve directly than to have a separate key, as is often done.

Other rules for clear graphs have been suggested by Court (1987). Repeated identical values should be indicated by a cluster of symbols, or one of larger size. Equal quantities should be represented by equal areas, not equal lengths, on any polar diagram, e.g. a 'wind-rose', mentioned in Chapter 6.

Equations

An equation is an abstract but concise way of summarising data. It can be manipulated mathematically to reveal useful features such as the average value, the maximum, and the dependent value corresponding to particular conditions of the independent variable, between or beyond those of the

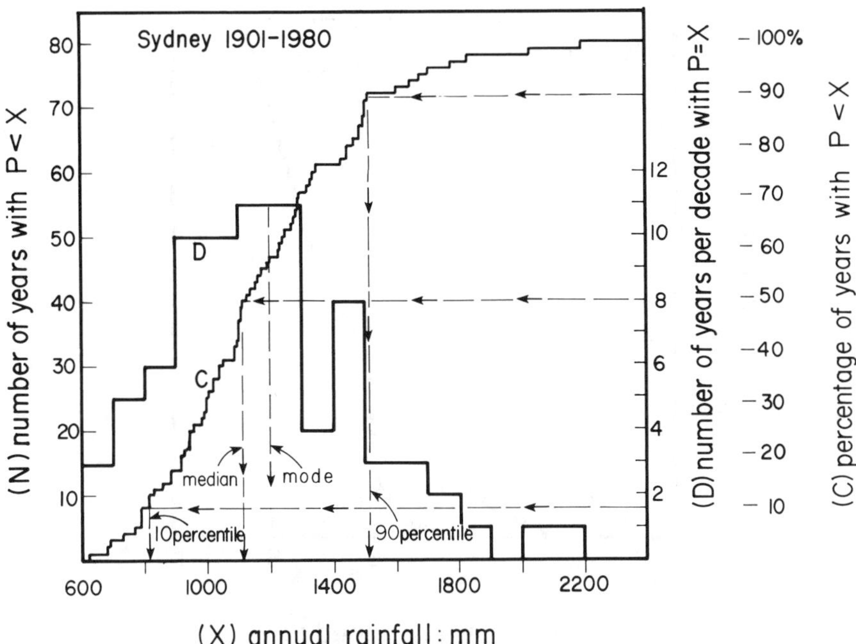

Figure 4.5 Histogram (D) and cumulative frequency curve (C) for annual rainfalls at Sydney during 1901–80

original measurements. The derivation of an equation is discussed in 'Scatter of values, pp. 126–34. The conversion of published equations to proper units was explained in Note 3, Chapter 1 pp. 312–3.

TYPICAL VALUES

Any long record of climate data creates the problem of handling a large set of numbers. This is overcome by concentrating the information, usually by replacing the set by two numbers representing the typical value and the spread of values from it, respectively. These two features are discussed in the present section and the next.

The typical value can be represented by either the 'mode', the 'median' or the arithmetic 'mean' value. The mode is the most commonly occurring value in a set. It is actually the mid-value of the most frequently occurring class of values, so it presupposes discrete data or the grouping of continuous data. The modal value of the dry-bulb temperature in Table 4.2 is 16.9°C, which is the average of 15.6°C and 18.2°C, the limits of the class with the highest column-total. The mode in Figure 4.5 is 1200 mm/a.

The mode can be derived only when there is a large number of data available, to allow reasonable numbers of cases to fall within each of several

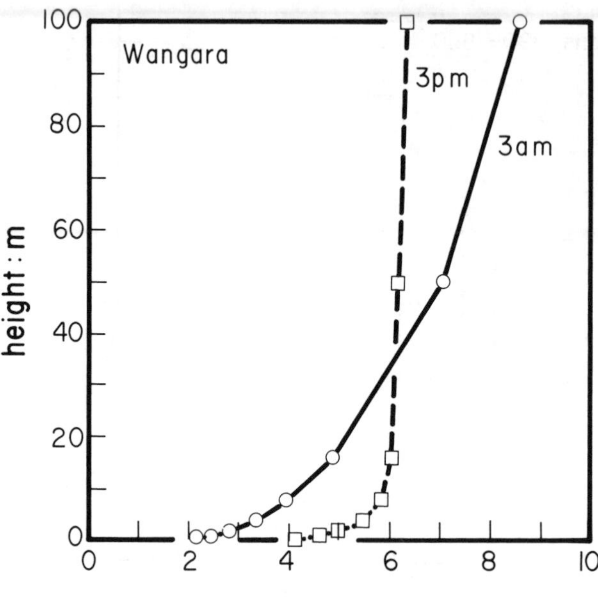

Figure 4.6 Profile of wind speed over flat land at Wangara, near Daley Waters, Australia (Clarke *et al.* 1971.)

categories. The number of categories should be as large as practicable, to permit each to be narrow, so that the mode is specified more precisely, as discussed in connection with Table 4.2. There is no single mode in a case like Figure 4.3. In general, the mode is not recommended as a measure of the typical value of a set, because it takes no account of all the numbers outside the most frequent class.

An alternative is the median, i.e. the 50-percentile (Figure 4.5). If there is an even number of values in the set, the median is the average of the two middle values. It has several advantages, best considered by comparing it with the mean.

Mean values

The arithmetic mean is the sum of all values in the set divided by the number of values. Unlike the median, the mean value of a set is affected by outlier values, whose very extremity may cast doubt on their validity. Thus the mean is said to be less 'robust' than the median, which is especially important when there are only a few values in the set.

Also, people are sometimes confused by the fact, for instance, that only 25 per cent of July months in California have a rainfall exceeding the mean.

(The explanation is that rainfalls in those few months are very high, balancing the small rains in the many other months.) From this point of view, the median seems more representative.

Another disadvantage of the mean is an occasional difficulty in calculating it. For example, the mean date of the last frost in the year cannot be calculated if there was no frost one year, since no date can be specified. Whereas the median is still relevant, even in that case.

However, the mean is usually a more 'efficient' indicator of a set's typical value. This signifies that using the mean of a sample to assess the mean of a normal population (from which the sample is drawn) requires a sample only two-thirds the size of that needed in using medians, for a particular accuracy. Another advantage is that the mean is readily calculated from a random sequence of data, and requires no preliminary arranging of the data into order of magnitude. Also, to update the mean when fresh data become available, it is not necessary to retain all previous values, but only their sum and the number of them. Most statistical procedures are based on this particular measure of the typical value.

Averaging over time

Averaging in climatology usually involves finding the mean of measurements taken over some particular *period*, as distinct from averaging simultaneous values measured at different places, for instance. The mean value over a number of years is useful for comparisons between places or periods, and for prediction. As regards the latter, one assumes that the typical value in the recent past represents a relatively stable climate, underlying short-term perturbations of weather due to cycles of variation and random fluctuations. This proves to be a simple and useful basis for forecasting in practice. From that point of view, a longer period of averaging is better, to even out perturbations more effectively.

Traditionally, the period of averaging to obtain the 'long-term mean' is 30 years, beginning in the first year of a decade. Dixon and Shulman (1984) found that a mean over the last 30 years is a better predictor of next year's temperatures overall than means over shorter periods. So it is unfortunate that the developing nations, where climate data are often needed for designing permanent structures such as dams, are particularly lacking in lengthy records. However, it is possible in some circumstances to compile a single homogeneous and complete record over a long period, by combining shorter, overlapping records from nearby stations.

A long record is not required, or even desirable, if there is any climate change, when the climate some decades ago may be unrelated to the decade to come, and the word 'normal' becomes inappropriate. When the climate is changing, a shorter, more recent period of averaging gives a better forecast of the immediate future, and several authors have concluded that a

Table 4.5 The effect of location and climatic element on the period appropriate for obtaining a long-term average for climatological forecasting

Place	*Time for averaging various elements*		
	Temperature	*Cloud*	*Rain*
between the Tropics			
Inland	5 years	2 years	30 years
Coast	8	3	40
Plain	10	4	40
Mountain	15	6	50
at higher latitudes			
Inland	10	4	25
Coast	15	4	30
Plain	15	8	40
Mountain	25	12	50

Source: Jagannathan *et al.* 1967: 13

period of 10–20 years gives the best prediction. Period lengths recommended for various climate elements and locations are given in Table 4.5. Whatever the period selected, any long-term mean should be accompanied by a statement of the dates involved.

The need to specify the averaging period applies equally to either a maximum, minimum or total value. As a result, it may be necessary to refer to the 'long-term maximum annual minimum monthly average of daily total rainfall', for instance. This unavoidable prolixity is better than the ambiguity which is common (see Note 2, p. 324). For clarity, it is important to *preface* the words 'mean', 'median', 'maximum', 'minimum', 'average', 'extreme' and 'total', with the period or area of the values being considered.

Sometimes the averaging is over a 'season'. This might be an astronomical season (e.g. summer) or a climatological season (e.g. the rainy season) or a phenological one (e.g. the season of flowering of some plant). 'Summer' in Canada begins when the daily maximum temperatures first exceed 18°C, and 'winter' when there is a snowfall of 25 mm. Bryson and Lahey (1958) in Wisconsin suggested unequal seasons of 140 days from 1 November, then 96 days, 57 days and 72 days. Landsberg (1982: 367) recommended starting the climatological year at the time of either a solstice or of an equinox (i.e. either 21 June and 21 December, or 21 March and 23 September). Trenberth (1983: 1276) argued for 3-month periods starting with 1 December, because this allows for the lag of temperatures on solar radiation (Figure 3.6). In the UK, the 'farmer's year' begins on the Sunday nearest to 1 March. Which is the best to use depends on the problem to be solved.

Cyclic fluctuations of climate occur daily and annually, and perhaps over longer periods too. In view of the 24-hour cycle, it is necessary to average

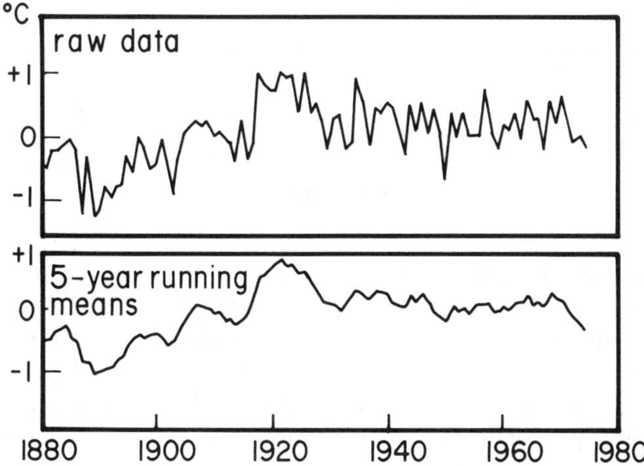

Figure 4.7 Deviation from the long-term average of the annual mean temperatures at Sydney during 1880–1974 and an equivalent graph of 5-year running means (Gibbs et al. 1978: 139). As an example, the 1930 value in the lower graph is the average of those for 1928, 1929, 1930, 1931 and 1932 in the upper graph. (By permission of Cambridge University Press)

over at least several days, or else a multiple of 24 hours, to avoid an effect due to unequal numbers of days and nights in the period. Averaging over a few random 15-hour periods, for example, would not lead to a description of the underlying stable climate since the values would depend on how much daytime was included.

Evidence of climate change is obscured by the scatter of short-term values. The underlying trend can be made more apparent by calculating 'running means' or 'moving averages'. For example, one may take a series of daily values x_1, x_2, x_3, etc. and form a series of 5-day averages, as follows: $(x_1 + x_2 + x_3 + x_4 + x_5)/5$, $(x_2 + x_3 + x_4 + x_5 + x_6)/5$, $(x_3 + x_4 + x_5 + x_6 + x_7)/5$, etc. The impact of any single anomalous daily value is reduced in such a series. An example is given in Figure 4.7, where 5-year running means are taken to filter out year-to-year variations.

Time-averaged means cannot be derived for ratios such as the relative humidity. An average of ratios of unspecified amounts is meaningless. To understand this, consider equal periods when the relative humidity is 67 per cent and 33 per cent respectively, the temperatures being 13°C and 24°C, respectively. The 'average' would be *50 per cent*, but of what is that a percentage? The vapour pressure was 10hPa in both periods, which is not 50 per cent of the saturation vapour pressure at the average temperature but *47 per cent*. Nor is 10hPa half of the average saturation vapour pressure, but *44 per cent*. Further consideration is given in Note 3, pp. 325–6.

Likewise, the non-linear relationship of saturation water-vapour pressure to temperature affects the averaging of dewpoints. Thus, a mixture of equal air volumes with dewpoints of 10°C and 20°C, respectively, does not have a mean dewpoint of 15.0°C, but one of 15.9°C. The latter is the dewpoint corresponding to a saturation vapour pressure of 17.9 hPa, this being the average of values for 10°C and 20°C, respectively.

Homogeneity

There can be a typical value of a set only if the set is 'homogeneous', as mentioned in 'Climate data', pp. 109–10. That is to say, climate measurements in a set must each have been obtained in the same manner and in comparable circumstances (Oliver 1981: 6). All the figures have to be of the same kind, not a mixture of rainfalls and wind speeds or of January daytime and July night-time temperatures, for instance. Heterogeneity can arise from a shift of the instruments, yielding distinct sets of measurements before and after the move. Mixtures like this yield bi-modal sets of data.

A climatological series can be tested for homogeneity by means of the Swed-Eisenhart Runs Test. First, count the runs of values which are respectively above and below the median. Then compare the number of these runs with the limits in Table 4.6. Too many runs would mean an oscillation, too few would imply a trend or shift of the mean, whilst an intermediate number of runs signifies possible homogeneity. Application of the Runs Test to more than 80 years' values of the annual mean temperature at ten first-rank climate stations in the USA revealed heterogeneity at eight of them. Of British weather stations, 34 per cent were

Table 4.6 Limits within which the number of runs[a] in a series must lie for a climatological series to be homogeneous (Langley 1979; and others)

Level of significance[b]		*Number of values in the series*						
		10	*20*	*30*	*40*	*60*	*80*	*100*
10%	minimum no. of runs[c]	4	8	12	16	26	35	45
	Maximum	7	13	19	25	36	47	57
5%	Minimum	2	6	10	14	24	33	42
	Maximum	9	16	22	28	37	48	59
1%	Minimum	2	5	9	13	22	30	39
	Maximum	9	16	22	28	39	51	62

Source: Langley 1979: 325, and others.

Notes: [a] A 'run' is a sequence of values all higher (or lower) than the median of the series. (Ignore any values equal to the median, apart from ending a run.)
[b] The 'significance level' is the likelihood that the number of runs could be outside the specified limits by chance alone.
[c] A number larger than the tabulated upper limit implies a regular fluctuation within the series, whilst a number below the lower limit implies a steady trend.

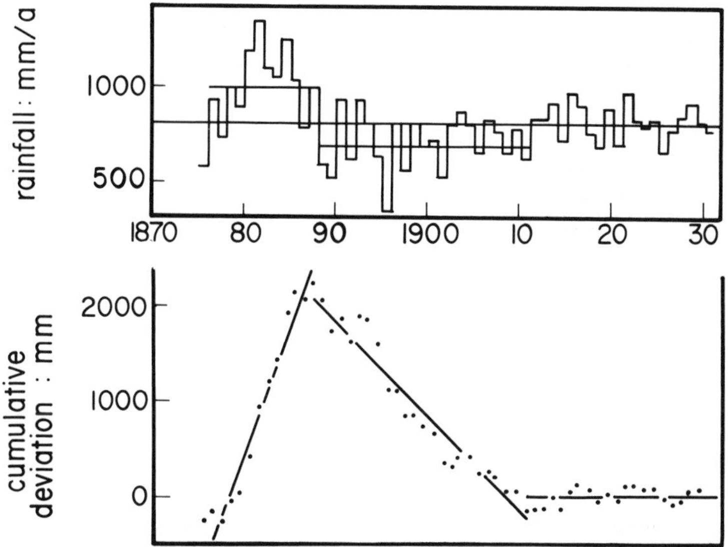

Figure 4.8 Rainfall data from Madison (Wisconsin) in terms of (a) a histogram of annual totals, and (b) the cumulative total of deviations (Dury 1980: 63). The cumulative total at any time is the sum, since the beginning of the record, of annual differences from the long-term mean. (By permission of the Institute of Australian Geographers)

found to have inhomogeneous temperature and sunshine records during 1951–80. Another example is given in Note 4, p. 326.

Non-homogeneity due to discontinuities of the record can also be detected by continuous comparisons with measurements at adjacent climate stations (see Figure 2.11) or by graphing 'cumulative deviations'. The latter involves plotting against time the cumulative total of the deviations of the values from the overall mean, as in Figure 4.8. The upper graph shows too much scatter for a clear appreciation of the changes of trend, whereas the summed deviations in the lower graph clearly reveal three distinct periods of change.

Obviously, the detection of global warming requires homogeneous records. Observations from any particular station can be used only if there is no change in their overall relationship to readings obtained at neighbouring stations, i.e. constant differences of temperature and constant ratios of rainfall.

Mean temperatures

Ideally, the daily mean temperature should be calculated by using a thermometer each fraction of a second, then averaging all the observations. But this is hardly practicable. Instead, the average of hourly measurements

Table 4.7 Estimation of the true daily mean temperature at Rehovot (Israel) by various means

Method of estimation		*30 days' average*[a]	*error: C°*
Average of 24 values, taken an hour apart, i.e. the true daily mean		15.7°C	
Average of 12 values, 2 hours apart		15.7	0
8	3 h	15.7	0
6	4 h	15.8	0.1
4	6 h	15.8	0.1
3	8 h	15.9	0.2
Average of the minimum value and values at 8 a.m., 2 p.m., and 8 p.m.		15.7	0
Average of values at 8 a.m., 2 p.m. and 8 p.m.		17.2	1.5
Average of the maximum and the minimum[b]		16.3	0.6

Source: Kalma 1968: 249

Notes: [a] Each temperature is the average of 30 daily values, obtained during December 1966, and March and June 1967

[b] The usual measure of the daily mean temperature

is called the 'true daily mean'. It is approximated by (T7 + T2 + 2 × T9)/4, for instance, where T7, T2 and T9 are readings taken at 7 a.m., 2 p.m. and 9 p.m., respectively. Similar empirical formulae apply if daily measurements are taken at other times. Alternatively, and more usually, one takes the average of the daily extremes, though this differs from the true daily mean by 0.2–0.9C°, for instance. The differences are usually less than 0.3C° in Britain. However, the average of the daily extreme temperatures can be an overestimate of the daily mean by 3C° in equatorial regions, because of only a brief hot period during the day.

Various methods of estimating the true daily mean are compared in Table 4.7. In that case, averaging measurements taken each six hours gave a better approximation to the true daily mean than the customary average of the daily maximum and minimum.

Annual means are usually taken as averages of the 12 monthly values, despite the variation of month length. However, Guttman and Plantico (1987) compared the annual average daily maximum temperature, calculated, firstly, from the 12 values of the monthly mean daily maximum, with, secondly, the figure derived from the 365 daily maxima There were significant differences in some places in the USA, attributed to one-day persistence, which is discussed later, see 'Persistence', pp. 135–8.

An alternative, approximate method of assessing the annual mean temperature is to measure the soil temperature at 0.3 m depth, which is deep enough for seasonal variations to be negligible.

Averaging non-scalar quantities

Different considerations apply in averaging ratios or vector quantities. For a ratio such as the relative humidity it is nonsensical to average a set of values, unless the temperature is constant (p. 121). As regards vectors, like winds there are two ways to average them. First, when the direction is unimportant, the speeds are averaged arithmetically. This would apply, for instance, if the average wind was needed for calculating the evaporation rate by means of the Dalton equation discussed in Note 2, Chapter 3. But other procedures are required when wind direction cannot be ignored, as in calculating the overall movement of air pollution, for example. A 4 m/s wind from the east for an hour and then one of 4 m/s from the west would average out as zero, not 4 m/s. The average wind in this case is calculated by adding vectors representing the various winds, and then dividing by the total time elapsed. The vector addition can be done either numerically or graphically. Note 5, p. 326 and Figure 4.9 give an example.

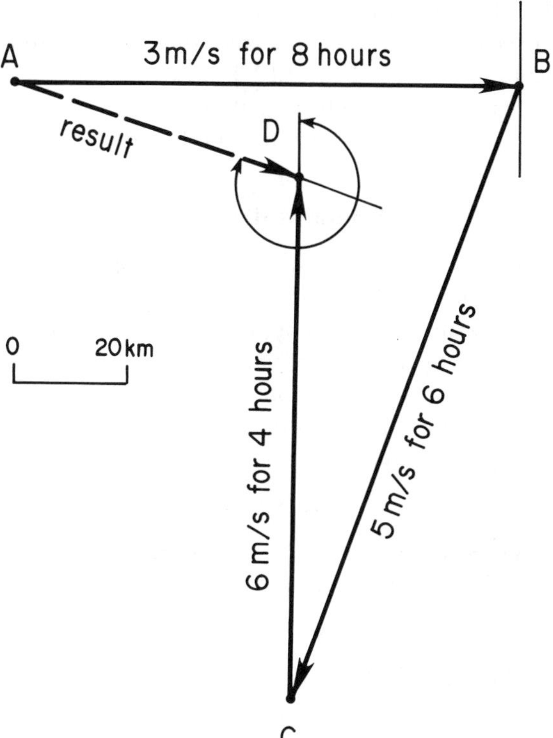

Figure 4.9 An example of vector addition, showing the graphical derivation of the resultant distance of travel (AD), after an initial west wind (AB), then NNE (BC) and then movement in a south wind (CD). The distance AD represents 47 km (in 18 hours), so the average wind is 0.73 m/s from a WNW direction.

SCATTER OF VALUES

To assess the accuracy of using typical values from the past for climatological forecasting, we assume that future differences from the norm will resemble the past scatter of values. This 'dispersion' can be derived from a long record of data in at least three ways – from either the 'range', or the average difference from the mean (called the 'mean deviation'), or the 'standard deviation'.

The range is the difference between the largest and smallest numbers in a set. It is easy to work out, but determined by only two outliers not by the main bulk of the data. This is unfortunate because the extreme values are those most likely to result from accident. A way round that difficulty is to ignore the top and bottom tenths of the ordered set. In other words, one takes the difference between the 90 percentile and the 10 percentile as a measure of dispersion. Better still, one divides that difference by the median value, obtaining a non-dimensional ratio, which can be compared with the dispersion of other sets with numbers of quite different magnitudes. The ratio is called the 'variability index'. Alternatively, one can use the difference between the 75 percentile and the 25 percentile – called the 'inter-quartile range' or 'mid-spread'. It also is unaffected by eccentric values in the set, so it is robust. A disadvantage of such measures of dispersion is their limited relevance to established statistical theory.

Standard deviation

Another method of describing the scatter of a set of N values is in terms of the difference of each value from the set's average. One takes the sum of the differences as a measure of dispersion, as in considering Figure 3.1. This is customary where the distribution of values is 'non-Gaussian', as explained later. For a Gaussian distribution, one takes the square of each difference, and then the sum of the squares is divided by N, to obtain what is called the 'variance' (s^2). The square root of that is the standard deviation (s):

$$s^2 = [\Sigma\ (x - \mathbf{x})^2]/N \qquad (4.3)$$

where $\Sigma\ (x - \mathbf{x})^2$ is the sum of the squares of the differences of x from the mean **x**. This is most easily calculated from the following version:

$$s^2 = [(\Sigma\ x^2) - (\Sigma\ x)^2/N]/N \qquad (4.4)$$

where $(\Sigma\ x)$ is the sum of all N values in the set, and $(\Sigma\ x^2)$ is the sum of their squares.

The 'coefficient of variation' (Cov) is a form of the standard deviation(s), which has the advantage of not involving units. It is a ratio, obtained by dividing s by the set's average (**x**):

$$\text{Cov} = 100\ s/\mathbf{x} \quad \text{per cent} \qquad (4.5)$$

It is not meaningful to obtain values of Cov for temperatures, since the magnitude of **x** depends on the units, Celsius or Fahrenheit.

Amongst various techniques for calculating the standard deviation approximately but quickly is that of Quenouille (1959: 5). His method involves dividing the set into groups, each of nine values, overlapping if necessary. To obtain the standard deviation, find the range of each group, average these ranges, and divide the average by three. A test with a set of 70 temperatures at Marsfield (whose true standard deviation was 0.38C°) yielded 0.32C° by Quenouille's method.

The 'normal' distribution

The histogram of values of annual rainfall (P) in Figure 4.10 shows the spread or 'distribution' of values. In this case the histogram is closely approximated by the bell-shaped curve which results from plotting values of the relative frequency of P, given by the following equation (4.6). It was first developed by Abraham de Moivre in 1733, although commonly named after Carl Gauss (1777–1855). The equation shows the proportion of times that x, the factor on the horizontal axis (i.e. P in Figure 4.10), lies within a small range dx:

$$p(x) = [39.9/s][\exp(-d^2/2)]\,dx \qquad \text{per cent} \qquad (4.6)$$

where d is shorthand for $[(x - \mathbf{x})/s]$, and s is the standard deviation of the set of values of x, **x** their overall average, and 'exp $(-d^2/2)$' means that the 'exponential number' (i.e. 2.7183) is multiplied by itself $(d^2/2)$ times and then divided into unity. (For example, if d is unity, so that the selected value differs from the mean by an amount equal to the standard deviation, the exponential term in Eqn (4.6) equals the inverse of the square root of 2.7183, i.e. 0.607.) The pattern of frequencies given by an equation is called a 'distribution' and Eqn (4.6) refers to a 'normal distribution' resulting from random variation.

The lower graph in Figure 4.10 is the cumulative equivalent of the upper graph, showing the percentage of years when the rainfall was *less* than the value shown on the horizontal axis. It is derived by adding the vertical values of the bars to the left in Figure 4.10(a), for each range of rainfall. It is notably more regular than the histogram and is matched well by the smooth line in the lower graph labelled C(P), obtained by mathematical integration of the right-hand side of Eqn (4.6).

A horizontal line from the 50 per cent value on the left of the lower graph intersects the curve at a point which represents the median rainfall on the bottom axis. That same value happens to correspond to the mode in the upper diagram. Eqn (4.6) shows that it also equals the mean rainfall in the case of a normal distribution.

Several climate quantities are distributed 'normally', ie according to Eqn

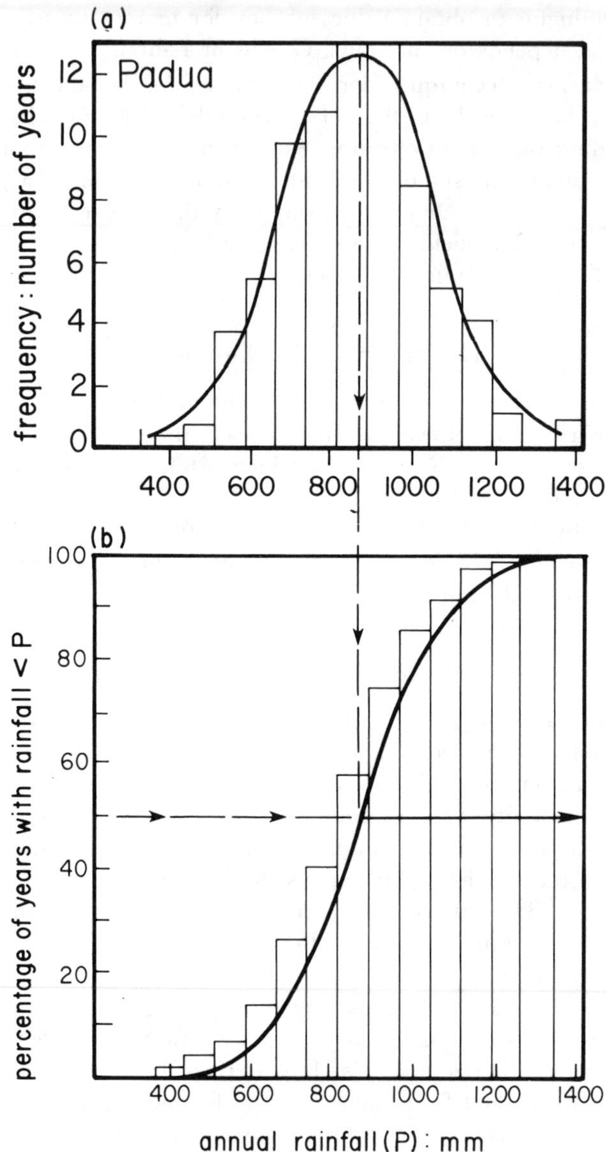

Figure 4.10 (a) Histogram of annual rainfalls at Padua (Italy), showing the relative number of years during 1725–1957 when the annual rainfall equalled P mm (Manning 1956: 462). This is an unusually long period of record. The histogram is accompanied by a bell-shaped curve from Eqn (4.6), representing the 'probability density distribution' p(P). (By permission of the Royal Society, London)

(b) The cumulative equivalent of (a), including the cumulative curve C(P), i.e. the mathematical integral of the right-hand side of Eqn (4.6)

Table 4.8 Proportion (Ed) of values which are beyond a certain number (d) of standard deviations from the mean, assuming a Gaussian distribution (i.e. Eqn 4.6)

d	0	0.1	0.3	0.6	0.6745	1.0	1.3	1.6	2.0	2.5	3.0
Ed:%	50	46	38	27[a]	25[b]	16	9.7	5.5	2.3	0.62	0.13

Notes: [a] 27 per cent of values lie above (**x** + 0.6 s), where **x** is the mean and s is the standard deviation of the set. Another 27 per cent of values lie below. So 46 per cent of values lie within 0.6 s of the mean.

[b] Exactly half the values lie within 0.6745 times the standard deviation from the mean. This is called the 'probable error' and sometimes used as a measure of scatter as an alternative to the standard deviation (Conrad 1946: 39, and others.).

4.6. They include the annual rainfall at places with a mean over 500 mm/a, summer rainfalls in northern China, and the dates of first frost. Even if a set of values proves not to be normal itself, the set of some function of the values may give a reasonable approximation. For instance, values of the *logarithm* of particularly intense rainfalls prove to be distributed normally. Likewise, a set of values of the *square root* of January rainfalls between 150–640 mm/mo at Rembang in Indonesia fits Eqn (4.6) better than do the raw rainfall data. Also, the square roots of annual rainfalls at places in eastern Australia, and the square roots of monthly mean wind speeds in New England (USA) yield normal distributions. The process of altering each value of a set (e.g. to its logarithm or square root) is called 'transforming the variable'.

The percentage of values beyond a certain number of standard deviations from the mean is given by Table 4.8. This shows that the standard deviation can be taken as approximately the difference either between the 16th percentile and the mean, or between the mean and the 84th percentile,

Table 4.9 Comparison between measured and estimated extreme values of daily maximum temperatures at various places in Australia.

Place	*Measured* *Mean*	*Measured* *86%*[a]	*Measured* *Extreme*[b]	*Estimated extreme*[c]	*Difference between extremes*
Tennant Creek	38.7°C	41.8	46.1	47.4	+ 1.3C°
Halls Creek	37.0	40.6	44.3	47.1	+ 2.8
Cairns	31.5	33.1	43.2	36.0	− 7.2
Parramatta	27.7	33.9	46.3	45.1	− 1.2
Albany	25.8	29.4	44.8	35.9	− 8.9

Notes: [a] The 86th percentile value.

[b] The highest of about 970 measured values of daily maximum temperature during January over about 30 years

[c] The extreme estimated as the sum of (a) the product of 2.8 times the difference between the 86-percentile and the mean, plus (b) the overall mean daily maximum.

or, better, the average of those differences. Furthermore, 14 per cent of values are more than 1.08 standard deviations above the mean, so daily values would exceed (**x** + 1.08 s) about once a week. In addition, it follows from Table 4.8 that 2.8 times the difference between the 86th percentile and the mean equals three times (i.e. 1.08 × 2.8) the standard deviation, and the excess above the mean is surpassed only once out of 770 occasions (i.e. 1/0.0013 from Table 4.8). So this approximates the expected maximum value. A comparison between values obtained by using this rule and maximum values measured in Australia is shown in Table 4.9. The disparity for Cairns and Albany suggests non-normal distributions there, perhaps due to different rainfall-producing processes at various times of the year, resulting in non-homogeneous data sets.

Tests for normality

A distribution can be tested for normality in several ways. Rigorous tests are used by statisticians, and suitable software can be obtained for computers. But three simple requirements of normality have been mentioned already: (a) the mode, median and mean are similar, (b) there is a symmetrical histogram, and (c) actual extreme values resemble those estimated on the assumption of normality. Another is that attributed to Alfred Cornu (1841–1902) – the standard deviation of a normal distribution equals 1.25 times the average difference of values from their mean. Or 68 per cent (about two-thirds) of values lie within one standard deviation from the mean (see Table 4.8). However, meeting these tests is necessary but not sufficient to prove normality.

A further test is based on the 'cumulative distribution function' represented by the smooth, S-shaped curve in Figure 4.10, the 'ogive'. This curve can be made straight if the vertical percentage scale is stretched appropriately, resulting in 'normal probability graph paper', or simply 'probability paper'. So, normality of measured values leads to their lying on a straight line when plotted on probability paper. The slope of the line yields the standard deviation, taken as one-third of the difference between the 93.3 and the 6.7 percentiles. Applying this test to July daily maximum temperatures at Frankfurt, for instance, shows that they are indeed distributed 'normally' about the mean given by the 50th percentile, but February values fit less satisfactorily.

Another example is given in Figure 4.11, using data from Table 4.10 and Note 6, p. 326. The graph shows good linearity, indicating normality of the data. Also, the median is seen to be 17.66°C, close to the mean of 17.56°C, and to the mode, shown in both Figure 4.2 and Table 4.10 to be in the range 17.40–17.59°C. Moreover, normality in the case of Sydney's annual mean temperatures is supported by the symmetry of the frequencies about the mode, seen in Figure 4.2. So it might well be concluded that variations

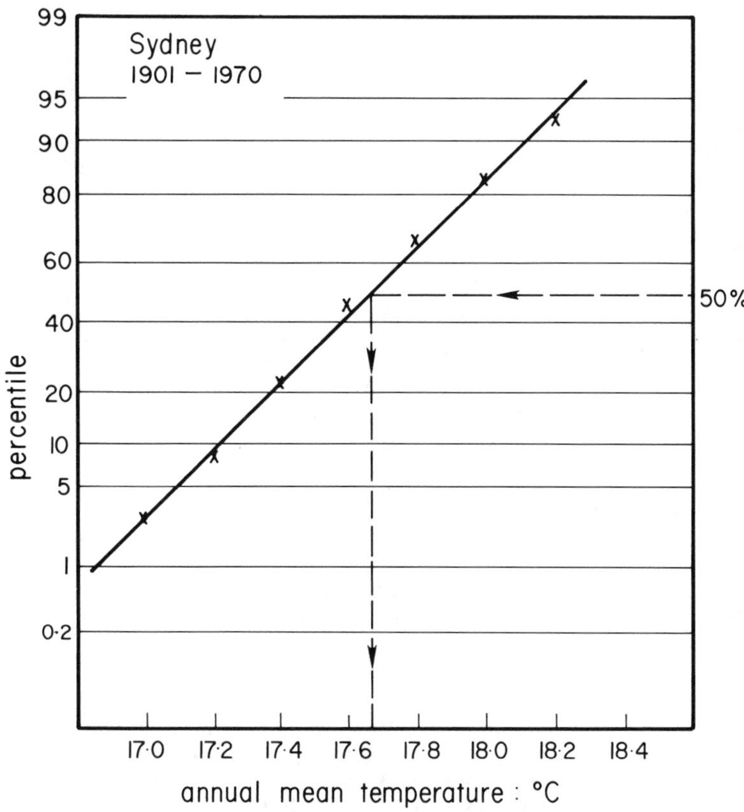

Figure 4.11 The cumulative frequency distribution of annual mean temperatures at Sydney during 1901–70, using data from Table 4.10 and Note 6.

Table 4.10 Annual mean temperatures at Sydney during 1901–70 (see Note 6, p. 326)

Range of values	*Number of years*	*Frequency p(T): %*	*Cumulative frequency C(T): %*
Below 17.0°C.	2	2.8	2.8
17.0–17.19	4	5.7	8.5
17.2–17.39	10	14.3[a]	22.8[b]
17.4–17.59	16	22.8	45.7
17.6–17.79	15	21.4	67.1
17.8–17.99	11	15.7	82.8
18.0–18.19	7	10.0	92.9
Over 18.19°C.	5	7.1	100.0%
Totals	70 years	100%	

Notes: [a] 10 years out of 70.
[b] 2.8 + 5.7 + 14.3, from the previous column.

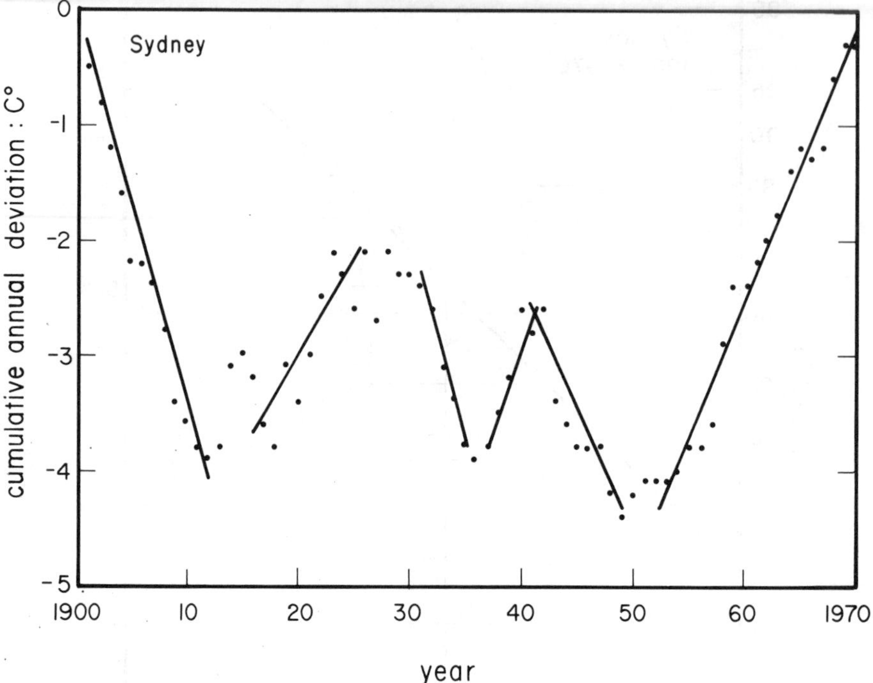

Figure 4.12 Variation of the cumulative deviation from the long-term mean of the annual mean temperature at Sydney, using data in Note 6. The long-term mean was 17.6°C, and the annual means were 17.1°C, 17.3°C, 17.2°C, etc. for 1901, 1902, 1903 etc., so the cumulative deviations were −0.5C°, −0.8C° (i.e. −0.5 + [−0.3]), −1.2C° (i.e. −0.8 + [−0.4]), etc.

of annual mean temperature have been simply random fluctuations from the average.

However, on the contrary, the more searching Runs Test reveals regular trends. It shows 24 runs with respect to the median annual mean temperature, which is fewer than the minimum of 26 for 70 homogeneous values, indicated in Table 4.6 at the 1 per cent level of significance. The pattern is clarified by plotting the cumulative annual deviations of temperature from 1901. Figure 4.12 shows relatively low temperatures in 1901–1911, 1928–1936 and 1941–1949, with warmer periods between. Slopes are the same for the three parts of the graph corresponding to warming, implying the same warm climate then, flipping to or from a fixed cool climate in 1914, 1927, 1936, 1941 and 1951. Subsequent data, plotted in the same way, indicate a yet warmer regime at Sydney from 1970 onwards.

If a logarithmic transformation of the variable is needed to obtain a normal distribution, the set gives a straight line on 'log-normal probability graph paper'. This has the horizontal axis proportional to the common logarithm of the variable, and a vertical axis as in Figure 4.11.

Other distributions

It is easy to overstate the universality of the normal distribution. It does not always fit the observations, e.g. those of daily mean wind speed. Monthly rainfalls rarely occur in a fashion that is Gaussian, because of seasonal changes. In addition, daily rainfall data include many days with zero precipitation, which prevents the histogram being bell-shaped. And an example has already been given of a bimodal curve (see Figure 4.3), which is quite different. Any equation to represent such data must involve an expression more complex than Eqn (4.6). This is dealt with in specialist textbooks.

One particular alternative to a normal distribution merits attention here. It is named after Simeon-Denis Poisson (1781–1840) and describes the frequency with which rare, independent events occur, like the number of hailstorms in a year at a certain place. The frequency of x events in a period is given by the following formula, when the event has a low probability of occurrence (p) on each of n occasions, n being a large number such as 365 days in a year:

$$p(x) = (n.p)^x/(\exp n.p).x! \tag{4.7}$$

where x! is the 'factorial function', equal to x times (x − 1) times (x − 2) etc. times 3 times 2 times 1. For instance, 4! has the value 24. The product n.p is the average number of events in a period. As an example, if there are 90 days with lightning in the course of 17 years, the average number is 5.3 lightning-days per year (and thus the chance (p) of any particular day having lightning is 5.3/365, i.e. 0.0145, or 1.45 per cent). So the chance of eight strikes in a particular year is given as follows:

$$p(8) = (5.3)^8/(\exp 5.3) \times 8 \times 7 \times 6 \times 5 \times 4 \times 3 \times 2 \times 1 = 7.7\% \tag{4.8}$$

The Poisson distribution applies only to discrete, random events and differs from Eqn (4.6) in giving an asymmetrical graph, unless the mean (n.p) exceeds about nine. The standard deviation of the Poisson distribution equals the square root of the mean.

Sampling

Often the problem arises of using an incomplete set of data to assess the properties of the whole set. For instance, there may be only a short period of record to be used in assessing the long-term mean value. So the questions arise – how best should one make this assessment, and how accurate is it likely to be? The first question is easy to answer – the mean of a complete set is best estimated as equal to the sample's mean. Then the second question is answered in terms of the 'standard error' (se), which is the standard deviation of the difference between the sample's mean and the set's mean:

$$se = s / N^{0.5} \tag{4.9}$$

where 'se' is in the same units as the sample's standard deviation (s), and N is the number of measurements in the sample. There is a 68 per cent likelihood that the set's mean lies between the sample's mean plus or minus the standard error, and a 95 per cent chance that it is within twice the standard error from the sample mean, so long as the sample contains at least 20 values (cf. Table 4.8). It follows directly from Eqn (4.9) that the number of values (N) in the sample, required for estimating the mean within a given standard error, is $(s/se)^2$. Also, the number of observations needed to determine the total set's mean within 1 per cent of the true value is $(100\ s/\mathbf{x})^2$ where **x** is the sample mean. That is the square of the coefficient of variation Cov, given by Eqn (4.5). An application of these ideas is demonstrated in Note 7, p. 327.

PERIODIC VARIATIONS

One cause of non-random scatter of climate elements is some regular or quasi-regular fluctuation. Daily and annual variations of temperature and radiation are obvious and usually linked with similar fluctuations of rainfall. There may also be more subtle rhythms, including the 26-month Quasi-Biennial Oscillation, the El Nino–Southern Oscillation, or a fluctuation associated with the 11-year cycle of sunspots which are discussed in 'Long-term, annual and seasonal rainfalls', pp. 276–83.

Various other rhythms also have been reported chiefly of temperature or precipitation, with periods of either about 2–3 years, 5–6 years, 18–24 years, 70–90 years or 190 years. Ye (1986: 89) found what might prove to be a useful 2–3 year cycle of annual mean temperature in a part of China. Mason (1976: 478) demonstrated rhythms of about 2.1 years and 23 years in the annual mean temperatures of central England during 1700–1950. Almost two centuries of monthly mean temperatures at three places in Germany show rhythms of 2.3 years and 3.5 years. And so on. There has been a constant search for regularities, as a contaminant of homogeneous data and as a possible guide in forecasting.

Any well-established periodicity helps in predicting the future. For example, if there is a clear rhythm of 11 years, say, then we may expect minimum conditions some 5–6 years ahead, when conditions are presently maximum. In this way, a subsequent dry period in New Zealand was successfully predicted in 1980 (Tomlinson 1987).

Subtle regularities imply only a small influence on the weather. Thus, Tyson (1981) discovered a rhythm of relatively wet and dry periods in South Africa, each cycle lasting 18 years or so, but the amplitude of this tendency is much less than the random variation from year to year (cf. Figure 1.3), so the effect is too small to be important in forecasting. This has proved generally to be the case. Few periodicities are large enough and certain enough to be of much practical value in long-range weather forecasting, contrary to the claims of some commercial weather forecasters. 'A search for dependable periodicities – inspired by the hope of finding some predictable element in the year-to-year and decade-to-decade, or longer, variations of climate – has long provoked effort out of all proportion to the results so far attained' (Lamb 1972: 214). 'Cyclicity is the bugaboo of climate studies. Many cycles are reported, few are confirmed' (Kerr 1984).

The variety of periods that have been discerned leads to the comment that some cycle can be found to fit most data, but the regularity seems to end as soon as it has been discovered (Kneen 1982: 4). An example is the 4-year rhythm of annual snowfall during 1956–72, illustrated by Linacre and Hobbs (1977: 94), which did not exist during 1935–46 and which disappeared from 1973 onwards. Such cycles are described as 'fugitive'.

The search for regularities of weather is beset with mathematical traps. Even random numbers can create the impression of regularity. For instance, a series of random values obtained from the last two digits of consecutive numbers in a telephone directory can yield a spurious rhythm. A simple method of testing for rhythms is described in Note 8, p. 327, based on the concept of 'persistence', discussed below.

One artificial regularity arises in taking running means of a climatological series (see p. 121). When running means are taken over n years, there is a tendency for a periodicity of 2n years to appear in the outcome. This is known as the Slutzky–Yule effect. 'Climatologists should use the technique of moving averages only with considerable care' (Landsberg 1957: 11).

PERSISTENCE

Persistence is a possible feature of a climatological series, preventing the individual data from being treated as statistically random. It is a tendency for conditions in one period to continue as before, like the common tendency for one wet day to be followed by another. It is due to a lag in changing from one kind of weather to another, which leads to a clustering of occasions with the same climate, i.e. settled weather. If it is much in evidence, the numbers in a set cannot be regarded as independent, having merely random variation from a mean, since persistence is neither wholly random nor entirely deterministic, but 'stochastic'. That means that the

Table 4.11 Evidence of persistence of monthly mean temperatures in Colorado (USA) during 1887–1957

June temperature	*July temperature*		
	Abnormal	*Normal*	*Subnormal*
Abnormal	14[a]	4	4
Normal	6	13	6
Subnormal	2	8	13

Source: Dickson 1967: 36.

Note: [a] 14 of the 71 years had both abnormal June and abnormal July mean temperatures.

sequence of events is partly determined by chance and partly by the frequency distribution.

Persistence occurs over the time taken by the dominant meteorological process. For instance, where the wet monsoon season lasts a few months, there is month-to-month persistence of rainfall in that period So persistence from one time to the next is more evident with short periods. On the other hand, annual values are unlikely to be governed by their predecessors, because a long period allows other influences to intrude on the inherited conditions.

Persistence over a long period tends to affect a large area, being due to some massive process (see Table 1.3). On a small scale, persistence is associated with the kind of lag discussed in the section on 'Measurement of temperature', pp. 36–7.

The weather forecaster welcomes any evidence of persistence, as another guide to future conditions. It has been widely reported to occur in various circumstances. For example, a warm spring often precedes a warm summer in the western USA. There is a good chance of a hot June in Colorado being followed by a hot July (see Table 4.11). Month-to-month persistence of temperature has been found at 100 places in the USA, occurring especially in summer. If monthly mean temperatures there are categorised equally into either abnormal or subnormal, month-to-month persistence (i.e. a repetition of the same kind of weather) gives correct forecasts 64 per cent of times against 50 per cent by chance. Month-to-month persistence of temperatures in winter is much more evident in the western and south-eastern USA than in the central states.

There are several ways of testing for persistence. One is the 'chi-squared test', see 'Relationships', on pp. 140, which may be applied to data like those in Tables 4.3 and 4.11. More directly, one can consider persistence in terms of 'initial probabilities' and 'conditional' (or 'transitional') probabilities. For example, consider the case of wet and dry days at Los Banos near Manila in the Philippines. There were 188 dry January days in 10 years

Table 4.12 Probabilities of dry and wet days[a] in each month at Los Banos (Philippines), based on 10 years of daily data

Month	$p(D)^b$	$p(W)^c$	$p(D/D)^d$	$p(W/W)^e$	$p(W/W) - p(W)$
January	0.606	0.394	0.717	0.566	0.172[f]
February	0.721[g]	0.279	0.794	0.468	0.189
March	0.819[g]	0.181	0.889	0.491	0.310
April	0.797[g]	0.203	0.826	0.328	0.125
May	0.442	0.558	0.690	0.610	0.052
June	0.370	0.630	0.500	0.707	0.077
July	0.297	0.703[h]	0.419	0.756	0.053
August	0.297	0.703[h]	0.489	0.732	0.029
September	0.267	0.733[h]	0.436	0.793	0.060
October	0.474	0.526	0.618	0.658	0.132
November	0.393	0.607	0.573	0.721	0.114
December	0.506	0.494	0.650	0.641	0.147

Source: Robertson 1970: 27.

Notes: [a] A 'wet day' is defined as one with at least 0.2 mm of precipitation,
[b] i.e the chance of a day in that month being dry, e.g. 0.606 in January,
[c] i.e. the chance of a day in that month being wet, e.g. 1−0.606 in January,
[d] i.e. the chance that a dry day is followed by another,
[e] i.e. the chance that a wet day is followed by another,
[f] i.e. the difference to the chance of a wet day, attributable to persistence,
[g] i.e. dry months.
[h] i.e. wet months.

(i.e. 188 out of 310), so the initial probability of a dry day p(D) was 188/310, i.e. 0.606. That is the overall chance of a dry day, irrespective of conditions beforehand. Consequently, the chance of a wet day p(W) was 0.394 (i.e. 1 − 0.606). Also, the 'return period' of a dry day was 1/0.606 (i.e. 1.65 days), the average time between dry days and therefore the average length of a wet spell. However, 135 of those 188 days followed immediately after a previous dry day, so the 'transitional probability' p(D/D) is 135/188, i.e. 0.717. In other words, 0.717 is the chance of a dry day, given that the day before had been dry. If, instead, the previous day had been one of the 122 wet days (which was the case on 53 dry days, i.e. 188 − 135), the transitional probability of this day being dry p(D/W) would be 53/122, i.e. 0.434. Hence one deduces the chance of a wet day after a wet day, denoted p(W/W), equal to [1 − p(D/W)], i.e. 0.566.

These figures imply some persistence of dry weather, because p(D/W), i.e. 0.434, is less than p(D), i.e. 0.606, which is less than p(D/D), i.e. 0.717. The difference between p(D/D) and p(D) is one measure of the amount of persistence. Data in Table 4.12 show that persistence of wet days is most likely in the early dry months, especially March. That is to say, wet days in the dry period tend to come in spells, rather than at random.

The term p(W/W) is called a 'first-order transitional probability', and

Table 4.13 The effect of the time of year and previous weather on the chance of a wet day in Brisbane

Previous weather	*Chance*	*January*	*September*
Wet or dry yesterday	p(W)	0.42	0.25
At least 1 wet day	p(W/W)	0.63	0.45
At least 2 wet days	p(W/WW)	0.67	0.43
At least 3 wet days	p(W/WWW)	0.67	0.39

Source: Brunt 1966: 533.

p(W/WW), which is the chance of a wet day after *two* wet days, is a second-order transitional probability. And so on. They are compared in Table 4.13 for wet days in Brisbane. The likelihood of another wet day declines after two wet days in September, so that is the length of the typical wet spell in spring, whereas wet spells last much longer in January, presumably because rainfall then has a different cause.

The procedure of estimating a sequence of events in terms of the initial and transitional probabilities is known as using a 'Markov chain model'. In a 'first-order' model, the transitional probabilities refer to conditions on the previous day alone [e.g. p(D/W)], whereas a 'second-order' model involves calculating the chances of 3-day sequences such as dry–dry–wet, like the third row in Table 4.13. Fortunately, the simpler first-order model usually fits observations quite adequately.

The following example shows the advantage of allowing for persistence. It is based on calculations like those in Note 9, pp. 327–8, yielding the figures in Table 4.14 for the frequency of runs of various length of dull days at Bangkok. The frequencies were estimated either by assuming no persistence (the calculations being based solely on the initial probabilities of dull and bright days), or by assuming first-order persistence. The latter gives results which better match the observed frequencies of runs.

Some climate elements exhibit more persistence than others, e.g. daily maximum temperature, compared with daily solar irradiance. The coefficient of correlation between the temperatures on one day and the next is 0.63, compared with only 0.22 for daily irradiance values in the USA. The reason is that thermal lag affects temperatures, but there is nothing equivalent for irradiance.

RELATIONSHIPS

It was pointed out in the section 'Estimating in general', pp. 64–7, that the estimation of one variable from the measurement of another depends on a strong link between them, i.e. on any alteration of the controlling (independent) variable A being always accompanied by an obvious change

Table 4.14 Test of alternative formulae for the number of runs of dull days annually at Bangkok. Compare the numbers of runs in italic, in particular

	Number of runs of n dull days in one year					
	Observed		*Estimated (see Note 9)*			
			No persistence		*With persistence*	
Length of run (n): Days	*Runs*	*Days*	*Runs*	*Days*	*Runs*	*Days*
1	32.0	32.0	46.5[b]	46.5	31.2[c]	31.2
2	9.4	18.8	9.2	18.4	10.7	21.4
3	3.4	10.2[a]	1.8	5.4	3.7	11.1
4	1.8	7.2	0.4	1.4	1.3	5.2
5	0.8	4.0	0.1	0.4	0.4	2.2
6	0	0	0	0	0.1	0.9
Annual totals		72.2		72.1		72.0

Source: Exell 1982: 6.

Notes: [a] 3×3.4; [b] 235×0.198, [c] $31.2 \times 0.343^{n-1}$, where n is unity in this case.

of the controlled (dependent) variable B. Of course, an association between changes of the two variables may not mean that A controls B. Instead, B may control A, or both A and B may be at least partly controlled by another variable, C. Likewise, an absence of a change of B when A alters, could mean that A controls both B and another factor C, where C in turn affects B in a way that compensates for the effect of A on B. In that case, B would not change when A alters, even though they may be strongly linked. So an association between changes of A and B is indicative, but not conclusive proof of the mechanisms at work. Relationships do not necessarily imply cause-and-effect.

The element C is called a 'confounding factor', which complicates what otherwise would be a straightforward connection between A and B. It is present, for instance, in the case of an increase of car accidents when air pressures are low. The association does not mean that low air-pressure causes accidents. The increase is probably due to low pressures occurring with the kind of weather which brings more cars on to the roads or makes roads slippery.

Another type of confounding occurs when both A and B values are multiplied by the same factor C. This produces an apparent association between the products A.C and B.C, even though A and B are mutually independent and C differs for each pair of A and B values. A similar spurious correlation arises if you relate A to either the sum (A + B) or the product A.B.

One way to test for an association between A and B makes use of the Runs Test (see p. 122). Pairs of A and B values are arranged in ascending order of A and then the number of runs where B is either above or below its

Table 4.15 The level of significance for various values of chi-squared

Chi-squared:	10.83	7.88	6.63	5.41	3.84	2.71	1.64	0.46
Significance: %	0.1	0.5	1	2	5	10[a]	20	50

Source: Lloyd 1984: 370.

Note: [a] If chi-squared is 2.71, the observed relationship would occur by chance on 10 per cent of occasions.

mean value is counted. If the number of runs falls below the range permitted in the table, there is indeed some connection.

Chi-squared test

A different method of checking for a relationship between two variables is based on the kind of contingency table shown in Tables 4.3 and 4.11. It is called the 'chi-squared test'. It hardly depends on the set of data being 'normal' or fitting any other known or assumed frequency distribution, so the test is described as effectively 'non-parametric'. In the simplest case, a dichotomous contingency table, one has a frequency 'a' in the top left, 'b' top right, 'c' bottom left and 'd' bottom right. The term 'a' in Table 4.3, for instance, is the number of cases when the conditions of the first column and of the first row were simultaneously met, i.e. the number of days of student attendance when temperatures exceeded 20°C. The term 'd' is the number of student absences at lower temperatures. One calculates chi-squared from Eqn (4.10) and compares the result with values in Table 4.15. This shows the 'significance' of any relationship, i.e. the likelihood of it occurring by chance. The larger is chi-squared, the more certain is any trend. A limitation on the chi-squared test is that each of the numbers a, b, c, d must exceed four.

$$\text{chi-squared} = [(a.d - b.c) - N/2]^2\, N/(a+b)(c+d)(a+c)(b+d) \qquad (4.10)$$

where (a.d − b.c) is a difference of products expressed as a positive number, and N is the sum (a+b+c+d).

Use of the chi-squared test prevents a particular common error in interpreting data. For example, a graph of the numbers of suicides in Houston, plotted against the day's temperature, shows a peak at 25°C. But this does not mean that there is a particular risk at that temperature, merely that it was the most common temperature during the study period. Applying the chi-squared test reveals that there is no significant dependence of suicide rate on temperature.

Determining the relationship

A chi-squared test indicates whether or not there is an association between two variables, but does not show what the relationship is. For that, it is common practice to plot the data on a graph and then to fit a line through the data. The equation corresponding to the line expresses the desired relationship.

As an example, consider the hypothesis that afternoon cloudiness is likely to be greater when the surface air is nearer saturation. The latter could be represented by the difference between the afternoon maximum temperature and the dewpoint temperature. To test the hypothesis, we assemble daily data on cloudiness and on (Tmax − Td), and then plot them as in Figure 4.13. (Ignore the lines and equations for the present.) Clearly it is not easy to draw a line through such a cloud of points, but it can be done, as follows.

Figure 4.13 has the axes correctly oriented, with the dependent variable as ordinate, if we assume that the degree of surface air saturation controls cloud, and not the other way round. Then, one method of obtaining a line summarising the dependence of the cloudiness on the humidity involves replacing all the points in each group of the independent variable by the column's median value. (The median is easier to find than the mean.) These medians are represented by the *triangles* in Figure 4.13. For estimating the unknown variable, the triangles can be joined by a line, the equation of which is as follows:

$$C = 10 - 0.5\,(\mathrm{Tmax} - \mathrm{Td}) \qquad \text{oktas} \qquad (4.11)$$

If, on the contrary, we were to treat (Tmax − Td) as dependent on the cloudiness, the data can be reduced by replacing each *row* of data by its median, shown by the *circles*. This yields a quite different line and equation. Rearrangement of that equation provides an estimate for the dewpoint temperature:

$$\mathrm{Td} = \mathrm{Tmax} + 0.8\,C - 14 \qquad {}^{\circ}\mathrm{C} \qquad (4.12)$$

The slope of the line in Figure 4.13 defined by Eqn (4.11), differs from that of the line defined by Eqn (4.12), because of the scatter of the points. The more the scatter, the bigger the difference.

The grouping of data in Figure 4.13 produces surprisingly well-defined lines, and the procedure is more useful than that which beginners in climatology usually adopt. They appeal to the reader to intuit a relationship from parallelism between graphs of the two variables against time. However, the dimension of time is usually irrelevant, and the graphs provide neither an equation for the relationship nor any quantitative measure of its certainty. To quote Gani (1975: 507): 'many of the arguments presented by climatologists are based on poor foundations – apparent similarities, parallel-looking curves and analogous trends do not constitute proof of a scientific hypothesis'.

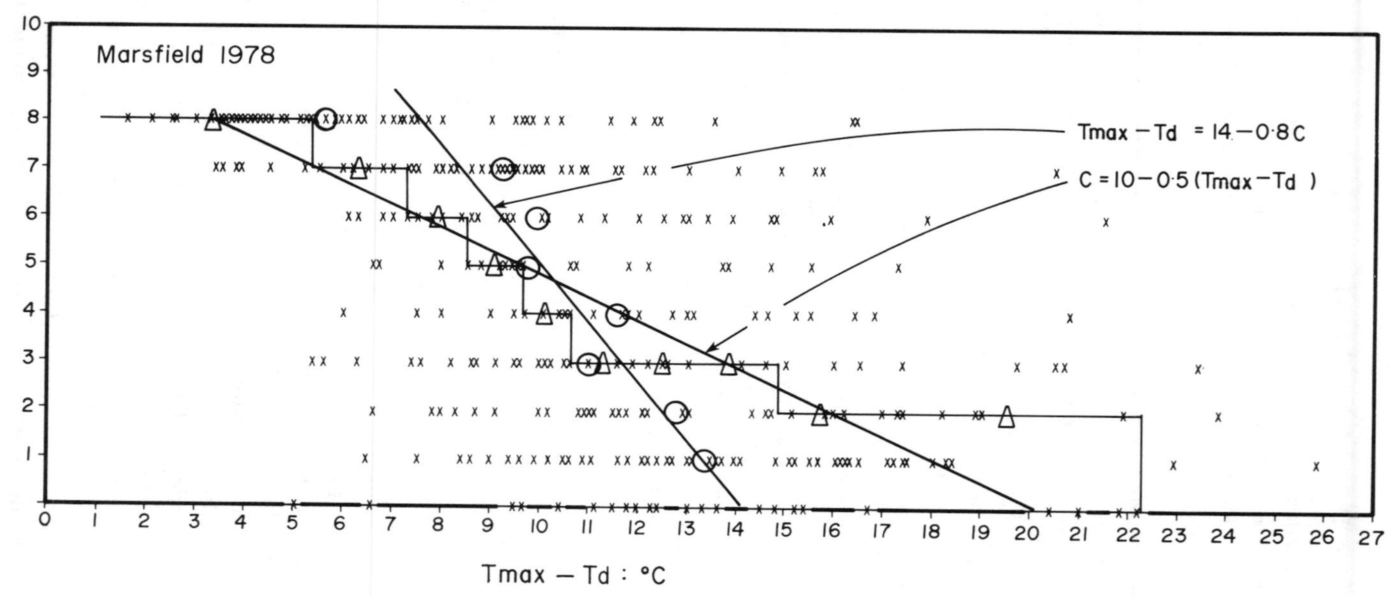

Figure 4.13 The effect of daily surface-air humidity conditions on the afternoon cloudiness at Marsfield in 1978. The proximity to atmospheric saturation is shown by the difference between the afternoon maximum and the dewpoint temperatures

The significance of the relationship in Figure 4.13 could be derived by means of a contingency table, constructed by counting the number of points in each quadrant formed when a vertical line and a horizontal line are drawn through the approximate centre of the cloud of points. Then one applies the chi-squared test.

James (1969: 103) described alternative, quick methods of 'regression analysis', i.e. of finding the appropriate line (called a 'regression line') through scattered points on a graph. One method is suitable if there are over seven points – divide them into three equally-sized groups from left to right, find the centroids of the outer groups and join them. (The 'centroid' of a group is the point whose ordinate is the average of the ordinates of the members of the group, and likewise for the abscissa.)

The usual procedure is the 'least-squares' method of estimating the regression line, is described in any statistics textbook. It shows an equation for the line, its level of significance and the 'correlation coefficient', the last being a measure of the strength of the relationship, discussed in Note 10, pp. 328–9.

It is good practice to use only some of the available data (the 'training set') to derive a regression equation, and then to use the remainder (the 'test set') for comparison with the equation's estimates of the dependent variable. This gives a realistic indication of the estimate errors.

Reporting

The final outcome of any analysis of climate data is a report. There is plenty of advice available on proper report writing (e.g. Day 1979). Here it is sufficient to add that most published papers would be easier to understand if there were stricter division into the proper parts. It is astonishing how many published papers lack crisp conclusions and how often the discussion is muddled with other sections. The proper content for each part is outlined in Note 11, pp. 329–30.

Part II

CLIMATE RESOURCES

5

SOLAR RADIATION

The aim in this chapter is to assess how much solar energy can be captured. This depends on what is emitted by the sun, its position in the sky, the attenuation of sunlight by clouds and by the atmosphere, and the slope of the receiving surface. We will take these in turn.

THE NATURE OF SOLAR RADIATION

The energy radiated from the sun is the same as that from a black surface at about 5760K. The energy consists principally of 'short-wave' radiation, which is mostly visible because the human eye has evolved to respond to what is chiefly available. In addition to light, with wavelengths of 300–700 nanometres (i.e. $3 - 7 \times 10^{-7}$ metres), short-wave radiation includes infra-red and ultra-violet components. However, the ozone layer at about 40 km elevation absorbs energy in the bands of 200–300 and 320–350 nanometres wavelength, and that will be considered no further. Nor shall we discuss long-wave radiation, which has wavelengths more than 3000 nm, since that was dealt with in the section 'Estimating net irradiance', pp. 95–7.

The sun's position

The sun's location in the sky is described in terms of two angles – the upwards 'zenith angle' (z), between the direct beam and the vertical from the ground, and, secondly, the sideways 'azimuth angle' (az). The horizontal azimuth angle is measured clockwise from the north (i.e. from north to east and onwards), as far as the vertical projection of the beam on to the ground. In other words, azimuth angles to the west are negative. The complement of the zenith angle (i.e. $90° - z$) is the 'solar altitude' (sa) or 'solar elevation', above the horizon.

The sun is highest at solar noon. At that moment its elevation above the *south* horizon is given by the following:

$$sa = 90 - A + d \qquad \text{degrees} \tag{5.1}$$

where A is the latitude and d the 'declination', both being positive in the northern hemisphere. Thus, for instance, at noon, if the declination is 20 degrees north, the solar elevation at 30°S is 140 degrees from the southern horizon (i.e. 90 + 30 + 20), which means 40° above the northern (i.e. 180 − 140).

The 'declination' is the seasonally varying latitude of the sun's path across the sky, north or south of the equator. At midsummer (which is 22 December in the southern hemisphere and 22 June in the northern) the sun passes overhead at noon at latitude 23° 27′, the Tropic. The value of the declination is given either by Figure 5.1 or by tables, or approximately by Eqn (5.2):

$$d = 23.45 \times \sin(0.986\ \mathrm{Nm}) \qquad \text{degrees} \qquad (5.2)$$

where Nm is the number of days since 21 March, the equinox, when day and night are of equal length (see Note 8, Chapter 3). The factor 0.986 is 360/365, the daily fraction of the annual rotation of the Earth about the sun.

Solar time

The sun does not reach its highest elevation precisely at noon in terms of the local clock time for three reasons. First, there is the convention that clock time is the same within each time zone of the Earth, even though places within a zone differ in longitude (normally by up to 15 degrees) and therefore have solar noon at times which differ by as much as an hour. Second, there is the effect of the 'Equation of Time' shown in Figure 5.1. This is the annual variation of noon by up to 16 minutes due to the non-circular orbit of the Earth around the sun and the tilt of the Earth's axis. Third, a displacement occurs with 'daylight saving' in summertime. This is an arbitrary shift forward of the clocks by law, for society's convenience at mid-latitudes, chiefly to reduce power consumption in lighting on summer evenings.

It is necessary to use the following concepts to relate local clock time to solar time:

1 *True Solar Time* (Tt) or Local Apparent Time is the time on a sun-dial set at 12 when the sun 'crosses the meridian', i.e. is exactly to the north or south.
2 *Local Mean Time* (Tlm) is the value of the Tt after subtraction of the Equation of Time correction.
3 *Universal Time* (Tu) or Greenwich Mean Time is the Tlm at zero degrees longitude.
4 *Local Standard Time* (Tls), or 'civil time' or Local Clock Time, is the Tlm at one of the standard meridians (Lo degrees of longitude), usually 15 degrees apart. It is the time shown on a normal clock.

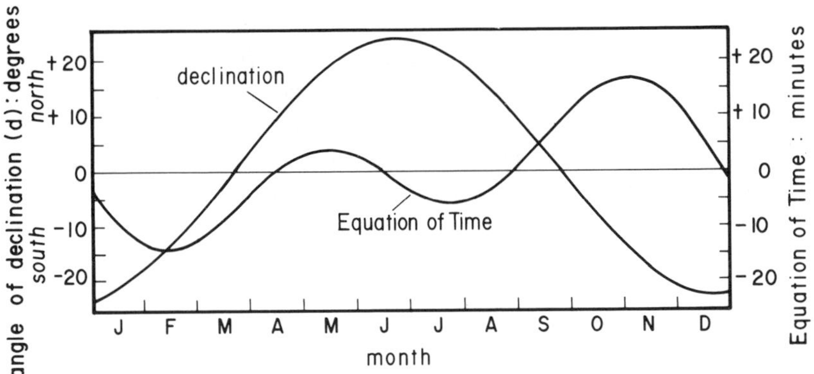

Figure 5.1 Annual variation of the sun's declination and the Equation of Time

The various times are related as follows:

$$Tt = Tu - L/15 + E/60 \quad \text{hours} \tag{5.3}$$
$$Tls = Tu - Lo/15 \quad \text{hours} \tag{5.4}$$

where E is the value of the Equation of Time (in minutes, from Figure 5.1), L is the longitude of the place (degrees) and Lo is the longitude of the standard meridian defining the local time zone. Longitudes west are counted as positive and east as negative. Eqns (5.3) and (5.4) and the definition of Local Mean Time show that it equals [Tls − (L − Lo)/15].

The clock is advanced an hour during the period of summer daylight saving, and that hour must be *subtracted* from the clock time to obtain local standard time (Tls).

An alternative measure of the time of day is the 'hour angle' (H), which is the difference of time from solar noon, expressed as 15 degrees per hour of difference. In other words, H equals 15 (Tt − 12) degrees, where Tt is derived from Eqn (5.3). Examples of its calculation are given in Note 1, p. 331.

Here we adopt the following conventions as regards signs, positive or negative – declination and latitude are positive *north* of the equator, longitude is positive *west* of Greenwich (in London), the hour angle is positive *after* noon, and the azimuth direction is positive *clockwise* from the north.

The hour angle (H) is used in calculating the sun's position at any time of the day. The solar elevation or solar altitude (sa) and azimuth (az) can be calculated from three factors – H, the latitude (A degrees, negative in the southern hemisphere), and the declination (d). The following equations are used (Hay and Davies 1980: 61):

$$\sin sa = \sin d \times \sin A + \cos d \times \cos A \times \cos H \tag{5.5}$$

$$\sin az = -\cos d \times \sin H/\cos sa \tag{5.6}$$

$$\cos az = (\sin d - \sin A \times \sin sa)/\cos A \times \cos sa \qquad (5.7)$$

The azimuth angle satisfies both Eqns (5.6) and (5.7). Eqn (5.6) is also satisfied by [180° − az] and Eqn (5.7) by [360° − az]. As an example, consider the sun's position when the hour angle is 30 degrees at 34°S, on the afternoon of 21 June. The declination is then +22 degrees, from Eqn (5.2), and the solar altitude is 27.1 degrees, from Eqn (5.5). So Eqn (5.6) gives the azimuth as either −31° or +211°, and Eqn (5.7) as either −31° or +31°. Therefore it is −31°, i.e. 31 degrees west of north.

Daylength

The elevation is zero at sunrise or sunset, when Eqn (5.5) becomes the following:

$$\cos Hs = - \tan A \times \tan d \qquad (5.8)$$

where Hs is the hour angle of the sun – positive at sunset, negative at sunrise. Eqn (5.6) gives the azimuth at that time, as the positive angle whose sine equals [− cos d × sin Hs]. For instance, at 34° on 1 June, the hour angle at sunrise is −74 degrees and the azimuth is +63 degrees, i.e. east of north. The daylength is given by Hs/7.5, which is ten hours in this example.

The times of sunrise and sunset are also given on the outer edge of the diagram in Figure 5.2, which graphically shows the sun's elevation and azimuth at any time (Tt) and on any date, for a particular latitude. Figure 5.2 applies to 32.5°S, but the equivalent for any other zone of latitude can usually be obtained from some government agency or publication.

Sunrise and sunset occur when the upper part of the sun's disc is exactly on the horizon of level ground or the sea, and the time between those occasions is usually taken as the daylength, N hours. This is practically the same as implied by Eqn (5.8), though not exactly. Some American climatologists subtract the periods after sunrise and before sunset when the sun's radiation intensity is too low to register on a sunshine recorder, even when the sky is clear. That applies when the sun is within about 5 degrees above the horizon. The advantage of the American practice is that it can yield values of unity for the ratio n/N, where n is the recorded duration of bright sunshine.

The 'civil twilight' periods before sunrise and after sunset are defined by the sun's centre being less than six degrees *below* the horizon. It is the time for activity by many wild animals. It lasts longer when the sun meets the horizon obliquely, being about 22 minutes at the equator but 108 minutes in winter at 60° latitude. 'Astronomical twilight' is the interval between the time when the sun's centre is 18° below the horizon and the moment of either sunrise or sunset.

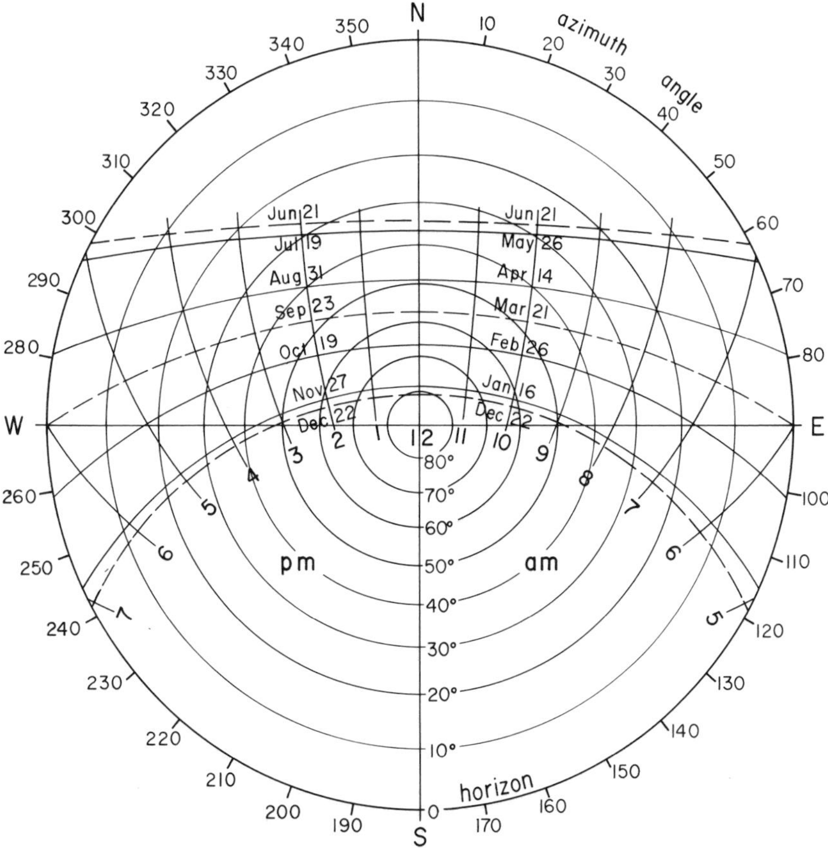

Figure 5.2 Sun-path diagram for 32° 30′S (Evans 1980: 172). A point on the curved date-hour grid can be referred to the concentric circles to obtain the solar altitude and to the outside 360 degree scale to obtain the azimuth angle. Thus, at 10 a.m. (True Solar Time) on 1 June, the sun is 27° above the horizon and 31° east of north

Radiation outside the Earth

Solar radiation takes about 8.3 minutes to reach us. The amount received varies during the year because of the elliptical orbit of the Earth around the sun. The radiant energy flux of the beam to the Earth is denoted by I (the 'solar parameter'), and has been calculated as over 1400 W/m^2 at the 'perihelion' near 3 January (when the sun is 147 million kilometres away) but only 1309 W/m^2 on 4 July when the Earth is at the 'aphelion', 152 million kilometres distant. The *annual average* irradiance (or 'insolation') of an area at the Earth's outer surface, facing the sun, is called the 'solar constant' (Io). The value of Io has changed as a result of improved measurements. During 1940–56 it was taken as 1323–1431 W/m^2. Then the NASA design standard of 1971 was 1353 W/m^2, and from 1977–81 the agreed value was 1371 W/m^2, until the World Meteorological Organization fixed on 1367 W/m^2 plus or minus 7 W/m^2. Recent satellite measurements indicate a cyclic variation by about 1 W/m^2 each decade or so.

The extra-terrestrial irradiance (Qa) is the radiant power which would be received at the ground if the atmosphere were absent. Values in various latitudes and months are given in Note 12, in Chapter 3, and represent the monthly average of [I.sin sa], where sa is the solar altitude. As an example, the tabulated monthly mean Qa is 311 W/m^2 at 30°S in April, so the daily 'exposure' is 2.9 megajoules per square metre (i.e. 311 × 3600 × 24). Also, the daylength is 11.5 hours, so the daylight mean of Qa is 649 W/m^2 (i.e. 311 × 24/11.5).

Diffuse radiation

The beam of extra-terrestrial radiant energy becomes reduced as it traverses the Earth's atmosphere, because part is reflected back into space and some is absorbed by the air, water vapour, particles and droplets within the air. Scattering of the radiation by the atmosphere increases the back-reflection and absorption. The combination of absorption, reflection and scattering amounts to 'attenuation' or 'extinction' of the beam.

Scattering leads to some of the radiant energy striking the ground from directions other than that of the beam direct from the sun. The sun's diameter (about 1.39 million km) means that direct radiation (Qd) comes within a cone whose half angle is about 0.27 degrees (i.e. half the angle whose tangent is 1.39/150), assuming no refraction. More widely scattered radiant power received at the ground comprises the diffuse component (D) of the total (or 'global') solar irradiance (Qs). Hence the following relationship involving the solar elevation (sa):

$$Qs = D + Qd.\sin sa \qquad W/m^2 \tag{5.9}$$

Figure 5.3 shows the proportions of D and Qd.sin sa in a typical case. The irradiance due to scattered radiation hardly varies when the sun is more than about 40° above the horizon in a clear sky. However, the ratio D/(Qd.sin sa) is more when the sun is low, which partly explains the greater importance of diffuse radiation in winter.

It is useful to distinguish two parts of the diffuse radiant energy. The 'sky-light' (Di) is 'isotropic', i.e. it comes more or less equally from all directions of the sky, whereas the 'circumsolar diffuse radiation' (Dc) comes from within about five degrees of the direct solar beam. The Dc is called the 'aureole' and typically is 5 per cent of Qd or about 30 per cent of Di when the sky is clear. However, the circumsolar irradiance can be 20–30 per cent

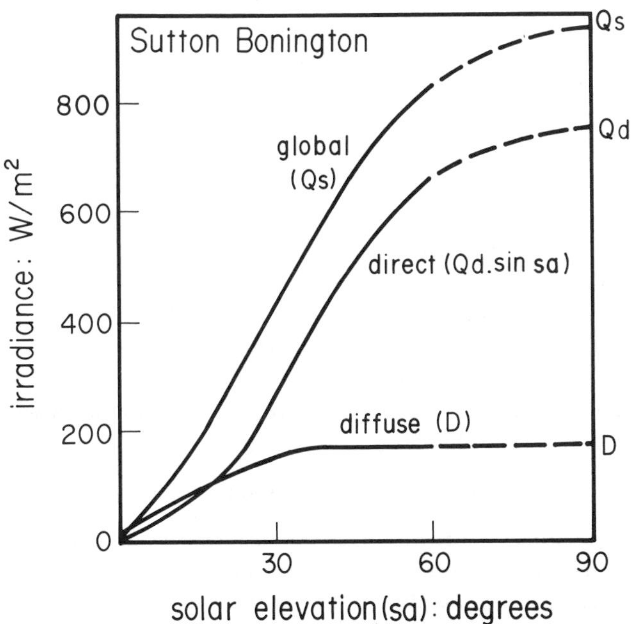

Figure 5.3 The effect of solar elevation (sa: degrees) on the diffuse (D), direct (Qd) and global (i.e. D + Qd.sin sa) irradiances, with clear skies at Sutton Bonington in England (Monteith 1973: 27)

of Qd in a hazy or turbid atmosphere. Naturally, it is less evident if there is cloud obscuring the sun, though the diffuse irradiance from right above remains about twice that from near the horizon, even when the sky is totally overcast, showing that there is still some departure from isotropy.

The non-isotropic nature of diffuse irradiance used to be ignored in considering the radiation on to a surface. But that can lead to appreciable underestimation of irradiance when the sun is high in the sky. Allowing for the circumsolar component in Canada reduces errors estimating Qs from as much as 28 W/m^2 to about 5 W/m^2.

The circumsolar diffuse component (Dc) may be taken as equal to D.Qd/I. So the isotropic part (Di) is given by the following:

$$\text{Di} = \text{D} - \text{Dc} = \text{D}\ (\text{I} - \text{Qd})/\text{I} \qquad \text{W/m}^2 \qquad (5.10)$$

The ratio of diffuse to other radiation fluxes

Whether solar irradiance is diffuse or direct is often important. The distinction affects the method of measuring the solar irradiance Qs (see pp. 164–70), the amount striking a sloping surface (see pp. 193–7) and the use of mirrors for collecting solar energy (see pp. 198–9). The amount of diffuse radiation is commonly expressed as a ratio of some larger flux, as follows.

D/Qo

This ratio (where Qo is the *clear-sky* solar irradiance of the ground) depends chiefly on the angle of the sun. Measurements in Belgium show that the ratio is unity when the sun is on the horizon, and then declines to about 0.6 as the sun rises 5°. At most places, D/Qo is about 0.3 when the sun is 30° above the horizon in a clear sky, and 0.17 when 60° above.

The ratio is less on high mountains because of cleaner and drier air. For instance, a relationship given by Klein (1948: 126) is as follows when the solar elevation is 65° in a cloudless sky in Switzerland:

$$D/Qo = 0.33 - 0.074 \log h \tag{5.11}$$

where h is the elevation (metres). Hence the ratio is 0.18 and 0.11 at 100 m and 2000 m, respectively. A similar expression for the sun at an elevation of 19° in a clear sky gives ratios of 0.29 and 0.14, respectively.

D/Qs

When there is cloud, the ratio equivalent to that in Eqn (5.11) is D/Qs. This depends on the solar elevation, the atmosphere's clarity and the amount of the cloud. The solar elevation and cloudiness are affected by the latitude, so the ratio is around 0.5 at the equator, about 0.26 at 20°S and 30°N (e.g. 0.23 in India), and 0.65 at 70° latitude.

The ratio varies also with the month of the year, ranging in Stockholm from about 0.23 in June to 0.93 in December. The ratio at Melbourne averages 0.31 in February and 0.54 in winter, for all days, cloudy and clear, instead of 0.14 for cloudless days alone.

Urban air pollution increases the ratio D/Qs by enhancing D and reducing the direct component of the global irradiance. Pollution at London was found to reduce Qd by 10 per cent, while D was increased by 8 per cent. Likewise, summertime global irradiance at the Australian city of Newcastle is 8 per cent less than in rural areas nearby, the direct component having been reduced by about 19 per cent and the diffuse increased by about 28 per cent.

Cloudiness both increases D/Qs and reduces the ratio of global irradiance to extra-terrestrial irradiance (Qs/Qa), so that the two ratios can be related to each other. This is convenient because Qs/Qa can be estimated fairly easily, as discussed later.

Published relationships between the ratios are shown in Figure 5.4. The scatter shown in the diagram may be partly due to differences of latitude. A

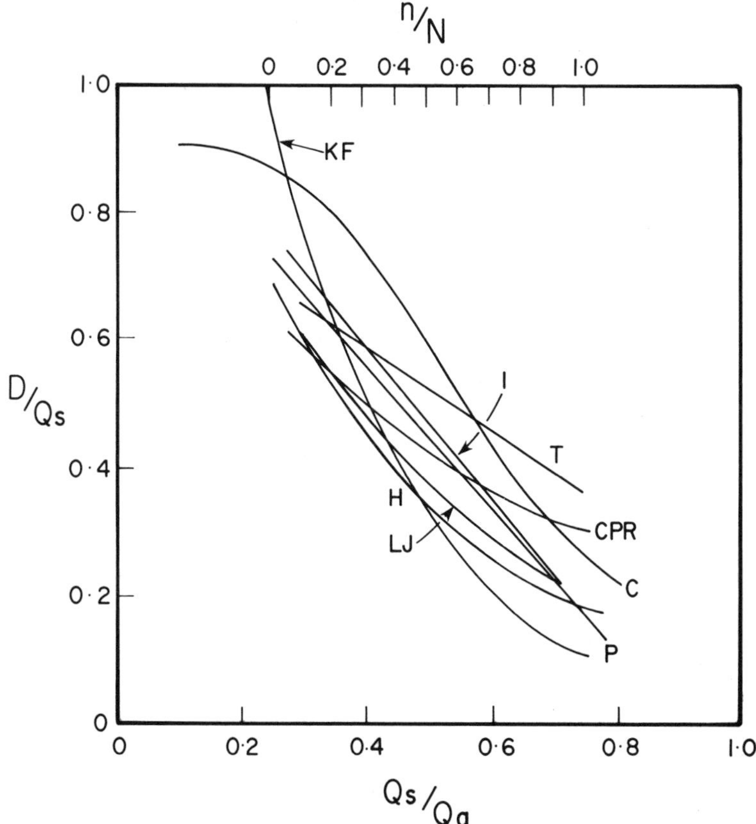

Figure 5.4 Published empirical relationships between D/Qs and Qs/Qa, where Qs is the measured global solar irradiance and Qa is the extra-terrestrial irradiance: LJ – Liu and Jordan (1960), confirmed by Atwater and Ball (1978): P – Page, confirmed by Neuwirth (1980: 424); C – Choudhury (1963); KF – Kalma and Fleming (1972), with a value of 0.25 for their function K; T – Tuller (1976); H – Hay (1976); I – Iqbal (1979); CPR – Collares-Pereira and Rabl (1979). The scale of n/N at the top of the diagram results from a later discussion of the relationship between Qs/Qa and the ratio of the number of hours of bright sunshine to the daylength (i.e. Prescott's formula, see pp. 177–9)

consensus can be derived from the median values of D/Qs for widespread values of Qs/Qa, e.g. the medians are 0.66, 0.44 and 0.23 for Qs/Qa equal to 0.3, 0.5 and 0.7, respectively, and these values define the following equation:

$$D/Qs = 1.0 - 1.2\ Qs/Qa + 0.13\ (Qs/Qa)^2 \tag{5.12}$$

A particularly simple equivalent was derived by Kalma and Fleming (1972), assuming that a quarter of the absorbed part of the solar beam becomes downwards diffuse radiation:

$$3\ D/Qs = Qa/Qs - 1 \qquad (5.13)$$

Eqns (5.9) and (5.13) together permit calculation of the global and constituent irradiances (see Note 2, pp. 331–2).

D/Qa

The following equation gives the fraction of the extra-terrestrial radiation which reaches the ground as diffuse radiation at Delhi (Choudhury 1963: 48):

$$D/Qa = 0.27 - 0.29\ Qd\ (\sin sa)/Qa \qquad (5.14)$$

That implies that D/Qa is 0.27 when no direct radiation reaches the ground, which occurs in overcast conditions. If the sun is overhead and the direct solar beam is reduced by a third, say, in passing through the atmosphere, then D/Qa is 0.07.

Similar values can be derived by combining an equation for D/Qs with Prescott's equation for Qs/Qa, considered later (see pp. 177–9). The equation gives Qs/Qa as 0.25 when the sky is overcast (when D/Qs equals unity), which fits Eqn (5.13). Prescott's equation also gives Qs/Qa as 0.75 when the sky is clear, and then Eqn (5.13) gives D/Qa as 0.08. This resembles the 0.07 derived in the previous paragraph.

D/Qd

A ratio involving the direct irradiance Qd was studied by Peterson and Dirmhirn (1981: 826), for the case of clear skies in rural Utah. They found that the proportion of diffuse to direct components of the solar irradiance hardly varies from hour-to-hour on any particular day, despite the variation of solar elevation. However, it does change from day-to-day, presumably because of differences of atmospheric humidity. Of course, the ratio is increased by cloud.

ATTENUATION OF SOLAR RADIATION

Both direct and diffuse irradiance are reduced as the result of absorption and reflection along the light beam's path through the atmosphere to the ground. How much the beam is reduced depends on the path length, the amount of air traversed by the beam, i.e. the air-mass. The ratio of the air-mass in particular circumstances, to the value when the sun is overhead, is known as the 'relative air-mass' (m). It depends on the air pressure (p) and solar altitude (sa), as follows:

$$m = p/(1013.\sin sa) \qquad (5.15)$$

where p is the atmospheric pressure (hPa) at the elevation of the climate station, and sa is the solar altitude. The relationship between the atmospheric pressure (p) and elevation (h kilometres) is as follows:

$$p = 1013 \exp - (g.h/R.\mathbf{T})$$
$$= 1013 \exp (-0.12\ h) \text{ approximately} \quad \text{hPa} \qquad (5.16)$$

where g is the gravitational acceleration, R is the gas constant for air (287.0 J/kg.C°) and **T** is the mean temperature (K) of the layer from sea level to height h. The acceleration g at sea level equals 9.78 m/s^2 at 0° latitude, 9.81 at 45° and 9.83 at the poles, and is 0.28 m/s^2 less at 900 m elevation. Hence, for example, the relative air-mass is 1.57 if the sun is 30° above the horizon and the altitude is 2 km, when the temperature at 1 km is 10°C, i.e. 283K.

However, Eqn (5.15) applies only to the direct irradiance. For *diffuse* irradiance, the appropriate air-mass is 1.66 (Davies and McKay 1982: 58). Also, Eqn (5.15) applies only to instantaneous conditions, with the sun at a particular elevation. The daytime average air-mass at various latitudes and times of the year is given in Figure 5.5.

Aerosols

The ratio (Qs/Qa) is one measure of radiation attenuation, which is controlled by:

1 the amount of cloud;
2 the optical air-mass (m, or 'optical path');
3 the amount of water vapour in the air;
4 the amount of 'aerosols' present.

Aerosols are droplets and particles so small as to remain airborne indefinitely. Particles and droplets smaller than 0.2 microns (i.e. 2×10^{-7} metres) are called Aitken aerosols, after John Aitken (1839–1919). They may result from chemical reactions of gases, often requiring high degrees of relative humidity. 'Large aerosols' are about 0.2–2 microns in size, and 'giant aerosols' above 2 microns. Urban air, and winds from inland, contain aerosols chiefly about 0.1 micron in size, whilst sea spray and forest-fire smoke consists of giant aerosols of about 10 microns. The latter arise in dust storms also.

What is called 'Rayleigh scattering' occurs when aerosols or gas molecules are of a size similar to the wavelength λ of the radiation. The amount of dispersion is then proportional to λ^{-4}, so that blue light of 0.47 microns is scattered almost four times as much as red light of 0.65 microns, i.e. scattered light is more blue than the remaining direct beam. That is why the sky is blue. However, there is 'Mie' scattering when the aerosols

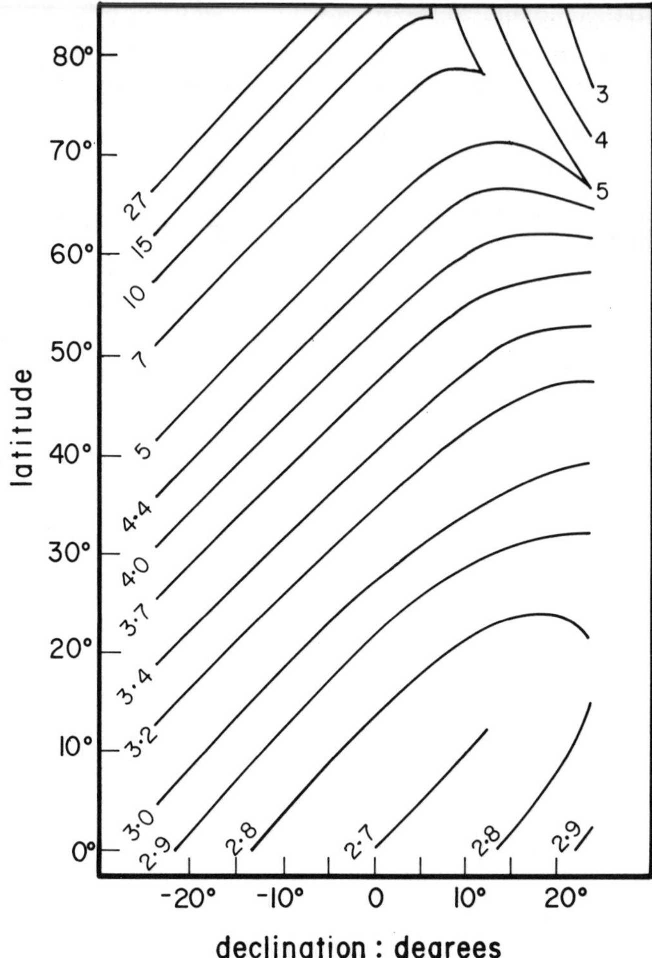

Figure 5.5 Effect of the latitude and time of year (i.e. solar declination) on the daytime average optical air-mass, shown by the numbers against the curves (Kennedy 1940: 302)

are much larger than λ, as with air pollution. This causes equal dispersion of all wavelengths, which creates a milky white haze

The atmosphere carries a particularly heavy load of aerosols after any large volcanic eruption, such as that of Krakatoa in 1883. This may reduce solar irradiance for several months over the whole globe. For instance, an eruption of El Chichon in Mexico in April 1982 affected the turbidity as far away as Alaska. At the time of maximum effect at Vancouver in Canada (which was in the following January), there was a quadrupling of the aerosols, leading to a 25 per cent reduction of the direct irradiance (Hay and Darby 1984). However, this was largely offset by a 76 per cent increase of the diffuse irradiance (D), so that altogether the global irradiance (Qs) fell by only 5 per cent at Fairbanks.

A similar change occurred in Australia after the eruption of Mt Agung in Bali in 1963. Also, there was 5.3 per cent less solar irradiance at the South Pole, and at Hawaii a reduction which lasted for seven years. The greatest attenuation at Hawaii and in Arizona occurred two years after the eruption.

The global warming that occurred early in the 20th century may have been due partly to fewer eruptions than usual (Jones *et al.* 1990).

Attenuation formulae

Attenuation of the direct solar beam in passing through a homogeneous, cloudless atmosphere can be described by the following formula, often named after Beer, though first formulated by Pierre Bouguer in 1760:

$$Qd = I.q^{m} \qquad W/m^2 \tag{5.17}$$

where I is the solar parameter, q is the 'transmission coefficient', and m is the relative air-mass. To clarify a confused terminology in the literature, one distinguishes the transmission coefficient q, from the 'transmittance' t and the 'transmission'. Each of these depends on the radiation's wavelength, though average values for the whole spectrum are frequently used for simplicity.

1 The *transmission coefficient* q can be estimated from measurements of Qd (the direct radiation measured at ground level), the solar elevation and the atmospheric pressure, which govern m. The value of q is about 0.93 on a high mountain in Hawaii, for example, and 0.76–0.80 in the north-western USA.

It is sometimes convenient to expand Eqn (5.17) as follows, to distinguish its components:

$$Qd/I = q^{m} = (qg \times qw \times qs)^{m} \tag{5.18}$$

where qg is the transmission coefficient for air which is both dry and clean, qw is the coefficient for the water vapour, and qs for the aerosols. The coefficient qw is 0.92 for an atmosphere containing sufficient water vapour to form a layer of liquid 10 mm deep, 0.89 if 30 mm, 0.86 if 100 mm. The aerosol's coefficient qs is 0.94 in Madison and 0.92 in Washington, for instance.

2 The '*transmittance*' (or 'transmissivity') t is the ratio Qd/I, which equals q^m in Eqn (5.17). Clearly, it depends on the solar elevation, as well as the factors governing q. Calculated values range from 0.76 when the sun is overhead, to 0.74 when 30 degrees away, and 0.66 when 30 degrees from the horizon.

Another expansion of Eqn (5.17), is as follows:

$$Qd/I = t = to \times tg \times tw \times ts \qquad (5.19)$$

where the terms to, tg, tw and ts are the separate transmittances of ozone, the rest of the air, the water vapour and the aerosols, respectively. However, a deeper consideration replaces the righthand side of Eqn (5.19) by this: [ts (to × tg + tw − 1)] (Davies and Hay 1980: 35). The term (1 − tw) is sometimes called the water vapour's 'absorptance'. The product (to × tg) is the fraction absorbed neither by ozone nor, subsequently, by the air's other gas molecules. Evaluation of Eqn (5.19) is considered in Note 3, pp. 332–3.

Typical values of the various transmittances in Samoa, Switzerland and Washington are 0.93 for the aerosols, 0.91 for the ozone and air together, and 0.86 for the water vapour. It can be seen that water vapour has most effect. The three values give an overall transmittance of 0.73.

As regards the aerosol transmittance, the air over oceans is relatively clean, so the absorptance by aerosols in particular is typically only 1 per cent at 350 m on the Canary Islands, i.e. the aerosol transmittance is 0.99. In Canada it is about 0.91. On the other hand, the absorptance by aerosols may be 8 per cent at Madison, 12 per cent at Washington or 15 per cent at Los Angeles, i.e. air pollution can greatly reduce the transmittance.

Knowledge of t allows calculation of the direct component of the global irradiance of the ground. If the relative air-mass is 2 at a particular time of the day, and the transmission coefficient 0.8, for example, the instantaneous transmittance is 0.64 (i.e. 0.8^2). So the beam intensity Qd would be given by Eqn (5.17) as 1367 × 0.64, i.e. 875 W/m^2. A relative air-mass of 2 implies a solar elevation equal to 30°, since 1/(sin 30) equals 2. Therefore, the direct irradiance of the ground would be 875.sin 30, i.e. 437 W/m^2 at that moment. In addition, the ground would receive diffuse irradiance, whose division into circumsolar and isotropic radiation is discussed in Note 4, p. 333.

3 The *transmission* is the ratio Qo/Qa, the clear-sky solar irradiance of level ground as a fraction of the irradiance at the top of the atmosphere. The ratio differs from Qd/I (i.e. the transmittance) in Eqn (5.18), because Qo includes diffuse radiation, unlike Qd, and both Qo and Qa refer to horizontal surfaces, whereas Qd and I are measured whilst facing the sun. However, values of the transmission and transmittance are similar in practice. The solar-noon value of transmission is 0.74–0.79 at Boston, for instance, but 0.82 at Albuquerque. It is 0.75 in winter and 0.60 in summer in Wisconsin. The transmission is 0.73–0.81 in Phoenix (Arizona) and about 0.70 in Kuwait. These values resemble transmittances quoted earlier. At Wellington in New Zealand, annual mean values of Qo/Qa are around 0.80, i.e. only about 3 per cent less than the transmittances (de Lisle 1966: 1004).

The transmission is greater on high mountains because of the thinner layer of atmosphere above, the reduced amount of water vapour in the cold air and the absence of dust. It is 0.67 at 300 m, 0.77 at 1000 m, 0.85 at 3000 m and 0.93 at 7500 m, for example (Klein 1948: 125).

Turbidity coefficients

An alternative to Eqn (5.17) is the following, discussed in Note 3, pp. 323–3:

$$Qd = I.\exp(-\mathbf{a}.m) \qquad W/m^2 \tag{5.20}$$

where **a** is the 'extinction coefficient' (or 'abortion coefficient') and m is the relative air-mass (Eqn (5.15)). The coefficient can be divided into parts, as previously:

$$\mathbf{a} = ag + aw + as \qquad \text{metre}^{-1} \tag{5.21}$$

where ag, aw and as refer to dry clean air, water vapour and aerosols, respectively. The coefficient ag is 1.4×10^{-4} p; where p is the atmospheric pressure in hectopascals, and is usually comparatively small. So that **a** in Eqn (5.21) depends chiefly on the water vapour and the aerosols (see Note 3). The term aw may be split into two parts which are multiplied together, representing the amount of water vapour and the attenuation due to unit amount. Likewise for as.

The attenuation of solar radiation by aerosols (the third component of Eqn (5.21)) is commonly quantified by one of three turbidity factors, named after Linke, Angstrom and Schuepp, respectively. These will be considered in turn.

Linke's turbidity factor, T

Linke's factor (T) is the ratio (**a**/ag), which is the number of air-masses of pure dry air with the same absorption as the given atmosphere. Insertion of the definition of T into Eqn (5.20) gives the following:

$$Qd = I.\exp(-ag.T.m) \qquad W/m^2 \tag{5.22}$$

Hence:

$$T = (\ln I - \ln Qd)/m.ag \tag{5.23}$$

Eqn (5.23) yields T from measurements of the direct irradiance Qd, because the product (m.ag) can be obtained from tables – when m is 0.5, the product is 0.023; when m is 1.0 it is 0.043; when 2 it is 0.078; when 4 it is 0.132; and when 8 it is 0.208.

The value of I in Eqn (5.23) can be calculated from the solar constant (Io), after adjustment for the time of year (which affects the distance between the sun and the Earth). Or it can be found by measuring the direct-beam irradiance Qd when the sun is at various elevations, corresponding to different values of the air-mass. Measurements are typically at 9 a.m., noon and 3 p.m. Then the results are extrapolated to find Qd for zero air-mass, which is I.

Monthly mean values of T are normally about 3 for temperate rural climates, about 4 between the tropics and 5 for humid polluted atmospheres. Linke's factor in Beijing varies between 2.8 in January and 3.8 in July. At Tashkent the factor is between 2.4–3.1, when the optical air-mass is in the range 1.5–2.0. The value is about 2 at 3000 m in Switzerland, but 5 at London in summer.

Unfortunately, the Linke turbidity factor is difficult to determine accurately and fails to distinguish the effects of water vapour from the absorption due to aerosols. Also, it is not truly independent of the optical air-mass and varies diurnally even with an unchanging atmosphere.

Angstrom's turbidity coefficient, b

Instead of Linke's factor, one may use a coefficient devised by Angstrom in 1929. This is denoted 'b' and defined in terms of the aerosol absorption coefficient as mentioned in Eqn (5.21):

$$b = as.\lambda^{n} \tag{5.24}$$

where λ is the optical wavelength (in micrometres) and n is an exponent, often taken as 1.3. In fact, n is zero for large particles (e.g. fog) but approaches 4 with a sky pure and clean enough for Rayleigh scattering.

The value of the Angstrom coefficient in particular circumstances is found by measuring the direct-beam irradiance Qd at various times of the day, for each of which the relative optical air-mass is given by Eqn (5.15). Insertion of values of Qd and m into Eqn (5.20) gives both I and **a**. Then Eqn (5.21) yields as, after subtracting the components due to dry air and water vapour, discussed in Note 3. Finally, Eqn (5.24) gives the Angstrom coefficient.

An assumed value of the exponent n can be avoided by taking measurements of Qd with each of two filters (e.g. for wavelengths of 380 and 500 nm respectively), to obtain two values of the aerosol absorption coefficient, as_1 and as_2, respectively. Then the proper exponent equals $[\log as_1 - \log as_2] / [\log 0.38 - \log 0.50]$.

Schuepp's turbidity coefficient, B

This is an improvement on Angstrom's turbidity coefficient, because it refers more specifically to visible radiation of 0.5 micrometres wavelength. Also, it differs in relating to common logarithmic attenuation:

$$Qd_5 = I_5 .\, 10^{-s} \tag{5.25}$$

Table 5.1 Representative values of the Angstrom or Schuepp turbidity coefficient

Atmosphere	*Circumstances*	*Turbidity coefficient*
Very blue sky	Low humidity, mountains, winter	0.02
Blue sky	Cold weather, visibility beyond 24 km	0.05
Clean sky	Average humidity, rural areas, fine weather	0.10
Urban	Only slight pollution	0.15
City	Visibility of 3–16 km	0.20
Tropical	Hot, humid	0.25
Polluted	Thick smog	1.00

Source: Iqbal 1983: 119, and others.

where Qd_5 is the direct-beam irradiance of 500 nm wavelength at ground level, I_5 is the equivalent radiation power above the Earth's atmosphere, and the exponent s equals (c + B).m, where m is the relative optical air-mass, B is Schuepp's coefficient and c is almost constant at about 0.067, representing the attenuation by clean dry air and ozone. As before, the value of I_5 is found by measuring Qd_5 for different solar elevations and then extrapolating to a zero optical air-mass.

The coefficient may vary with the season. Monthly mean values in Bangkok range between 0.05–0.11. At New Delhi it is 0.05 in winter and 0.3 in July. In Madrid it is about 0.10, rising to 0.13 in November. In Washington and Europe it is usually highest in summer.

The Schuepp coefficient (B) equals 1.07 times Angstrom's b, if the exponent in Eqn (5.24) is 1.3. Assuming this similarity, values of both have been gathered in Table 5.1. However, the values can differ by a factor of two on occasion. Also, both are cumbersome to derive and the derivation is subject to large errors because of differences between similar magnitudes involved in calculating the coefficient from the basic measurements (see 'Errors of estimation', p. 68).

Visibility

The clarity of the atmosphere can be described in terms of the 'visibility' (V), also called the 'standard visibility' or 'visual range'. It is the horizontal distance over which one can clearly discern an object against a contrasting background. The object should be black and either well ahead of the background or on the horizon. It should subtend 0.5–5.0 degrees to the observer, that is about 10 m at 1 kilometre or 8 mm at arm's length. In clean air one can see an adult person at a kilometre, a car at 3 km and a horse at 8 km.

The visibility is approximately related to the coefficient **a** (Eqn (5.20)) by the following:

$$V = 3.9/\mathbf{a} \qquad \text{km} \tag{5.26}$$

Alternatively, the visibility can be related approximately to the Angstrom turbidity factor (b) as follows:

$$b = 0.044\ (1/V - 0.003)(V + 41) \tag{5.27}$$

Accordingly, b has a value of about 0.23 if the visibility V is 10 km, and the coefficient **a** would then equal 0.39, from Eqn (5.26). The absorption coefficient of absolutely clean air at sea level is about 0.015, implying a visual range of over 250 km. The coefficient of what is called the 'US standard atmosphere' is 0.16, implying a visibility of 25 km.

MEASURING SOLAR IRRADIANCE

The global short-wave irradiance of the ground (Qs) is measured at only a small minority of climate stations. A Climatology Commission of the World Meteorological Organization recommended in 1969 that there should be a radiation instrument at every 500 km, which means about 40 in a country the size of Australia, Brazil, China or the USA. In fact, there were 21 in Australia in 1976, and around 40 in the USA in 1977. There were 52 in Canada in 1979. But in many other places there is a great need for more measurements, in view of the considerable spatial variation of irradiance discussed in 'Values of solar irradiance, pp. 184–93. The lack is especially great in developing countries.

Many of the values that have been published were obtained with inferior instruments. Even in the USA, the National Weather Service stopped publishing irradiance data in 1972 because of errors of up to 30 per cent in previous measurements, due to calibration inaccuracy and discoloration of the irradiated surfaces of the instruments. The network was upgraded by 1977.

Instruments for measuring irradiance are called 'radiometers' or 'solarimeters'. Those directed vertically upwards are 'pyranometers', as distinct from 'pyrheliometers' (or 'actinometers') which face the sun to measure the direct beam Qd. The instruments in general use involve either (i) the distortion of bimetallic strips heated by the radiation, (ii) the thermoelectric effect, (iii) distillation, or (iv) photo-electricity. These will be described. We shall use the approved SI units of W/m^2, rather than traditional units such as the langley (i.e. one calorie/cm^2, or 41.87 kJ/m^2) or the cal/cm^2.min (i.e. 697.5 W/m^2) or Btu/ft^2.min (i.e. 189.3 W/m^2).

Robitzsch pyranometer

The Robitzsch pyranometer was introduced in 1932 and in common use until recently. It is based on the bending undergone by a horizontal, blackened, bimetallic strip, when it is heated by the sun's rays. Only radiation with a wavelength less than about 2.8 microns penetrates the glass dome which protects the bimetal strip from the wind's temperature. Flexure of the strip is amplified by levers to move an ink pen up and down, across a chart wrapped around a drum which rotates on a vertical axis. Thus the record is a graph of instantaneous values and it is necessary to use a planimeter, which is laborious, to find the hourly or daily exposure and hence the mean irradiance.

Accuracy is affected by friction of the pen on the chart and by changes of the ambient temperature. Errors are caused also by the sluggish response of the instrument to any rapid changes of irradiance. The lag time is 5–8 minutes, so that it takes 10–15 minutes to respond by 98 per cent to a sudden change. As a result, there are errors of up to 10–20 per cent in measuring the daily exposure, even with recalibration every month. Measurements from side-by-side instruments can differ by 10 per cent, and seven instruments used together for a year were found to change calibration by amounts between −15 per cent and +26 per cent. Consequently the instrument is no longer recommended.

Thermopile pyranometers

More accurate are the 'Moll–Gorczynski' radiometers made by Kipp in Holland and by Eppley in the USA, and the Yanishevskii instrument of the USSR. In each case, the principle is that radiant heating of one surface of a thermopile generates an electrical voltage which is then measured. (A 'thermopile' is an electrical loop of alternating lengths of wire of two suitable alloys, where each length reaches from a surface warmed by the radiation being measured to a surface held cool. It is a pile of thermocouples.) Typically, 100 W/m^2 creates 1–3 millivolts, which can be recorded either continuously as a graph or digitally at intervals, or most conveniently by the integration of a current generated by the voltage. The hot junctions of the thermopile are coated with a black paint to absorb all the incident solar radiation and are protected from wind and from long-wave irradiance by a glass dome or sphere.

The instrument made by Kipp (see Figure 5.6) has two concentric glass domes of 30 mm and 50 mm diameter to reduce heat loss from the 18-mm diameter blackened surface of the thermopile. The pile consists of 14 thermo-junctions in series, and its electrical output achieves 98 per cent of a change within 30 seconds of suddenly altering the irradiance intensity. Errors overall are less than 5–10 per cent. The most recent design is

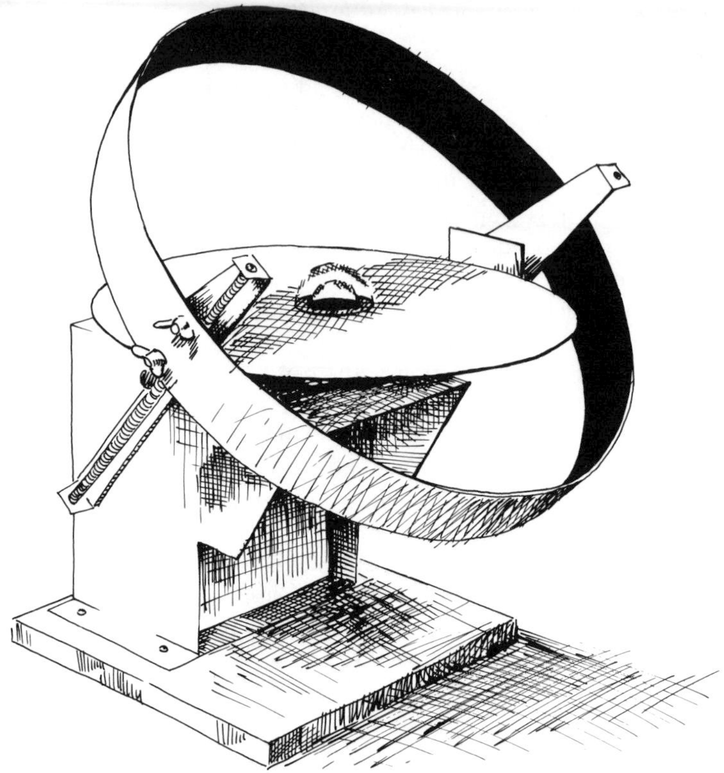

Figure 5.6 A Kipp solarimeter. In this case it is fitted with an occulting band, to measure only the diffuse solar irradiance

perhaps the best pyranometer available, with a precision of about 2 per cent.

The latest Eppley version is similar to the Kipp instrument and is standard in the USA and Australia. It has six alternately white and black sectors of a 3-cm diameter circle containing 48 copper-constantan thermocouple junctions, which provide 11 microvolts per W/m^2.

Figure 5.6 shows a thermopile pyranometer adapted for measuring diffuse irradiance. The sensor is shaded from the direct beam by an 'occulting band'. The band is typically 76–81 mm wide and 640–780 mm in diameter. The axis is tilted at the angle of the local latitude, and the position of the band is adjusted each 2–3 weeks to allow for changes of declination. The inside of the band should be repainted black each year.

The instrument reading has to be increased to allow for the fraction of the sky which is obscured by the occulting band. There are seasonal and latitude effects. The correction is 3 per cent in winter but 9 per cent in summer, in the case of an overcast sky at Utah. At 40°S, it is 8 per cent in winter and 21 per cent in summer. Unfortunately, there is no international standard for the size of the band nor for the procedure of correcting for it. The usual assumption of the same amount of diffuse radiation from each part of the sky may give a slight underestimate, because of the aureole (see 'Nature of solar radiation', p. 152).

Distillometer

A distillometer is a cheap radiometer whose accuracy is between those of the Robitzsch and thermopile instruments. The best-known is Gunn's version of the instrument invented by Bellani in 1836, and illustrated in Figure 5.7. One measures the amount of liquid condensed in the lower tube, distilled from the upper black sphere, which is heated by the sun. The lower tube is relatively cool because it is buried in the ground.

The liquid was originally specified as alcohol, but that distils too rapidly for convenience if the irradiance exceeds 200 W/m^2; so water is now preferred, especially at low latitudes, and shows less scatter of results. The amount distilled since the last time of resetting is a measure of the exposure meantime. Resetting the instrument consists simply of momentarily inverting it, usually each night – though this is inconvenient.

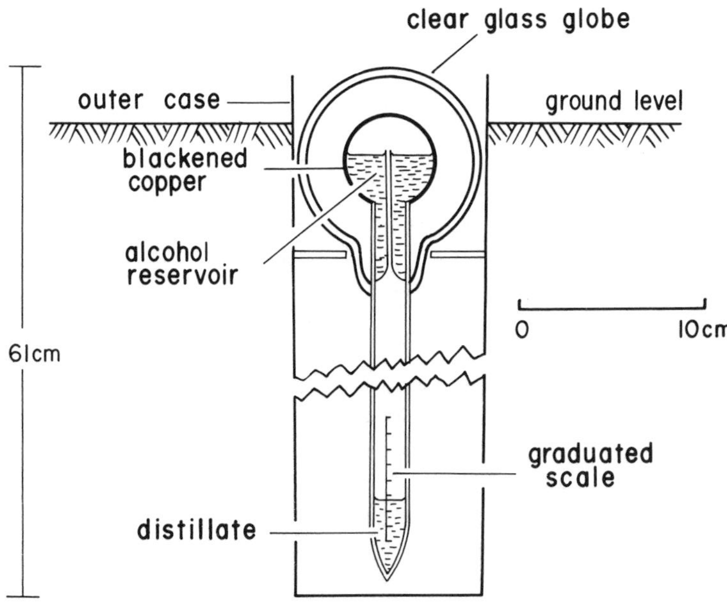

Figure 5.7 The Gunn–Bellani radiometer

The rate of alcohol distillation in a Gunn–Bellani radiometer in England has been found related to the solar irradiance as follows:

$$Qs = 0.8\ Dt - 40 \qquad W/m^2 \tag{5.28}$$

where Dt is the rate of distillation (mL/week). The equation implies a threshold of 40 W/m^2 before any distillation occurs. The value is higher if water is used, and it increases at low temperatures.

There are three disadvantages with the Gunn–Bellani radiometer. First, it measures the sum of the direct and diffuse irradiance of a sphere [D + Qd], not that of flat ground [D + Qd.sin sa]. The difference is greatest at high latitudes, where the solar elevation is small. Second, the instrument is unsuitable in temperate climates, where the threshold is an appreciable fraction of the daily irradiance. Third, the thermal capacity of the liquid leads to a slow response to sudden changes of irradiance: the distillation rate reaches 90 per cent of the new rate only after a delay of about 45 minutes. As a result of these three features, the instrument is most suited to equatorial latitudes, where the daily readings correlate well with those from thermopile radiometers and errors are typically only 7 per cent.

A different kind of distillometer is the 'wig-wag', which consists of two glass bulbs, joined by a pivoted, almost horizontal, glass tube part-filled with alcohol. The lower bulb is exposed to the sunlight, and alcohol distils into the other higher and hidden bulb until eventually the increasing weight in that tips it instead into the sunlight. Distillation then occurs from the second bulb into the first until the device tips back. The rate of rocking indicates the irradiance and the number of tippings gives the exposure. The instrument is said to be cheap, accurate, easily maintained and convenient, and it requires no power.

Photocells

Several authors have suggested that solar irradiance could be measured simply, by means of photocells. However, Brodie (1964) found that the calibration of at least one commercial instrument was unsteady, and other authors have pointed out that a photocell's response to solar irradiance is complicated by effects due to the solar altitude and the colour of the light received. As regards the latter, a photocell measures only a narrow sample of the wavelengths present in solar irradiance, and gives a bigger voltage for the same irradiance when the sky is more overcast because the light is then more blue. Fortunately, errors due to solar angle and wavelength susceptibility tend to cancel, so the overall accuracy may be adequate. The instrument responds to changing radiation within a fraction of a second and gives a conveniently large voltage output. A typical silicon solar cell of 2 cm^2 area produces 0.1 volt from 200 W/m^2.

Errors of radiometers

One aim in measuring solar irradiance is to limit the error to less than 5 per cent of the hourly or daily average, and less than 25 W/m^2. This is possible only with careful work. The error depends on the choice of instrument, the period of averaging, care in using the radiometer and the accuracy of its calibration. The maximum error of the Kipp and Eppley instruments can be reduced to about 8 per cent for an hourly average or 5 per cent for a daily average. The errors for the daily exposure may be less than 7 per cent on 95 per cent of days, and the error of annual mean irradiance may be only 2.5 per cent. But larger errors occur occasionally.

There are several causes of inaccuracy, including damage to the black paint which coats the sensor surface of the radiometer, condensation on the glass dome, the instrument not being level, inexactness of the assumption of a fixed proportionality between irradiance and the instrument's output, errors of the standard instrument used for calibration, and a temperature effect. For instance, Moll–Gorczynski instruments give 0.2 per cent more volts when the air temperature falls by 1C°, even when the global irradiance is unchanged. But such a temperature sensitivity is only about a quarter of that of the Robitzsch instrument, and the effect can be compensated for by suitable electrical circuitry between the thermopile and the voltmeter. In addition, there is some error due to variations of cloudiness. Also, dew and frost on the sensor reduce its sensitivity. Finally there may be a 'cosine error', the change of calibration caused by alteration of the solar altitude. The error can be 5–10 per cent when the sun is less than 30° from the horizon, though it is possible to allow for this cosine error, mathematically. Or the effect can be reduced by covering the cell with a shaped plastic diffiuser.

Operation

Several factors must be considered in using a pyranometer. Obviously, it must be mounted clear of shadow. There should be no obstruction higher than 5 degrees above the level nor above 3 degrees in the directions of sunrise and sunset to the east and west, and not more than 3 per cent of the sky should be obscured above a horizontal plane through the instrument. One should describe the climate station in terms of the angle of the horizon above the horizontal in each direction, shown on a circular diagram, with superimposed lines showing the track of the sun across the sky in each season.

The instrument should not be near any surface which is highly reflecting, and must be accurately level. It should be inspected after tropical dust storms or heavy rain, to clean the glass and check for leaks or condensation. The electrical resistance of the device should be checked annually for constancy, and the instrument's calibration should be tested regularly.

It is important that any radiometer be properly calibrated. Users are advised to be sceptical of the calibration constant supplied by the radiometer manufacturer, since its accuracy depends on the climate near the factory, the time of storage and the history of the instrument since calibration. In practice, many radiometers are poorly calibrated, leading to underestimates of the irradiance. Also, there may be a change of calibration: Heerman *et al.* (1985) reported changes of standard instruments in the USA by 10 per cent in five years. Aslyng and Nielsen (1962: 343) mentioned an *increase* of sensitivity by a few per cent when the radiometer is in constant use, whilst Szeicz (1968: 126) observed a steady *decrease* of calibration by 1 per cent over 11 years. Regular recalibration is the answer.

Calibrations carried out in different decades would not have given quite the same reading for the same irradiance, because of slight changes of the international standard. The Angstrom scale used since 1905 gave readings 1.5 per cent below those of the Smithsonian scale, which was introduced in 1913, whilst the International Pyrheliometric Scale since 1956 gives values 2 per cent above. The World Radiometric Reference Scale, used since 1980, gives values 2.2 per cent higher than those based on the 1956 scale.

It is possible to calibrate locally, but this should be confirmed by comparisons with side-by-side readings from a national standard instrument at least once each five years (see 'Measurement accuracy'). Preferably, the comparison should involve hourly averages taken over a few days.

Bright sunshine

An alternative is to measure the duration of 'bright sunshine'. This is the total time during the day when the sun is unobscured by cloud, apart perhaps from thin cirrus. The procedure is cheap, convenient and often sufficiently accurate for estimating daily mean solar irradiance. As a result, there are several times as many climate stations with bright-sunshine recorders as there are with radiometers.

The simplest and most common equipment in temperate climates for measuring the hours of bright sunshine in a day is the 'heliograph' developed by Campbell and Stokes during 1853–80. It was a standard instrument during 1964–81, recommended by the World Meteorological Organization. It consists of a glass sphere backed by a concentric frame which holds a strip of cardboard, as shown in Figure 5.8. The 100-mm sphere focuses direct radiation from the sun into a spot on the cardboard, burning a hole if the irradiance is sufficient, i.e. when there is no cloud in front of the sun. The hole becomes a slot as the sun moves across the sky, and the slot's length indicates the duration of bright sunshine. The time indicated on the chart by the position of the focused light is sun-dial time (Tlm, p. 148) and not clock time.

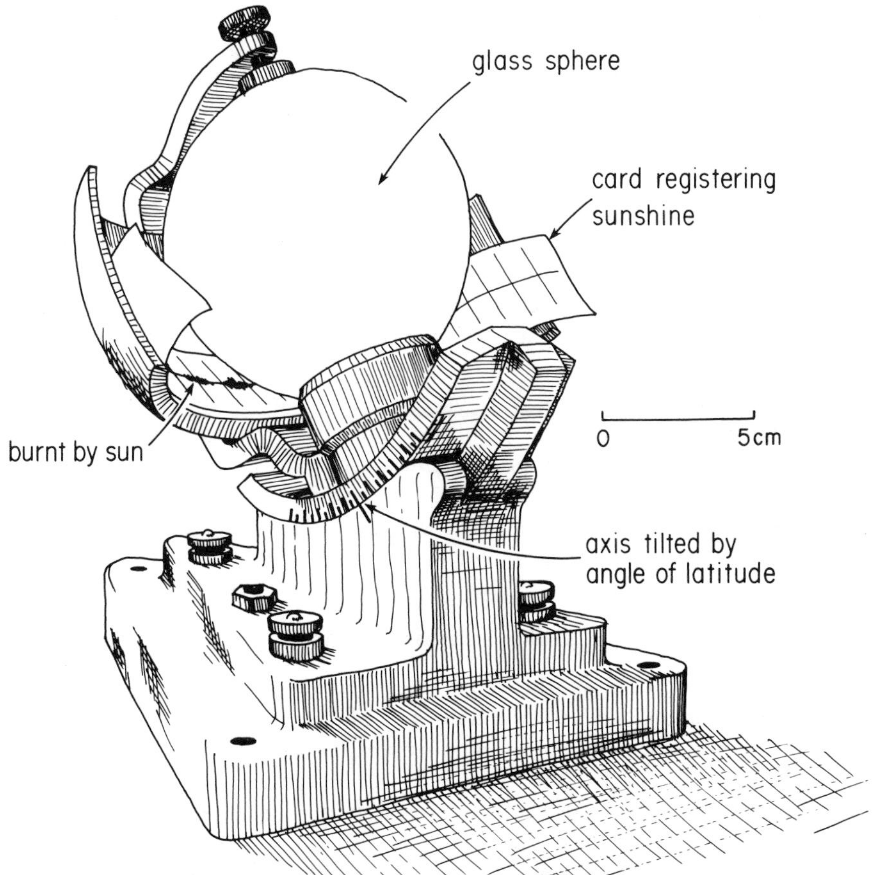

Figure 5.8 A Campbell–Stokes sunshine recorder

The device is robust and without moving parts, but tedious to use, since a new card must be fitted each day before early dawn, and the old one laboriously examined to derive the total duration of bright sunshine (see Note 5, p. 334). A home-made equivalent can be made from a spherical glass flask filled with 500 mL water, though damage may occur on frosty nights.

The Campbell–Stokes instrument should be mounted firmly on a brick pillar at least 2 m high, as with all instruments for measuring irradiance. Radiometers and sunshine recorders are usually mounted on the flat roof of a high building so that no shadows fall across the instrument once the sun is above 5 degrees from the horizon. When the sun is below that, the radiation is too weak to register and as a result there is no 'bright sunshine' for about 15–30 minutes after sunrise and for the same duration before sunset.

There are four fundamental difficulties with the Campbell–Stokes recorder. The main problem is that the minimum solar irradiance required to burn a hole in the card varies, on account of the card's wetness and of the effect of atmospheric turbidity in reducing the ratio of direct to total global irradiance (see 'The nature of solar radiation', pp. 154–6). The threshold value of global irradiance is nominally about 200 W/m^2, but in practice may be below 100 W/m^2, or above 230 W/m^2. Painter (1981) quoted a range of 106–85 W/m^2. The World Meteorological Organization recommended that the threshold be regarded as relating to a *direct* irradiance Qd equal to 120 W/m^2. But the uncertainty of the threshold value was the reason for the WMO eventually abandoning the instrument as an interim standard.

The second problem is that the device measures only the direct beam irradiance and is unaffected by diffuse radiation, even when the latter may be appreciable. Third, the instrument is useless at high latitudes in winter when the clear-sky direct irradiance is below the threshold for registration. Fourth, it is difficult to measure the daily duration of bright sunshine accurately when there is frequent intermittent cloud.

Other sunshine recorders

An instrument used in the USA in preference to the Campbell–Stokes recorder is associated with the names of Maring and Marvin. It is a glass thermometer whose mercury column is expanded by solar heating of the blackened bulb, so that the mercury closes an electrical switch. An electric clock measures the total duration of switch closure when the irradiance is above a threshold of about 260 W/m^2 measured facing the sun. The advantages of the instrument include its convenience, the possibility of automatic and remote recording and the absence of any need for daily attention. However, the device is less robust than the Campbell–Stokes instrument and less sensitive. Also, it is affected by temperature, so that an adjustment has to be made in autumn and spring. As a result, it is no longer made and has been replaced by the Foster sunshine switch.

The latter has a pair of silicon photocells, one shaded to measure only the diffuse irradiance, and the other measuring both the direct and diffuse irradiance. The electrical output of the first is subtracted from that of second to determine the direct radiation alone. The device is stable, reliable and responsive.

A particularly simple and trouble-free alternative is the device shown in Figure 5.9. Four pairs of parallel leaves of bimetal strip are arranged as part of a hemisphere so that the sun shines on the outside leaf of at least one pair. The heating of that leaf causes it to bend and touch the parallel inner leaf, thus closing a switch, provided that the irradiance facing the sun exceeds 200 W/m^2. All four pairs of leaves are switches in parallel. So, if

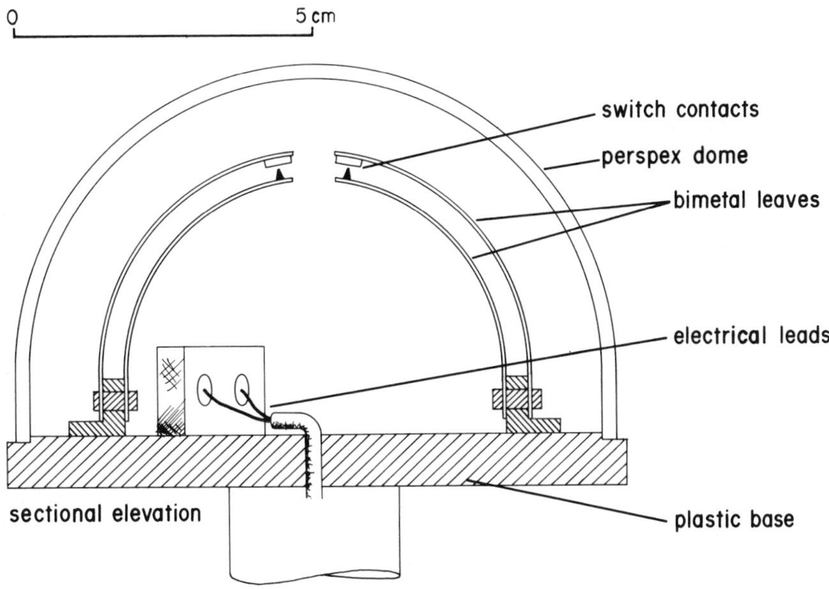

Figure 5.9 A recorder of the duration of bright sunshine (Sumner 1966)

any pair is closed, a small constant-speed electric motor drives a revolution counter, the number of revolutions being a direct indication of the duration of bright sunshine. No switch closes simply because temperatures rise, as both the parallel leaves of each pair will then bend, remaining apart. The instrument is cheap, automatic, robust and registers all periods of bright sunshine lasting longer than about 20 seconds. Daily readings of a Sumner instrument at Griffith (New South Wales) showed 0.4 hours more sunshine, on average, than was indicated by a Campbell–Stokes recorder.

Cloudiness

Clouds are an important influence on both diffuse and direct irradiances. Consequently, it is customary to assess how much of the sky is covered by cloud in the course of taking daily measurements at a normal climate station. One should average separate assessments of the fraction of the sky covered by cloud in each quarter of the compass. On days of strong wind there should be repeated assessments, averaged once more, to iron out cloudiness fluctuations. However, the judgement is difficult because of the irregular coverage, the various heights of different kinds of cloud and their various thicknesses.

The observed cloudiness is commonly quantified in terms of tenths of the sky's area, e.g. C tenths. However, the International Meteorological

Organization (the forerunner of the World Meteorological Organization) recommended in 1947 that the unit should be the 'okta', an eighth of the sky's dome, e.g. *C* oktas. Oktas are easier to estimate and they describe an overcast sky by means of a single digit. Zero tenths or oktas signifies a cloudless sky, 8 oktas or 10 tenths an overcast sky, and 9 oktas a sky obscured by fog.

The assessment of the fractional cloudiness of the sky, seen from a point on the ground, gives what may be termed the 'point cloudiness' (Cp). It involves an oblique view mostly. Even an experienced observer tends to overestimate the amount of cloud near the horizon, but underestimate the amount when the cloud is uniform across the sky. Also, comparison with satellite photographs shows that ground observers tend to overestimate low clouds and underestimate high clouds. Further, there is a tendency to moderate the extremes, so that the cloudiness is overestimated by surface observers if the sky is less than half covered, and vice versa.

Another figure (Ca) is obtained from a satellite looking vertically downwards. The difference of cloudiness estimates is particularly great if the sky has isolated tall clouds near the horizon. So the difference is greatest at low latitudes where clouds tend to be taller than the stratus which prevails at high latitudes.

A third estimate of cloudiness (Cs) is given by a recorder of the number of hours of bright sunshine (n) in a daylength of N hours:

$$\mathrm{Cs} = 10\,(\mathrm{N} - \mathrm{n})/\mathrm{N} \qquad \text{tenths} \tag{5.29}$$

Daily average estimates of Cs equal those from satellites (Ca), if the cloudiness is even across the sky and constant during a day. In general, instrument values of Cs are preferred to the subjective estimates (Cp), derived from visual observation. Raju and Kumar (1982: 497) found them related as follows, using monthly mean values from 33 climate stations in India:

$$\mathrm{Cs} = 0.88\,\mathrm{Cp} - 0.3 \qquad \text{tenths} \tag{5.30}$$

Eqn (5.30) indicates that Cp is an overestimate by about 0.7 tenths when the sky is a third clouded, but by 1.3 tenths with 8 tenths of cloud. The difference is likely to be less at higher latitudes, e.g. about 0.6 tenths in Europe, but zero in the USA. Both possibilities can be accommodated within the considerable scatter of the measurements.

Effect of snow and cloud on the measured irradiance

Any irradiance measurement at high latitudes is affected by either snow or ice on the ground, if there is also partial cloud, because of additional short-wave radiation reflected down from the cloud. When radiant energy reaches snow-covered ground, the high albedo leads to substantial reflection upwards, and this in turn is partly reflected down again by any cloud, augmenting the measured irradiance. In fact, there is a train of reflections between the sky and the ground, so that the total measured downwards short-wave radiation Qs is related to what it would be in the absence of ground and sky reflection (Qsa) as follows:

$$Qs/Qsa = 1 + \alpha.\alpha s + (\alpha.\alpha s)^2 + (\alpha.\alpha s)^3 + \text{etc.} \qquad (5.31)$$

where α is the albedo of the ground and αs is the reflectivity of the sky. The latter depends on the elevation of the sun and the amount of cloud.

A formula of Hay and Won (1980: 37) for the albedo for a clear sky gives 26 per cent when the sun is 10° above the horizon, 9 per cent when at 20°, and 4 per cent when at 60° or overhead. The albedo of cloud depends on its thickness: it is 40 per cent for a cloud which is 30 m thick, 70 per cent for 100 m and 90 per cent for 1 km. Also, the albedo of a broken layer varies with the square of the fractional cloudiness. On the whole, the sky's albedo is increased by roughly 3.5 per cent for each tenth of cloud, with about 60 per cent when the heavens are totally overcast.

The ratio Qs/Qsa in Eqn (5.31) is 1.39 if the albedo of a sky is 40 per cent and there is snow on the ground (with an albedo of 70 per cent). On the other hand, the ratio is only 1.01 with grass and the sun overhead in a clear sky. So the augmentation of global irradiance by multiple reflections is significant only at high latitudes in winter.

ESTIMATING THE SOLAR IRRADIANCE OF LEVEL GROUND

The shortage of radiometers in routine use, especially in developing countries, makes it necessary on occasion for the global irradiance to be estimated instead of measured. Estimated values supplement available measurements and are useful in deducing other climatic factors such as the net irradiance (see pp. 89–103), which in turn governs the evaporation rate and hence the soil-moisture content and the growth of crops. Furthermore, estimates of solar irradiance (Qs) are relevant to house design, human comfort and the use of solar energy.

One method of estimating Qs was mentioned in 'Estimating screen temperature', pp. 72–85 – the so-called 'loop method' (see pp. 78–80), which gives monthly mean irradiance and temperature figures. When this method is applied to eight different places, estimates and actual measurements compare as shown in Figure 5.10, the average error being 18 W/m^2 and the root-mean-square-error 26 W/m^2, or about 13 per cent of the average measurement.

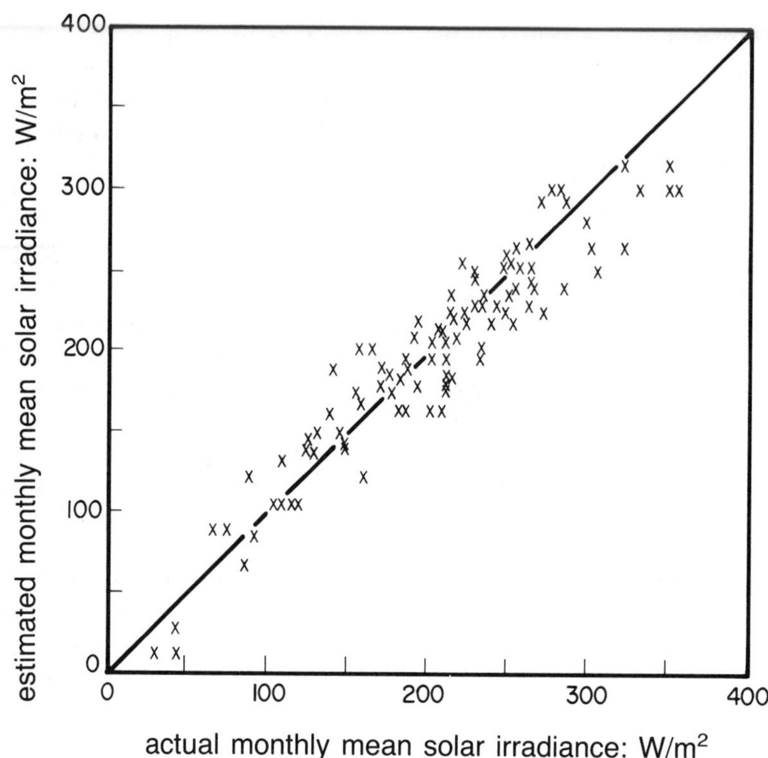

Figure 5.10 Comparison of estimates of monthly mean solar irradiance Qs (estimated by the loop method) with one year's measured values from eight places – Williamtown in Australia, Cape Town, Edmonton, Nairobi, Dar-es-Salaam, Port Moresby, Albuquerque and Athens

The Angstrom formula

In estimating Qs, one seeks a proxy which is closely related. The closest relationship found by Lund (1968), in a study of nine years' monthly mean data from Massachusetts, was with the duration of bright sunshine – the correlation coefficient exceeded +0.93. This was more than the −0.76 for the correlation between Qs and cloudiness or +0.54 between Qs and temperature. The strong relationship with sunshine duration supports the suggestion by Angstrom (1924) that the following equation be used:

$$Qs = Qo\,[A + (1 - A)\,n/N] \qquad W/m^2 \qquad (5.32)$$

where Qo is the daily mean solar irradiance when the sky is cloudless, and n is the number of hours of bright sunshine during a daylength of N hours (see Note 8, Chapter 3). The factor A in Eqn (5.32) was found by

Angstrom to be 0.24 for Stockholm, but it appears to vary with location and season. A value of 0.37 was obtained for places in the USA and 0.34 in Australia. It is 0.36 in summer in Melbourne but 0.40 in winter. Values around 0.36 are typical, resulting in Eqn (5.33):

$$Qs = Qo\ (0.36 + 0.64\ n/N) \qquad W/m^2 \qquad (5.33)$$

One difficulty with this is that the relationship between Qs/Qo and n/N is not in fact quite linear, so that the formula overestimates Qs when the sky is either cloud-free or totally overcast. The reason for the non-linearity is that turbidity of the air becomes important in completely clear skies, whilst cloud thickness affects the relationship in totally overcast conditions. And there are two other problems with Eqn (5.32). Inaccuracy arises if there is any variation of cloudiness during the day, since cloud at noon is not equal, in its effect on Qs, to cloud occurring when the sun is low. Also, there is the problem of determining the Qo value in places or at times of the year when skies are rarely without some cloud.

The Prescott formula

The problem of determining the clear-sky global irradiance (Qo) as a preliminary to using Eqn (5.32) is avoided by the stratagem of Prescott's group, who suggested using values of the extra-terrestrial radiation intensity (Qa) already given in Note 12, Chapter 3:

$$Qs = Qa\ (a + b.n/N) \qquad W/m^2 \qquad (5.34)$$

where a and b are factors to be determined empirically. The sky is overcast when the ratio n/N is zero, so [a.Qa] represents the diffuse irradiance, whilst the term [b.Qa.n/N] is a measure of the direct irradiance.

The straightness of the line in Figure 5.11 implies that the factors a and b in Eqn (5.34) are constants, i.e. 0.20 and 0.51 respectively. Similar figures have been reported by many authors, including those mentioned in Table 5.2 from a wide range of places. Most figures lie near 0.25 and 0.50, respectively, chosen as convenient fractions of unity. These particular constants are in the ratio of two-to-one, as in Eqn (5.33), approximately. Eqn (5.34) thus becomes the following:

$$Qs = Qa\ (0.25 + 0.5\ n/N) \qquad W/m^2 \qquad (5.35)$$

The constants add up to 0.75, and then refer to the value of Qs/Qa when there is a clear sky all day, i.e. to Qo/Qa. This is the daily mean transmission, found earlier to be around 0.75 (see p. 60, p. 333).

The scatter of values in Table 5.2 is presumably due to differences of solar elevation (which governs the distance that the direct radiation travels through the atmosphere), of altitude, turbidity and to a non-linear relationship between radiation and cloudiness. In addition, Rietveld (1978)

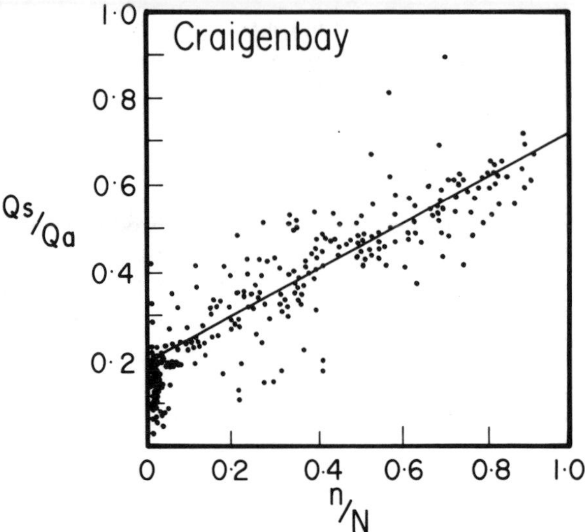

Figure 5.11 Relationship for Craigenbay (Scotland) during 1965 and 1966, between (a) the ratio of the daily mean measured solar irradiance (Qs) with respect to the extra-terrestrial irradiance (Qa), and (b) the ratio of the bright-sunshine duration (n) with respect to the daylength N (Nkemdirim 1970: 134, by permission of Springer-Verlag Inc. and Dr. L. Nkemdirim)

suggested that the factor a in Eqn (5.34) depends on cloud height, whilst b may be affected by cloud thickness. It has been suggested that the constants depend on latitude, but closer examination shows no significant effect. Seasonal increases of turbidity may reduce the constants slightly. Elevation does not affect the constants (at least not in Austria).

Small differences of the values hardly alter the estimated solar radiation. For instance, a measured annual mean solar irradiance in Greece of 182 W/m^2 was closely approximated by estimates from Eqn (5.34), whether using 0.20 and 0.51 for a and b (i.e. 182 W/m^2) or 0.31 and 0.41, i.e. 184 W/m^2. So it is satisfactory to use Eqn (5.35), quite generally.

Errors found in using Prescott's formula for estimating values of the *daily* mean solar radiation have been quoted by various authors as either 10 per cent, 15–20 per cent or up to 35 per cent of the values. Estimates of *monthly* mean errors have been found to be either about the same as for daily means, or, as one would expect, only a quarter or a third as much. As a result, Eqn (5.35) is preferably not used for periods of less than 10 days. However, errors are less than those incurred by actual measurements made with a nearby Robitzsch radiometer, or with a superior radiometer more than 30 km distant.

One proper use for the formula is in supplementing irradiance measurements, as in a survey of conditions in India where radiometer

Table 5.2 Values of the constants in the Prescott formula (Eqn 5.34), for estimating global radiation from observations of sunshine duration

Reference	*Place*	*Typical factors* a	b	$a+b$
Prescott 1940	Canberra	0.25	0.54	0.79
Penman 1948	England	0.18	0.55	0.73
and thirty other sets of values				
Rao and Bradley 1983	Oregon	0.23	0.52	0.75
Kotoda 1986: 15	Japan	0.21	0.53	0.74
S. Turton 1987: priv.comm.	Humid tropics	0.28	0.47	0.75
Median		0.23	0.50	0.75
Mean		0.25	0.50	0.74
Standard deviation		0.05	0.05	0.05

values from 16 climate stations were augmented by estimates based on sunshine-duration measurements at 121 other places (Mani and Rangarajan 1982).

Estimation from cloudiness

Where there is no instrument to measure the *duration* of bright sunshine, it is necessary to replace n/N in Eqn (5.35) by an equivalent expression which depends on the observed *degree* of cloudiness. This is because there are even more climate stations where daily observations of cloudiness are made than there are places with sunshine recorders. Unfortunately, there is some subjectivity about the estimation of sky cloudiness (see 'Measuring solar irradiance', pp. 173–4) and a non-linear relationship between n/N and cloudiness, as seen in Figure 5.12. Similar non-linearity has been observed in Thailand, i.e. n/N is 0.8 for 3 oktas, 0.7 for 5, and 0.4 for 7 (Oldeman 1987: 104). The scatter in Figure 5.12 is due partly to the subjective observation of cloudiness, and also to the comparison of a time average with a space average. The averages are equivalent only if the degree of cloudiness is constant as the pattern moves across the sky.

In view of the connection between cloudiness and the duration of bright sunshine in Figure 5.12, Angstrom's formula can be replaced by an equivalent involving the cloudiness (C tenths), such as the following:

$$Qs = Qo\ (1 - 0.07\ C) \text{ approximately} \qquad W/m^2 \qquad (5.36)$$

One weakness of Eqn (5.36) is that no allowance is made for the effect of cloud density on the solar irradiance, only cloud extent. Another is the need to determine Qo, which can be side-stepped by using an equation equivalent to Prescott's formula involving the extra-terrestrial irradiance

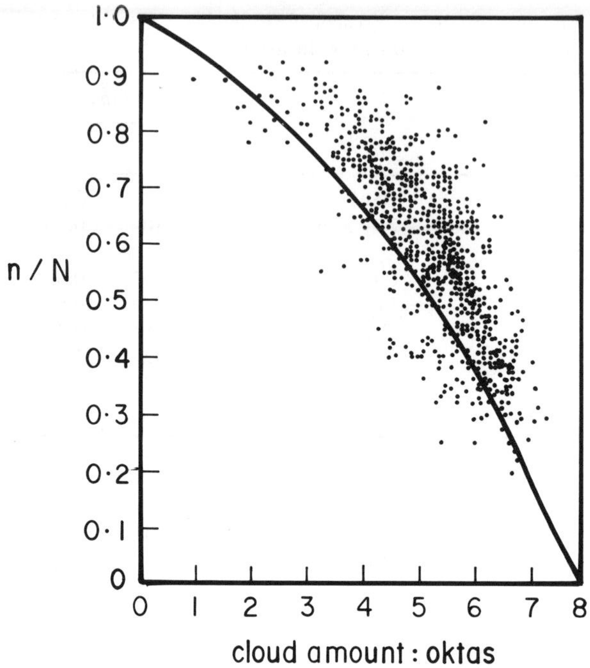

Figure 5.12 Relationship between (a) the sunshine-duration/daylength ratio and (b) the daily mean cloudiness, in East Africa (Chang 1971: 23). The curved line comes from an equation of Satterlund and Means (1978: 368).

(Qa) instead. Examples are illustrated in Figure 5.13. The median values of the Qs/Qa ratio for cloudiness values of 2.5, 5 and 7.5 tenths are 0.71, 0.58 and 0.47, respectively, which yield the following consensus equation:

$$Qs = Qa\ (0.85 - 0.047\ C) \qquad W/m^2 \tag{5.37}$$

Similar expressions have been derived empirically in Holland, Japan, Thailand and Sydney. However, Eqn (5.35) is preferable, since it avoids the assumption that clouds of various kinds have the same effect on solar irradiance.

Estimation from temperature data

Temperature is an alternative to either geographical data, sunshine duration or cloudiness, as a surrogate for measurements of irradiance. After all, the weather is warm when the sunshine is strong. Bagnall (1982) pointed out that the dividend of monthly mean values Qs/T in Australian cities is about 30 $W/m^2.C°$. Similarly, the sum of daytime hourly temperatures at Nairobi is roughly proportional to the solar irradiance.

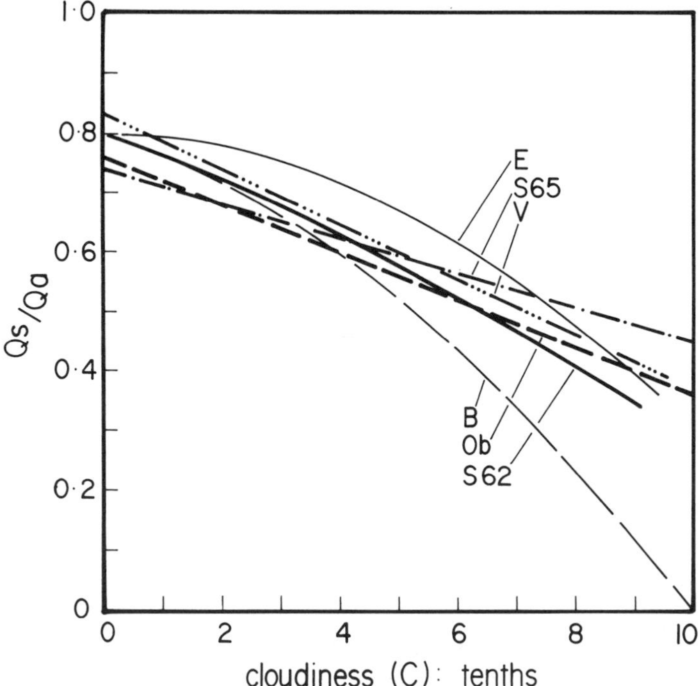

Figure 5.13 Effect of cloudiness on the ratio of global irradiance to extra-terrestrial irradiance. 'E' signifies data from Exell (1982); 'S65' values come from Stanhill (1965: 56); 'V' from de Vries (1958); 'B' from Black (1956, see Gates 1962: 55), Stringer (1972: 215) and Rosenberg (1974: 16); 'Ob' from Obasi and Rao (1977); and 'S62' from Stanhill (1962)

However, simplistic correlations ignore the effects of altitude and latitude and are inferior to the following, based on Figure 5.14:

$$Qs = 600\ Tc/(84 - A) \qquad W/m^2 \tag{5.38}$$

where Qs is the annual mean solar irradiance, Tc is the annual mean equivalent sea-level temperature (see p. 73) and A is the latitude in degrees. At a latitude of 45°, representative of the USA, the equation implies warming by 0.07C° for each extra one W/m^2 of global irradiance.

Attempts to develop relationships like Eqn (5.38) for monthly mean instead of annual mean values are complicated by the fact that the temperature lags about a month behind the annual rhythm of radiation. The month's lag of temperature means that monthly mean irradiance estimated from Eqn (5.38) is best derived from the *following month*'s mean temperatures give an average error of 68 W/m^2. Likewise, using amended temperatures in Eqn (3.31) to estimate evaporation reduces the mean error

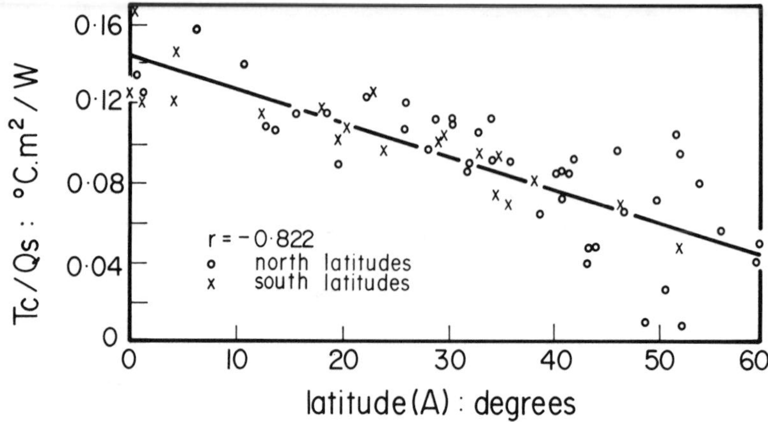

Figure 5.14 Relationship between (a) the ratio of the annual means of temperature and global irradiance, and (b) the latitude (Linacre 1969a: 8, by permission of Springer-Verlag Inc.)

from 0.65 mm/d to 0.63 mm/d, in the case discussed in 'Estimating evaporation', pp. 105–6. The lag explains the tendency in Figure 3.18 for estimated evaporation values to be to the right of the line joining measurements.

The three connections, (a) between the daily range of temperature (Rd) and the ground-level humidity (see 'Estimating dewpoint temperature', p. 88), (b) between the latter and cloudiness (see Figure 4.13), and (c) between cloudiness and Qs in Eqn (5.36), suggest an indirect dependence of Qs on Rd. This was evaluated statistically by Cengiz *et al.* (1981: 1271) from daily measurements at Columbia (Missouri) in terms of the clear-sky irradiance (Qo), as follows:

$$Qs = 0.048\ Qo - 3.5\ Rd + 0.029.Qo.Rd + 24 \qquad W/m^2 \qquad (5.39)$$

Disadvantages of such an equation are the uncertain physical meaning of the product (Qo.Rd) and the equation's unknown applicability outside Missouri.

Estimation from other proxies

There have been several suggestions that the global irradiance can be estimated from measurements of relative humidity, but empirical relationships at particular places have proved to be inaccurate elsewhere. Alternatively, a rough estimate of the solar irradiance can be obtained from the rainfall, since the clouds which bring rain also attenuate the sunshine.

A more direct surrogate for Qs is the rate of evaporation from water. McWhorter and Brooks (1965) reported the following:

$$Qs = 32\ Ep + 65 \qquad W/m^2 \tag{5.40}$$

where Ep is the rate of evaporation from a US Class-A pan evaporimeter (mm/d). A test of a similar empirical equation against daily mean data from Marsfield in 1981 showed an average error of only 13 W/m^2 in estimating Qs, i.e. 8 per cent.

A statistical correlation of measurements of Qs with the latitude, longitude and the amount of precipitation, provides a useful method of estimating Qs at intermediate locations. Regressions of monthly mean values of Qs in China against elevation, the number of sunshine hours, and the number of days with dust or smoke give estimates which are within 4–10 per cent of measured values.

Alternatively, the long-term mean solar irradiance can be estimated by interpolation between nearby measurements. Interpolations between 130 radiometers in Canada lead to errors of less than 1 per cent. Monthly mean global irradiances at places in Australia have been estimated with a root-mean-square-error of less than 14 W/m^2 in summer, or 9 W/m^2 in winter. This was based on data from 44 radiometer records and measurements of bright sunshine at 102 places.

More complex methods

So far we have been concerned with estimates for periods of a week or more. For times of an hour or less it is necessary to deal with matters in more detail, allowing for solar elevation, ozone content of the air, the properties and hourly amounts of cloud at various levels, the atmospheric turbidity, and estimation of the two kinds of diffuse radiation. However, this generally requires considerable input data and is beyond the scope of this book.

A recent development now completely alters the estimation of ground-level solar irradiance over short periods as well as other meteorological elements. This involves the remote sensing of cloud and of surface temperatures by means of satellites. For instance, Nunez (1988) described a simple method of estimating daily irradiance using an empirical correlation with satellite measurements of the Earth's reflectance, the standard estimation error being 30 W/m^2. However, once again, this is outside the capability of an observer at a normal climate station. In any case, one can usually avoid such sophistication and still achieve an accuracy sufficient for most practical purposes.

VALUES OF SOLAR IRRADIANCE

The global distribution of Qs is seen in Figure 5.15. This shows over 300 W/m^2 in Australia and South Africa in January, and in the Middle East in July. There is more than 360 W/m^2 in June in north-west Africa. Average irradiance in the USA is 140–240 W/m^2. At the South Pole, the mid-summer value is only about 200 W/m^2, even though the extra-terrestrial irradiance (Qa) exceeds 540 W/m^2 on account of the 24-hour day and the fact that summer in the south coincides with the perihelion. The intensity at parts of the equator is greatly reduced by cloud.

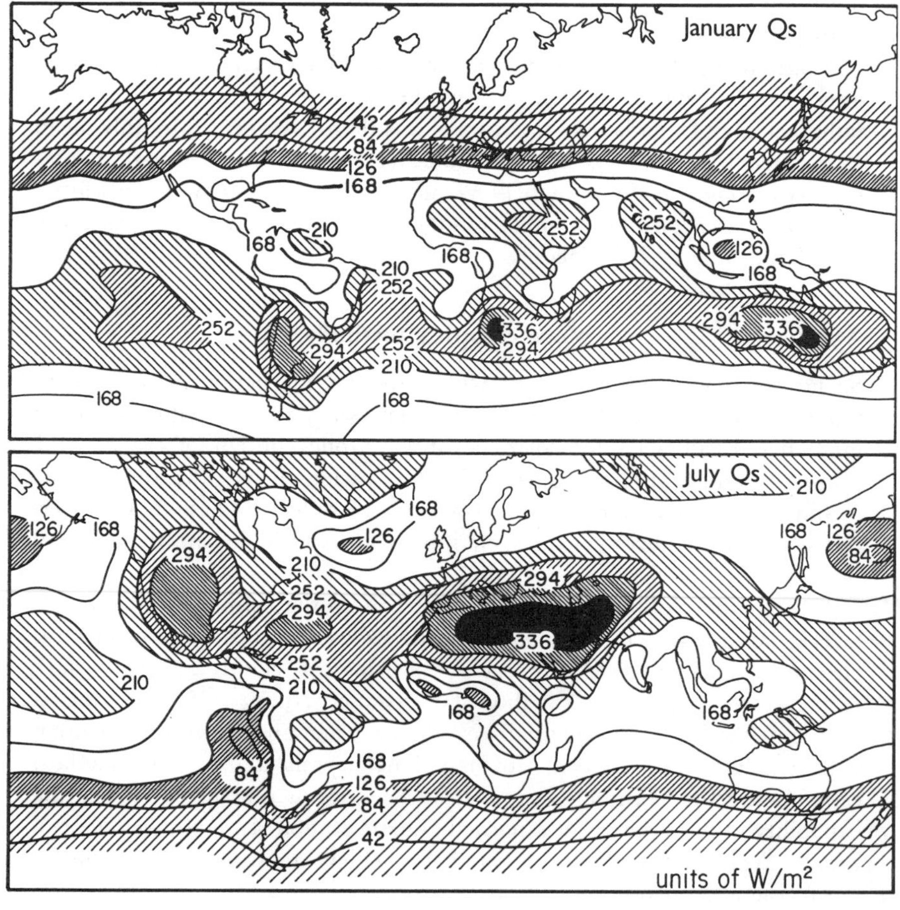

Figure 5.15 Global patterns of monthly mean solar irradiance in January and July. The numbers show the irradiance in units of W/m^2.

The irregularity in Figure 5.15 is partly due to the spatial variation of cloudiness, seen in Figure 5.16. Other reasons are differences of altitude and atmospheric turbidity, and the shorter daylength in north–south valleys. Fig. 5.16 shows that there is slightly more cloud in the southern hemisphere than in the north, presumably due to greater expanse of ocean.

The duration of bright sunshine varies widely from place to place. There are about 1600 hours annually at London, 2000 at Melbourne, 2400 at Sydney, 2800 at Miami, 3200 at Pretoria, 3600 at Phoenix (Arizona) and 4000 at Aswan in Egypt, for example.

Effect of elevation

Solar irradiance increases by 2–14 per cent per km of height up a mountain, assuming the same degree of cloudiness. It is 7 per cent per km when the sky is clear in the European Alps, despite a decrease of diffuse irradiance from 49 W/m^2 at 200 m elevation to 30 W/m^2 at 3 km. When the sky is overcast, diffuse irradiance *increases* with elevation, e.g. from 75 W/m^2 at 200 m to 196 W/m^2 at 3 km.

The amount of cloud generally varies with elevation. The sunshine-duration/daylength ratio in Java falls from 0.67 at 100 m, to 0.43 at 1600 m and to 0.38 at 3100 m. Those figures imply a reduction of 0.70 hours of sunshine annually for each metre ascent between 100 m and 1600 m, and 0.15 h/a.m between 1600 m and 3100 m. Likewise, figures for Switzerland show 0.20 h/a.m between 11 m and 2500 m, and 0.17 h/a.m between 2500 m and 3460 m.

The variation of fog with elevation up mountains in Japan leads to a minimum of irradiance at a height of 1500 metres, for instance. However, the height of maximum fog and least irradiance depends on the locality and macroclimate.

Effect of spacing on irradiance differences

Variations of cloudiness, altitude and turbidity over quite short distances make it unsafe to assume that solar irradiance measurements at one place are relevant elsewhere. The effect of distance on the change of irradiance can be discussed in terms of either the difference, or the correlation coefficient, relating the values at two separate places. We will consider these.

Figure 5.17 shows how the difference between irradiance readings at separate places increases with the distance apart. The variation is more or less similar in the USSR and western Canada, but is affected by the season. Monthly mean values at places 50–100 km apart in the USSR, Wisconsin and Connecticut differ by about 8 per cent, 10 per cent and 19 per cent, respectively.

January

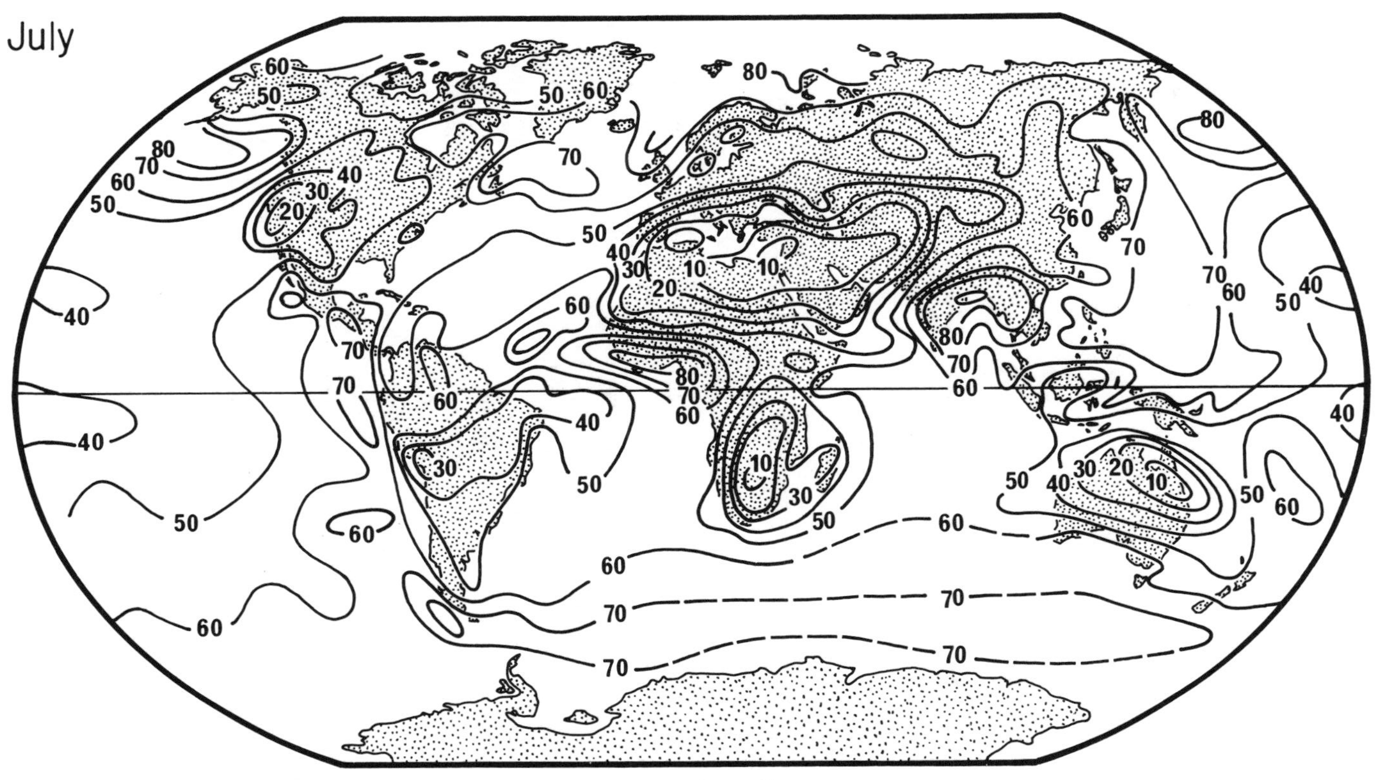

Figure 5.16 Global patterns of monthly mean cloudiness in January and July (Okolowicz 1976: 325). The numbers show the percentages of the sky obscured by cloud, i.e. 100 per cent equals 10 tenths or 8 oktas. (By permission of PWN – Polish Scientific Publishers)

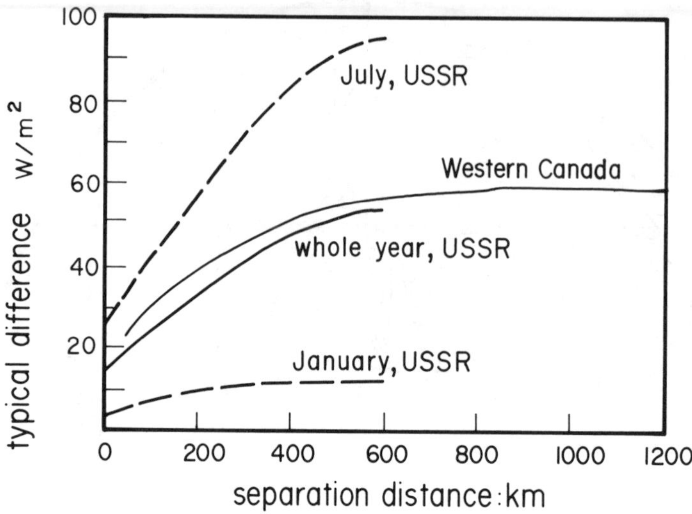

Figure 5.17 Effect of the distance between measurements of global irradiance on the mean square of the difference between daily mean values (WMO 1981b: 192, by permission of the World Meteorological Organization)

Another method of relating irradiances at separate places involves the correlation coefficient, linking simultaneous hourly, daily or monthly mean values. The coefficient declines with increasing distance apart, even within short distances of separation: Hay (1984: 430) quoted 0.95 for the correlation of daily mean Qs at places 10 km apart in Vancouver, 0.87 for 20 km separation and 0.85 for 40 km.

The correlation coefficient of irradiance measurements is higher for long periods of averaging. For *annual* mean values of irradiance, it may be 0.90 for 300 km separation, compared with 0.86 for *monthly* sunshine durations. *Daily* values of global irradiance in the north-central USA correlate with coefficients which are 0.88 for places 60 km from each other and only 0.30 if the spacing is beyond 550 km. For *hourly* sunshine duration, the coefficient in Switzerland is only 0.55 for 180 km. The effect of the period of averaging on the correlation coefficient can also be inferred from an example given by Hay and Hanson (1985: 152): for a separation of 250 km in western Canada the coefficient is 0.96 for annual averages of Qs, 0.92 for monthly values, 0.71 for daily, and only 0.40 for hourly means.

Figure 5.18 shows the effect of the distance of separation, the period of averaging, and also of the time of year on the correlation coefficient. The coefficient is higher in summer than in winter, presumably because of seasonal variations of cloudiness.

The solar energy at some distance from where Qs is measured depends on the separation, the period of averaging irradiance, the cloudiness, and

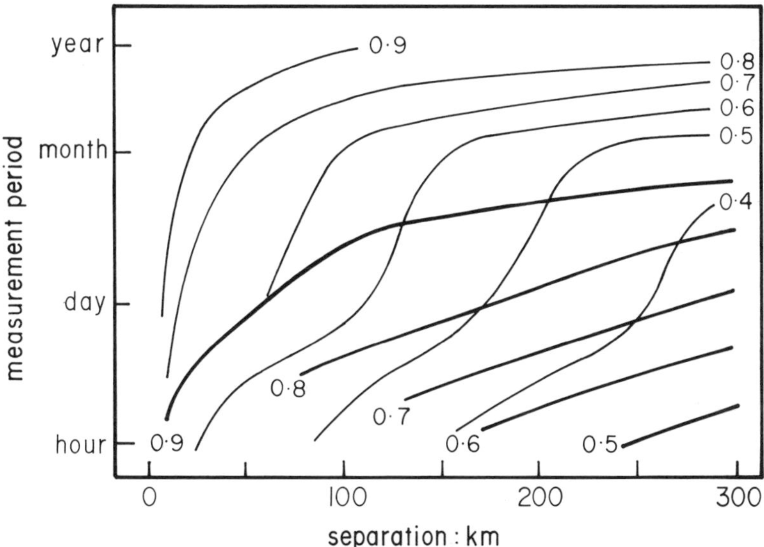

Figure 5.18 Effects of the time of year, the period over which measurements were taken, and the separation distance, on the coefficient of correlation of values of sunshine duration at pairs of places in Switzerland (WMO 1981b: 241). The fat lines refer to July conditions and the thin lines to those in December. For instance, the correlation coefficient is about 0.7 for values of the daily sunshine duration at pairs of places 270 km apart in July, or 60 km in December (By permission of the World Meteorological Organization)

the season. For instance, daily values in various parts of Wisconsin may differ by 25 per cent, and monthly values by 10 per cent. Daily values differ by over 15 per cent when places in Labrador are over 60 km apart. Monthly irradiance values, at places separated by 200 km in the southern USA, differ by as much as 10 per cent on 90 per cent of occasions. To limit the error of daily values within 10 per cent on 95 per cent of occasions, the distance would need to be less than 100 km in Washington State. Such figures indicate the instrument spacing required for particular degrees of accuracy.

The separation of radiometers in a network has to be a compromise between a close spacing for accuracy and a wide one for economy. Won and Truhlar (1980: 1.11) suggested as suitable whatever distance achieves a correlation coefficient of 0.7 for monthly values. This distance would depend on the terrain, being less in mountainous country, near the coast and in humid climates, where clouds are more prevalent (see Table 2.1, p. 28).

Table 5.3 Range of monthly means of measurements of solar irradiance for particular months during runs of a few years

Place	Number of years	Range of monthly mean Q_s W/m^2	
		January	*July*
Nairobi	5	68	31
Dar-es-Salaam	4 or 3	69	24
Albuquerque	4	26	34
Cape Town	5	42	24
Williamtown	4	82	11
Edmonton	4 or 5	7	40
Means		48	47

Source: Berliand 1970.

Variation in time

Solar radiation varies in time also. The tilt of the Earth's axis and the elliptical orbit lead to variations of the sun's declination (see Figure 5.1) and to the minimum annual range occurring not at the equator, but at 3.4°N.

There seems to have been a long-term change of cloudiness. Changnon (1984: 147) reported a steady rise in the number of cloudy days at six places in Illinois, from 1120 days during 1901–10, to 1710 days in 1971–80. Over the USA as a whole, average cloudiness rose from 5.0 tenths in 1900 to 6.1 in 1980: only one of 77 climate stations did not report an increase. The biggest increase occurred during 1930–50, which was before much high-level air traffic, so it was not due to condensation trails. Between 1940–77, more cloud in the USA was accompanied by a very small increase of turbidity, which also would reduce irradiance.

Considerable changes of monthly mean irradiance from year to year are indicated in Table 5.3. The consequence is that there is no need for extreme accuracy of measurement or estimation, for many problems. For example, even fairly large observational errors hardly alter the general pattern of data needed for designing solar-energy collectors.

Daily irradiance can range appreciably within a month, as exemplified in Table 5.4. For this selection of Australian cities and towns, the variability index (p. 126) varies from 20 per cent at Darwin in July to 89 per cent at Newcastle in January. The median value is 62 per cent, with a tendency for more variability at higher latitudes. The table also shows that the variation with latitude in summertime (i.e. January) is remarkably small, because longer days at high latitudes compensate for lower solar elevations. This is important in growing crops.

Table 5.4 Values of daily mean solar irradiance at Australian cities and towns which exceeded the values on 10%, 50% or 90% of days during 1968–1974

Latitude		*January: W/m²*			*July: W/m²*		
(°S)	*Place*	*10%*	*50%*	*90%*	*10%*	*50%*	*90%*
12	Darwin	93	252	326	198	229	244
20	Port Hedland	227	337	361	172	208	221
23	Longreach	168	328	369	160	193	208
24	Alice Springs	208	332	365	128	202	217
29	Geraldton	262	355	381[a]	83	147	169
31	Forrest	215	243	372	63	139	164
31	Woomera	180	348	385	75	136	159
33	Newcastle	108	298	372	65	134	156
38	Melbourne	137	295	358	41	73	105
43	Hobart	146	278	349	35	68	92

Source: Gaffney 1975: 9.

Note: [a] For example, at Geraldton, which is at 29°S, the daily mean solar irradiance was *less* than 381 W/m^2 on 90 per cent of days in January.

The risk of a long run of cloudy days, considered on pp. 135–9 for the case of Bangkok, is important in collecting solar energy. Data from Lisbon, for instance, show that there were 38 days with less than 145 W/m^2 during the 279 days of nine Octobers, and six runs of three such days. Consecutive daily mean data from Wisconsin showed that one day's exposure differed from the next by 18 per cent on average.

Hourly variation

The solar irradiance may alter greatly during the day, as shown in Figures 5.19 and 5.20. The latter indicates how the ratio of the hourly total to the daily total irradiance depends on the daylength and on the time of day. It can be deduced that 90 per cent of the day's exposure occurs during the middle two-thirds of the daytime, and when irradiance is more than 40 per cent of the midday value.

Considerable hour-to-hour fluctuations of cloudiness are illustrated in Table 5.5. In this case, there are fewer clear skies in summer than in winter, and cloudiness is greatest in the afternoon, showing the importance of thermal convection in cloud formation at 34°S.

A feature of cloud observations shown in Fig. 4.3 is the non-normal distribution of the frequency of various degrees of cloudiness. The

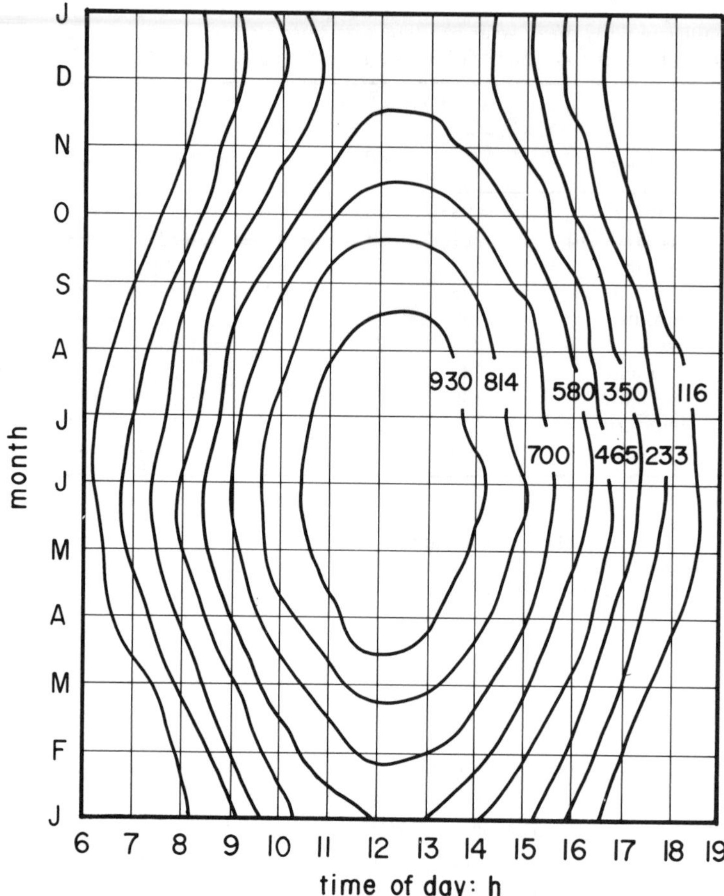

Figure 5.19 Hourly and monthly variations of solar irradiance measured at Albuquerque at 35°N (Stewart and Mohnin 1977: 364, by permission of the World Meteorological Organization). Units are W/m^2.

Table 5.5 Number of days per month with clear skies at Sydney at three times of the day in 1970

	Jan	*Feb*	*Mar*	*Apr*	*May*	*Jun*	*Jly*	*Aug*	*Sep*	*Oct*	*Nov*	*Dec*	*Total*
9 a.m.	3	2	4	5	2	4	19	9	2	5	0	3	58
3 p.m.	0	2	1	1	2	1	11	4	2	1	1	0	26
9 p.m.	3	3	6	11	8	10	22	10	8	9	3	1	94

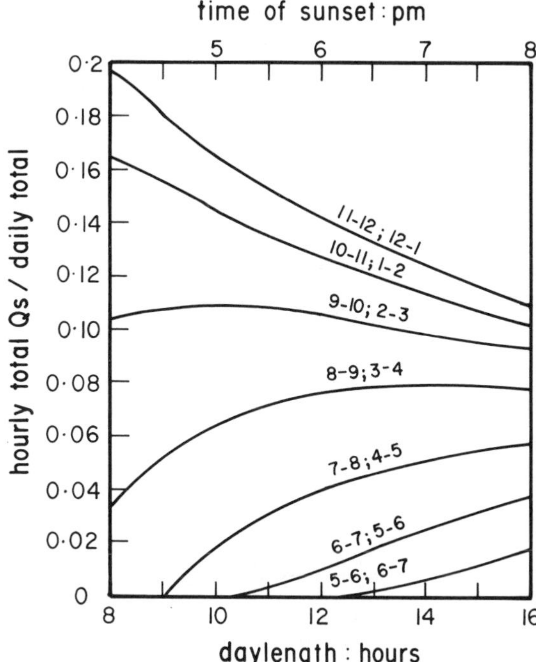

Figure 5.20 Relationship of the hourly total to the daily total solar irradiance for various hours of the day. (By permission of Pergamon Press)

distribution tends to be bimodal, with relatively few occasions of moderate cloudiness.

SOLAR IRRADIANCE OF SLOPING SURFACES

Previous sections have dealt with solar irradiance of a flat horizontal surface. But it may be necessary to know the irradiance of a sloping surface, such as a tilted flat-plate collector of solar energy, the side of a building, a standing person, or a field on a hillside. The variation of irradiance is considerable even for the degrees of slope likely in terrain used for agriculture, as in Table 5.6. The effect is sufficient to alter surface temperatures and hence crop growth. So this section deals with the radiant energy on to a surface tilted at angle b to face a direction with azimuth angle sz. The arrangement of the sloping surface, the surrounding horizontal ground and the sun's direct beam is shown in Figure 5.21.

The solar irradiance of a sloping surface consists of four elements, each to be calculated separately and then added. They are (i) the direct component (Qd), (ii) the circumsolar diffuse irradiance (Dc), (iii) the isotropic diffuse irradiance (Di), and (iv) the diffuse radiation reflected on to the sloping

Table 5.6 Effect of slope and aspect of land at 49°N on the mean solar irradiance received during the growing season (April–October)

Slope (degrees)	*Seasonal mean radiation when the slope faces*		
	North	*East or West*	*South*
0	215 W/m^2	215 W/m^2	215 W/m^2
20	173	212	234
40	118	204	234

Source: BC Grape Growers 1984.

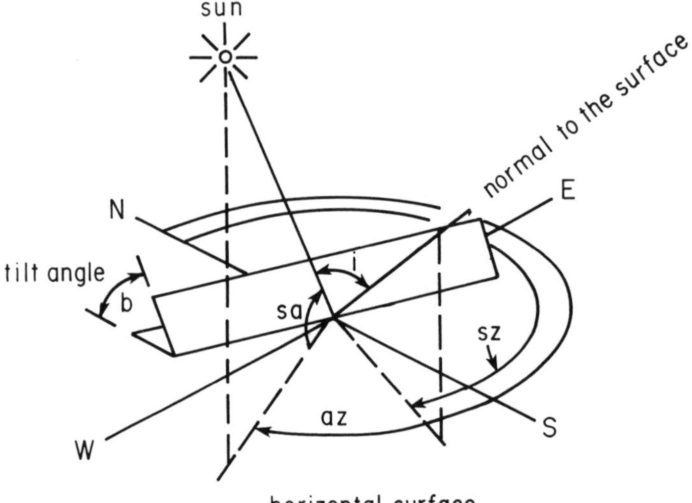

Figure 5.21 Arrangement of the sloping surface considered in this section, showing the normal to the surface ON, its azimuth angle sz with respect to north (in the south hemisphere), the slope's angle b, and the solar elevation sa

surface by the surrounding level area (Dr). We will consider these in turn. It may be assumed that we already know the values of global irradiance of a horizontal surface (Qs) and its diffuse component (D), the latter being divided in turn into circumsolar and isotropic parts.

Direct irradiance of a sloping surface

The parts of the solar irradiance which come either directly from the sun or as circumsolar diffuse radiation, depend on the solar elevation and the slope's angle to the horizontal. The direct irradiance of the sloping surface (Qds) is as follows:

$$Qds = Qd \cdot \cos i \qquad W/m^2 \qquad (5.41)$$

where Qd is the solar-beam irradiance on to a surface facing the sun (calculated from Eqn (5.17) in the case of a clear sky). The term i is the angle of incidence of the beam on to the surface, i.e. the angle between the beam and a perpendicular from the sloping surface (see Figure 5.21), given by the following formula:

$$\cos i = \cos b \cdot \sin sa + \sin b \cdot \cos sa \cdot \cos (az - sz) \qquad (5.42)$$

where sa is the sun's elevation above the horizon (see Eqn (5.5)) and az is the sun's azimuth angle from the north (Eqns (5.6) and (5.7)).

Diffuse irradiance of a sloping surface

The simplest procedure for calculating the total solar irradiance of a sloping surface involves considering only two components – Qd and Ds, where the latter is the diffuse radiant power regarded as essentially isotropic. Schulze (1975: 56) derived the following empirically, for calculating Ds from cloudless skies in South Africa:

$$Ds = 94 (\sin sa)^2 \cdot [\cos (b/2)]^2 \qquad W/m^2 \qquad (5.43)$$

The factor $[\cos (b/2)]^2$ allows for the reduction of the amount of sky affecting the slope, caused by the tilt of the surface. Sometimes the factor is expressed instead as $(1 + \cos b)/2$, which is mathematically the same.

Lunde (1980: 77) proposed calculating the diffuse irradiance of a slope by multiplying $(1 + \cos b)/2$ into the irradiance of horizontal ground, but this assumes that all the diffuse irradiance is isotropic, which can lead to errors of 40 per cent in estimating the solar irradiance of a sloping surface (WMO 1981b: 146). A better procedure involves separate consideration of the three components of (Ds) – the isotropic, the circumsolar, and the reflected parts. The isotropic part to a sloping surface (Dis) is given thus (Hay 1986: 18):

$$Dis = D (1 - Qd/I) (1 + \cos b)/2 \qquad W/m^2 \qquad (5.44)$$

where D is the diffuse irradiance of a horizontal surface and I is the solar parameter (on average equal to the solar constant Io). The circumsolar irradiance of a sloping surface (Dcs) is given thus:

$$Dc = D \cdot Qd \cdot \cos i/I.\sin a \qquad W/m^2 \qquad (5.45)$$

The irradiance (Dr), reflected on to the sloping surface from a horizontal environment with an albedo α, is given by the following:

$$Dr = \alpha.Qs (1 - \cos b)/2 = \alpha.Qs.\sin^2 (b/2) \qquad W/m^2 \qquad (5.46)$$

Eqns (5.41) and (5.44)–(5.46) together yield the instantaneous value for the total irradiance of a sloping surface. The daily exposure may be calculated by totalling the rates at each hour.

Irradiance of a vertical wall

The procedure outlined above can be applied to the case of a vertical wall, for which the slope angle (b) equals 90°. In this case, the instantaneous total radiation (Qv) is given as follows:

$$Qv = Qd\,[\cos i + D\,(\cos i/\sin sa - 0.5)/I] + \alpha.Qs/2 + D/2 \qquad W/m^2 \qquad (5.47)$$

Eqn (5.47) implies that the ratio Qv/Qs (the irradiances of vertical and horizontal surfaces) depends on cos i, which in turn is governed by sz (the

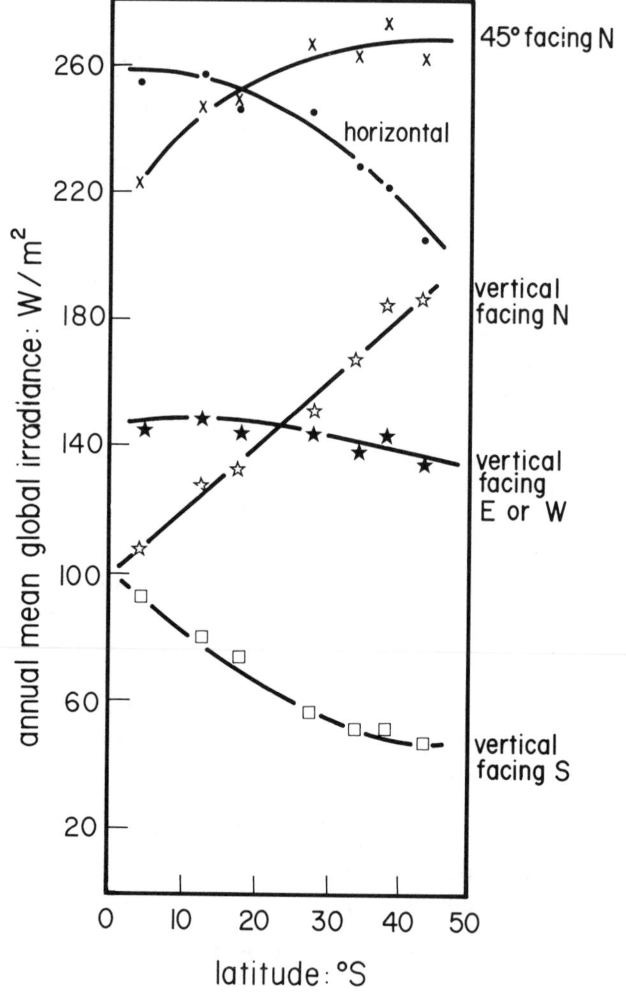

Figure 5.22 Effect of latitude, orientation and slope of a surface on the annual mean short-wave irradiance of a surface on cloudless days (Martin 1976).

azimuth angle of the vertical surface – Eqn (5.42)). The average ratio in Russia varies from about 1.8 for a wall facing south to 0.4 for one facing east or west, and 0.2 for one facing north. For the southern hemisphere, the effect of the wall's orientation, as well as its slope and latitude, is shown in Figure 5.22. It can be seen that the irradiance of east and west walls is little affected by latitude.

SOLAR ENERGY

The solar constant is so large that the solar irradiance of the surface of the whole Earth is many thousand times the world's present consumption of electricity. The solar energy *each year* on to the Arabian Peninsula alone is over twice the world's total oil reserves. It is natural, then to, consider using solar energy instead of either the usual fuels (with their wastes and air pollution) or nuclear fuels (with their link to nuclear weapons). Solar energy is clean, quiet, hardly affects the thermal economy of the world, and it is everlasting. On the other hand, solar energy requires considerable space and is capital-intensive; this makes it still relatively expensive.

The average irradiance of the ground is far less than the solar constant on account of night-time, latitude, season, cloud and atmospheric attenuation. Average values of around 250 W/m^2 are normal and would necessitate a collecting area of several square metres to provide 2.5 kW continuously, the average energy use of an Australian household, for instance. A much larger area would be needed if solar energy were also to replace not only the family's but also their share of the needs of industry, transport etc.

Difficulties with large-scale solar power are as follows:

1 Radiant energy is dilute in space, so that collectors must be large in area, which is expensive even with the cheapest materials. These areas should be near cities to reduce transmission losses, yet land is expensive there and cities are frequently near the coast where cloud reduces irradiance.
2 The radiation is intermittent, whereas the demand for power is relatively constant. Cloud and night interrupt the supply, and the greatest availability at noon does not coincide with peak demands.
3 The mismatch of supply and demand necessitates either the additional expense of ample energy storage or the extra cost of a reserve energy source, such as diesel generators.
4 The greatest demand for electricity is in gloomy climates, where and when irradiance is least. Solar power systems are most promising at latitudes less than 35°, though passive use of irradiance in heating suitably designed houses, for instance, is practicable up to 55°.

Collecting solar energy

In the first instance, solar power will become significant for particular purposes, e.g. electricity production in remote and sunny places, and for domestic hot water. At present, photocells are best suited to small-scale applications in isolated places. However, the falling price of photovoltaic power and the rising cost of alternatives suggest a crossover early next century, for wider use.

Solar energy is often collected on a flat, black plate, which gathers both direct and diffuse radiation and is protected from the wind by a pane of glass. The collector is usually tilted towards the equator at about the angle of latitude to capture the most solar energy. A steeper angle would increase irradiance in winter and reduce the accumulation of an obscuring layer of dust in a dry area. But a less steep angle increases diffuse irradiance, which would be useful in a cloudy place.

The collected heat is carried away in water flowing through the plate. If the flow rate is zero the water may boil and damage the equipment, whereas a fast flow rate reduces the warming of the water. Typically, about 40 per cent of the solar irradiance can be collected in the warm water, but as little as 15 per cent if the hottest water is required. Current research will improve matters.

Daily weather reports in some cities in the USA include an index of the expected available solar energy. What is called the 'Solar Index' is the percentage of the power needed for 360 litres of hot water that could be supplied by the expected solar energy, assuming about 7 m^2 of flat-plate collector.

Use of mirrors

Irradiance can be increased by focusing the solar beam on to a small area by means of mirrors. This was done at the World Fair in Paris in 1878 to create steam and thus drive a printing press. In 1983, an array of 1818 steerable mirrors at Barstow in California began generating 10 MW of electricity in the daytime. However, the electricity generated by this pilot plant cost several times what was generally paid for grid electricity.

A fixed mirror of 56 m diameter – along with 63 smaller, steered mirrors trained on the sun – has been used for high-temperature chemical reactions in the French Pyrenees. Also, a pilot installation at White Cliffs, a remote town in New South Wales, has generated electricity since 1983 from steam created at the focus of each of 14 parabolic mirrors of 5 m diameter. The arrangement produces about 180 kW of steam and thence up to 30 kW of electricity, on average. If solar-power generation is interrupted, battery cells can maintain supply for 10 hours; thereafter a backup diesel engine takes up the load. During rainfall, the dishes are parked in a horizontal position and become washed by the rain. The electricity produced is expensive.

Mirrors collect only the direct beam, not the diffuse radiant energy. Also, the mirrors become dirty or tarnish, and must be continually adjusted to face the sun.

Collecting solar energy into water

Water is a cheaper collecting surface than a black plate or a mirror. Also, it has a low albedo (see Table 3.3) and therefore a high absorption, and provides convenient storage of the collected energy on account of water's high specific heat. Three kinds of collector are water bags, solar ponds and water stills.

A cheap Israeli device involves water in long bags which are 0.1 m deep, made of clear plastic on top and black plastic beneath. A typical design involves a bag 38 m long and 3.5 m wide, resting on rock-wool insulation. The bag is filled each morning and drained in the late afternoon.

A solar pond has an open surface and is about 2–3 m deep, lined at the bottom with black plastic to absorb the solar radiation. The bottom layers of the pond are made heavy by dissolving common salt there (or else magnesium chloride), so that heat collected by the plastic is not convected to the surface and then lost to the atmosphere. Such a pond in Melbourne achieved bottom temperatures which were 38C° above the air temperature, but at Townsville (at 19°S) the bottom temperature would be nearly 100°C. The heat can be extracted in a flow of clean water through pipes laid within the bottom layers of the pond.

A pond at Alice Springs has an area of 2000 m^2 to drive a 20 kW generator. This implies an overall efficiency of only 4 per cent in transforming the solar energy, assuming an irradiance of 250 W/m^2. There are much bigger ponds in Israel generating up to 10 megawatts.

Ripples on the surface increase a pond's albedo and so reduce the energy collected. Algal growth similarly decreases the pond's effectiveness. Fortunately, such problems can be solved. More serious is the gradual decay of the gradient of salt concentration from bottom to top of the water, so that the density difference is eroded and eventually heat can escape by thermal convection to the surface. The deterioration of the salinity difference is promoted by surface evaporation and by stirring of the pond surface by wind, though these can be reduced by floating rings. The main problem of solar ponds is that replacing a single 660 MW coal-fired power station (and New South Wales, for instance, has six stations each of 1000 MW) would require a pond of 66 square kilometres or so.

Solar energy into water can be used for distillation to provide potable water from brackish. A solar still consists of a large area of impure water flowing over black rubber or plastic beneath a cover of clear plastic or glass.

Solar irradiance evaporates pure water, which condenses on the underside of the cover, runs to the bottom edge into a channel, and then is collected for drinking or other use. A typical installation was erected in 1967 at Coober Pedy in central Australia. The collecting area was 3800 m^2, delivering about 16 tonne/day of fresh water, i.e. about 4 litre/m^2.day in summer, though only about 1 L/m^2.d in winter. Much larger and much smaller units are in use in Israel.

A difficulty in operating a solar still is the possible formation of a white crust of salt which reflects radiation away. Also, the glass must be kept clean and panes damaged by birds have to be replaced. In addition, there must be no strong winds like those that wrecked the Coober Pedy installation.

Cooling

Solar energy can be used for cooling as well as heating, using either of two principles. With the first, the equipment is like a gas-operated refrigerator with the needed heat coming from the sun instead of a gas-flame. This boils a liquid, which later condenses elsewhere in the system and then evaporates, creating cooling. The alternative principle involves absorption refrigeration. Heat from the sun is used to boil ammonia out of water, the ammonia vapour is then condensed in a receiver at the far end of a tube and subsequent evaporation of the ammonia causes refrigeration. A unit with 1.44 m^2 of surface for collecting solar energy in Thailand can make 6 kg of ice daily. Buildings can be cooled in the same way. An advantage is that most solar energy is available for cooling at precisely the times that temperatures are highest. Unfortunately, the equipment is expensive.

Biomass energy

Relatively cheap collection of solar energy is achieved by vegetation, which absorbs the energy in photosynthesis and releases it on subsequent combustion. The biomass can be used to produce solid, liquid or gaseous fuels, either with simple technology suited to small decentralised installations or with large-scale industrial treatment, which would have an efficiency far above that of simply burning firewood, for example.

Only about 1 per cent of the solar irradiance is captured by most temperate crops. The figure for water hyacinth is 1.5 per cent, for rainforest 3.5 per cent, sugar cane in Hawaii 3.7 per cent, bulrush millet 4.2 per cent. On the other hand, an absorption efficiency of 25 per cent has been achieved experimentally with algae. The rate of biomass growth depends on the temperature and rainfall, so this form of capturing solar energy is best suited to equatorial regions.

Vegetation may be used either in the burning of wood or the combustion of some more convenient fuel derived from the biomass, like ethanol. Ethanol from sugar-cane powers about a third of Brazil's cars. About 22 tonnes of dry fuel grow each year in a hectare of rainforest, and about 90 km^2 of *Leucaena leucocephala* growing in a climate like that of the Philippines could supply a 75 MW power station.

A problem in growing biomass for fuel in a semi-arid country, where most of the land is already committed to food production, would be that of finding unused space to grow it on and water to nourish it

RADIATION AND CLIMATE CHANGE

Short-wave irradiance of the ground has probably lessened slightly during this century as a result of the observed increase of cloudiness (see pp. 190–1). More cloud is consistent with more evaporation from warmer oceans as part of a global change of climate. On the other hand, it will reduce future warming, i.e. cause negative feedback (see 'Climate change', Chapter 1).

Some rough idea of the effect of the extra cloud can be derived from the slope of a line through the points in Figure 5.12 (i.e. the ratio n/N decreases by about 0.15 for each extra tenth of cloud), and from Eqn (3.24), an expression for the net irradiance Qn. Differentiating the equation, and assuming an albedo of 0.22 and a temperature of 15°C, shows that 1.1 more tenths (see 'Values of solar irradiance', this chapter) reduces Qn by [0.064 Qa − 13] W/m^2, where Qa is the extra-terrestrial irradiance (see Table 8.3). So the reduction depends on the season and latitude. At the equator it is around 13 W/m^2, whilst at 40 degrees in winter the reduction is only 2 W/m^2. Such decreases are too small to be detected in past radiation measurements in view of the errors discussed in the section 'Measuring solar irradiance', pp. 164–70. However, they are of the same order as the increase of downwards long-wave radiation due to more carbon dioxide, and tend to offset it.

An attractive partial solution to the climate-change problem lies in the substitution of solar energy for fossil fuels to obtain perennial power without greenhouse gases. The problems mentioned earlier (see 'Solar energy', pp. 197–201) mean that solar power generally entails considerable capital investment and therefore is presently economic only for special purposes. However, vigorous research and development promise to improve the efficiency of photovoltaic generation, solar ponds, and mirror collectors, for instance. These will lead to gradual extension of the use of solar energy, particularly as oil costs increase.

6

WIND

Surface winds affect crops, the evaporation of water, air-pollution concentrations, city and house design, human comfort and many other aspects of our world. Also, wind can be a source of power.

In this chapter we first consider the causes of surface winds, since they explain spatial and time variations. These in turn govern the practicability and location of collecting useful amounts of wind energy.

CAUSES OF WIND

Winds near the ground, at 10 m height, say, have four possible causes:

1 Gradient winds blowing at a few hundred metres above the ground, which result from the patterns of pressure shown on the weather map.
2 Horizontal temperature differences over the Earth's surface, which lead to sea-breezes, for instance.
3 The slope of the land, which accounts for cold air draining downhill in the night-time and for daytime winds up hillsides facing the sun.
4 Instability of the atmosphere, which causes tropical cyclones and thunderstorms, for example.

There are no other causes, and any wind can be attributed to one or more of these four. Which are operative at any moment depends on the synoptic pressure pattern, the time of day, the terrain, the land-use and the character of the prevailing air-mass.

The gradient wind

The gradient wind is controlled by horizontal differences of pressure on a synoptic scale (see Table 1.3). It occurs high enough above the ground for surface roughness to be unimportant, typically beyond 250 m above the sea, 300 m over farmland, 400 m over a forest or suburbs, and 500 m over a city. The gradient wind controls the surface wind if two conditions are met: (a) the isobars are close together on the normal synoptic chart so that the gradient wind is strong, and (b) there is adequate vertical churning of the lower atmosphere, due to either thermal convection or turbulence, to link the gradient wind to the lowest levels of the atmosphere. Good connection is most likely in the daytime when heating of the ground by the sun promotes thermal convection. This is absent over land surfaces at night, when there may be an inversion within the lowest 200 m of the atmosphere decoupling the surface air from the gradient wind above. Hence the familiar calm of night-time. A similar detachment from the gradient wind results from other low-level inversion layers, like that above a sea-breeze or above the downhill flow of a katabatic wind.

The surface wind speed is only a fraction of the gradient wind speed (Vg). The fraction varies widely, but the wind measured on board ship in the North Sea is typically around 0.6 Vg, for instance. Winds at 10 m above a land surface are up to 0.7 Vg for strong winds, but less for light.

The ratio is most in unstable conditions (e.g. in daytime) or with strong wind which closely couple the layers together. Surface winds at Oklahoma City average 3.5 and 5.1 m/s at midnight and noon respectively, when winds at 260 m are 11 and 7 m/s – the ratios are therefore 0.32 at night and 0.71 in the daytime.

Surface roughness reduces the ratio. For instance, a 10-metre wind over flat grassland is only about 84 per cent that over the sea but 1.62 times that over woodland, the gradient wind being 50 m/s in each case.

The gradient wind tends to drag the surface wind in more or less the same direction, the difference between the gradient-wind and surface-wind directions being greater over a rough surface. In theory, the ends of vectors, representing the wind direction and speed at levels between the ground and the gradient wind, trace out a spiral called the 'Ekman spiral'. This is due to the Coriolis effect. The difference between the directions of light gradient winds and surface winds may be 45° inland, or 30° over the sea. Measurements over the North Sea, when gradient winds were 8–18 m/s, showed that the surface winds came from a direction relatively anticlockwise on 79 per cent of occasions, the average difference of direction being 14°. Observations in Canberra showed close alliance of gradient and surface wind directions in the daytime, but surface winds mainly roll downhill at night.

Sea-breezes

Sea-breezes are 'gravity currents' (or 'density' or 'buoyancy' currents), being primarily horizontal and due to a difference between the densities of adjacent fluids. In this case they are due to an inflow of maritime air, replacing air rising from coastal land which has been warmed by the sun. To a lesser extent, the opposite happens at night when the land cools to less than the sea-surface temperature. A wind similar to a sea-breeze occurs at night around a large city, which tends to be hotter than the suburbs. Likewise, there may be a pseudo-sea-breeze towards a heated plateau. In each case there is a surface wind towards the warmer area.

Sea-breezes occur particularly where there are extensive coastal plains and cool ocean currents offshore, at latitudes with strong sunshine to heat the land. These requirements tend to reduce their incidence at both high latitudes and the equatorial zone, though they have been observed in both Jakarta and Iceland.

For a sea-breeze to occur, there must be sufficient difference between the temperatures of the land and the adjacent sea. For instance, measurements on a Sydney beach in April showed that a sea-breeze started only when the sand surface was 2.5C° above the sea-surface temperature. As a result of the importance of the land/sea temperature difference, the frequency of sea-breezes is greatest in summer, unless there is more cloud in that season. This is shown in Figure 6.1, for example.

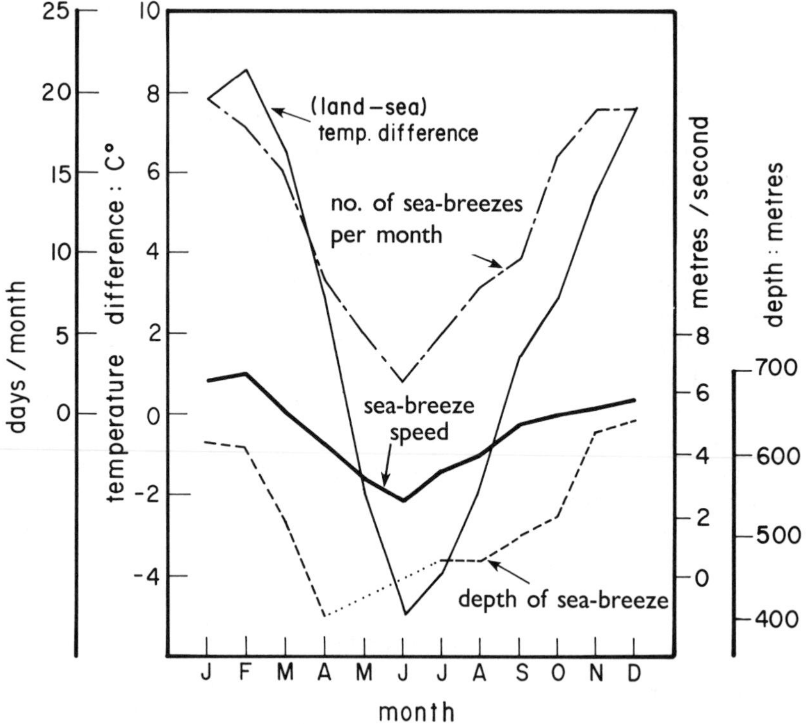

Figure 6.1 Effect of the season on sea-breezes at Perth during 1941–3 in terms of the number of days with a sea-breeze each month, the wind speed within the sea-breeze, the vertical depth of the breeze, and the associated difference between the daily maximum screen temperature over the land and the coastal water temperature (from tables of Hounam 1945).

The required temperature difference (D) seems to depend on the place, on the strength and direction of the gradient wind, and on atmospheric stability. As a result of so many factors, the evidence is confusing. First, there are reports that D must exceed either 0.17 Uo^2 or 0.65 Uo in England, or 0.06 Uo^2 in Hobart, where Uo is the offshore component of the gradient wind. Second, different evidence points to the need for a sufficiently small gradient wind, irrespective of its direction. Thus, sea-breezes happen in southern Western Australia when the onshore component is between −3 and +3 m/s and the daily maximum temperature exceeds 30°C. Likewise, sea-breezes at Sydney occur mostly when the gradient wind is northerly or north-east, ie parallel to the coast. Perhaps this reflects the need for anti-cyclonic conditions, which entail both light gradient winds and the clear skies which allow adequate heating of the land, to create a sea-breeze. Third, there is evidence from Wales that a considerable *offshore* component of the gradient wind may promote a sea-breeze, when it is sheltered between 250 m hills, by strengthening the return flow above the sea-breeze itself.

A sea-breeze begins out to sea and is first detected on land at any time between 7 a.m. and 7 p.m. The onset in England is at about 9 a.m. if the gradient wind is 2.5 m/s but 4 p.m. if it is 8 m/s. The breeze usually ends in the late afternoon or early evening but is prolonged by cloud in the evening, presumably by delaying cooling of the land. South of the equator its direction backs (i.e. shifts anticlockwise) during the day, because of the Earth's rotation.

Figure 6.1 indicates a linear relationship between the speed of the sea-breeze (Us) and the land/sea temperature difference (D). However, theory and other observations show the following kind of relationship:

$$Us = 3\ D^{0.5} \qquad \text{m/s} \qquad (6.1)$$

This gives 9 m/s for a land/sea difference of 9C°, for instance, which is within the range of published sea-breeze speeds, e.g. a typical 5 m/s on continental coasts, about 7 m/s at the Tropics, up to 12 m/s measured at Sydney, and as much as 14 m/s in the Namib desert. A similar breeze from the smaller area of a lake is only 2 m/s.

The inland extension of the sea-breeze is part of a cellular pattern. The inland edge of the cell constitutes the 'sea-breeze front', a boundary initially parallel to the coast, between cool, moist maritime air and the unaffected air ahead of the front. The cell enlarges during the day, i.e. the sea-breeze front moves inland as illustrated in Figure 6.2. Its arrival at a place is signalled by a change of wind direction and by an abrupt decrease of the wet-bulb depression.

The sea-breeze front moves at perhaps half the speed of the sea-breeze itself. The difference between the speeds accounts for the atmospheric convergence and ascent at the front. Speeds of the front quoted in the

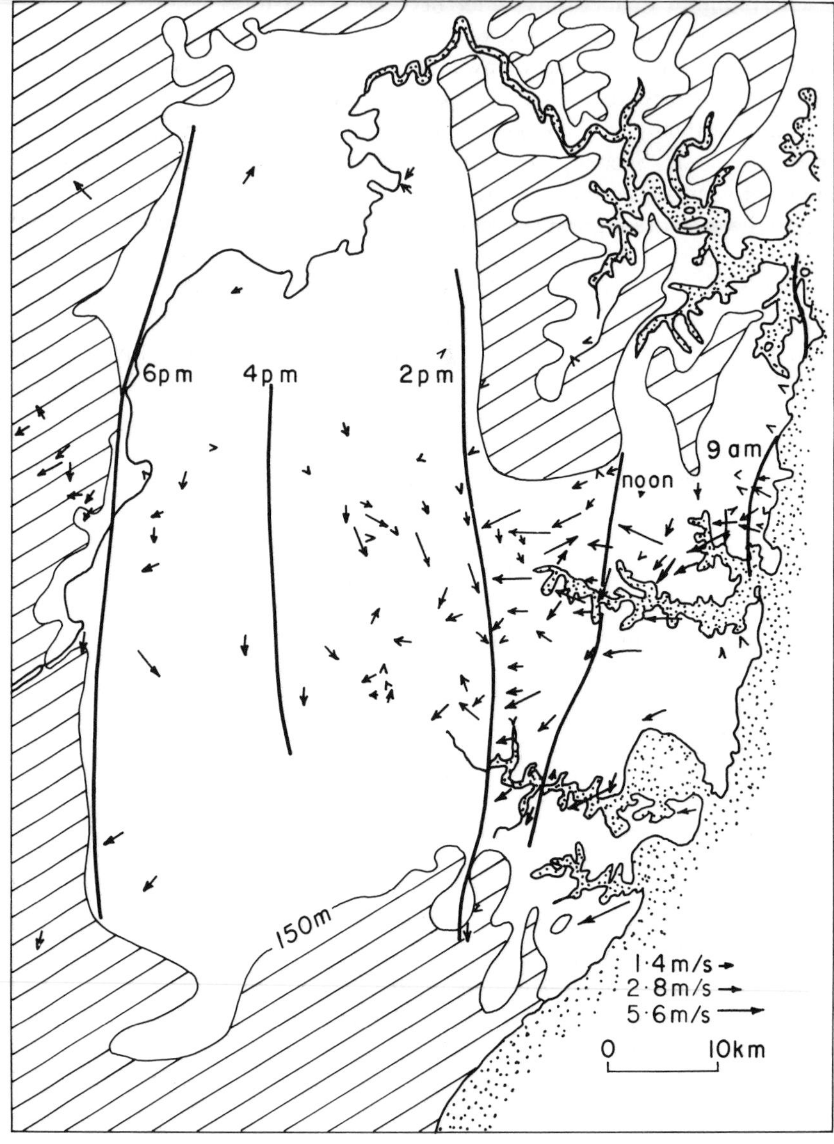

Figure 6.2 Winds at various places in the Sydney region at 2 p.m. on 1 March 1974, showing also the positions of the sea-breeze front at various times that day (Linacre and Barrero 1974).

literature include 2.2 m/s at Danzig, about 3 m/s in southern England and India, 4 m/s in Australia and 1.4–3.6 m/s in Massachusetts. Figure 6.2 implies about 1.2 m/s in the morning and 2.0 m/s after midday, which accords with observations elsewhere of an acceleration in the afternoon. Theoretical studies indicate, for instance, a steady advance at about 6 m/s after a sea-breeze at 30° latitude has reached 70 km inland. Low latitudes are associated with rapid frontal movement, but high ground obstructs the breeze, unless the slope faces the sun, in which case the extra warmth stimulates the wind.

The front moves inland to a distance which depends on the daylength and on the front's speed. Distances mentioned in the literature include beyond 40 km or more from Lake Ontario, 80 km in Egypt and Chile, over 100 km from the shores of Lake Victoria in Africa, 160 km in south-east England and the Namib desert, and 400 km across the coastal plain in the south of Western Australia. The distance is greatest across dry plains at low latitudes.

There is a similar but opposite land breeze at night. It is lighter, e.g. less than 3 m/s from the Namib desert.

Slope winds

Slope winds blow up a slope in the daytime and down at night, and are called 'anabatic' and 'katabatic', respectively. Sometimes katabatic winds are defined as originating on a slope, whereas 'drainage winds' come from a plateau higher up. However, the usage varies, and the two are hard to distinguish in practice.

Katabatic winds are unaffected by the roughness of the slope. They slide down the sides of hills and then may converge on the floor of a valley, where they contribute to deeper and slower 'valley winds' (or 'drainage flows') down the valley, at an angle to the initial slope winds. These various surface winds are often surmounted by a deeper layer flowing down the general slope of the whole region.

Katabatic winds blow when the land slants at more than about 2 degrees. In other words, cold air can drift even on land which is almost level; Brooks (1959: 137) mentioned flows developing slowly to 1.8 m/s across orchards on almost flat land in California. The speed may pulsate with a rhythm of either 10–30 minutes or some hours. Slope winds are commonly disrupted by mesoscale winds stronger than 5 m/s or so, especially near the top of a ridge.

The winds are remarkably shallow, e.g. less than 20 m deep at 1 km down a slope of 20 degrees. Measurements on summertime katabatic winds in Washington have shown a depth of only 6 m. In general, the depth is about 5 per cent of the vertical distance from the top of the slope.

The wind is strongest in a steeply sloping, deep and bare valley where there is a large daily swing of temperature. It can be shown theoretically to vary with the square root of the sine of the angle of slope, and of the distance down it. Usually, katabatic winds are only two or three metres per second, but they can blow at 40 m/s on the steep and high edge of Antarctica.

Anabatic winds are opposite to katabatic winds. They occur on steep slopes facing the sun, especially in deep valleys, and may be 2–3 times stronger than the nocturnal cold winds. Upslope winds may begin about 2 hours after sunrise, reaching a maximum at noon, whilst katabatic flows

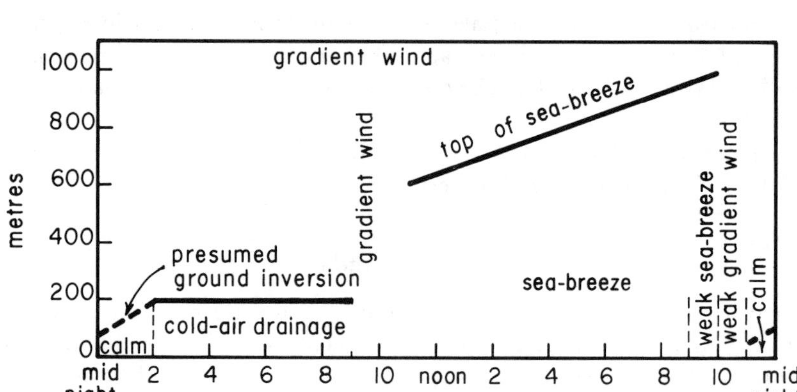

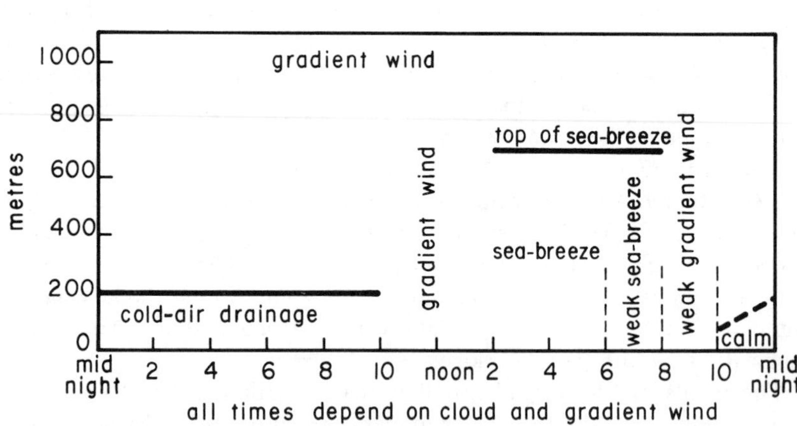

Figure 6.3 Typical sequences of surface winds and inversion layers on cloudless days in Sydney, in terms of the causes. The values used here depend partly on data from McGrath (1972) and four surveys of winds in Sydney (Linacre and Barrero 1974).

begin 2 hours before sunset. However, the exact time of an anabatic wind starting depends on the slope's orientation to the sun.

Anabatic winds are popular with glider pilots, and the winds up the mountains east of Los Angeles help to vent the city's air pollution.

Daily pattern

The diurnal variation of gradient winds, sea-breezes and slope winds makes it possible to determine the cause and hence the normal strength and direction of the surface wind at any time of the day. An example is illustrated in Figure 6.3. The early cold-air drainage is directed downhill and is slow and calm. Later there is a sea-breeze, which flows more briskly and tends to blow inland. The direction backs (i.e. the wind comes from a more anticlockwise direction) during the day in the southern hemisphere, on account of the Coriolis effect in the southern hemisphere. Then daytime convection links the surface air to the gradient wind, with a corresponding strength and direction.

The periods of dominance by each factor causing the surface wind vary with season, because of the longer day and more intense sunshine in summer. Also, cloud reduces not only daytime convection but also the heating which causes a sea-breeze and the cooling which promotes drainage winds. Another factor is elevation: winds are strongest on Chinese mountain tops at night, but in the valleys in the daytime. In addition, there may be disruption of the usual daily pattern by storms.

Storm winds

Storms influence the surface wind more or less at random caused by tropical cyclones, tornadoes, whirlwinds or thunderstorms, for instance. The most intense local wind-storms in Australia, for instance, occur when there is a cold front nearby, and a tornado-scale motion is imposed on a rising atmosphere of large-scale cyclonic character. In general, storm winds are due to the discharge of intense instability on the toposcale or mesoscale, and are associated with particular regions and times, as shown in Table 6.1

Tropical cyclones

These are great whirls of the atmosphere found between 20°S–35°N, usually at around 15° latitude, three-quarters of them in the northern hemisphere. Thirty-eight per cent occur east of China. None affect South America, because there is insufficient distance along the equator between Africa and South America for the ocean surface to warm enough.

They originate over oceans warmer than 26°C and involve winds which

Table 6.1 Summary of the typical characteristics of atmospheric events creating strong winds.

Event	*Preferred location*	*Preferred season*	*Preferred time of day*	*Diameter*	*Duration*	*Wind: m/s*
Tropical cyclone	Low-latitude coasts	Summer	Anytime	12 km	1–2 hours	32–60
Tornado	Flat inland	Spring	3–6 p.m.	100 m	15 mins	Up to 100
Whirlwind	Dry inland	Summer	1–2 p.m.	5 m	1 minute	–
Thunderstorm	Mountains	Summer	3–6 p.m.	200 km	2 hours	Up to 50
'bomb'[a]	Mid-latitudes	Winter	–	Meso-scale	–	Up to 50

Note: [a]An unusual intense storm, little understood.

sometimes exceed 60 m/s. There is a central eye of calm whose diameter is from a few kilometres to 80 km. Outside the eye, winds may exceed 34 m/s for 50–150 km and be more than 17 m/s for a further 300 km. The centre generally meanders westwards before curling away from the equator. It soon dies out when the eye crosses the coast and travels inland. The total life is up to 20 days, typically nine days near Australia, for example.

Tornadoes

Extremely intense thermal convection may lead to a huge black cloud from which a tapering pillar descends to the ground. At that point the tornado exerts great suction, inducing winds which can demolish buildings. The force of these winds indicates speeds above 90 m/s on occasion. As the point of contact moves across the ground at 10–27 m/s, an intermittent swathe of devastation is left behind, 20–1000 m wide. They occur notably in the mid-western states of the USA, though elsewhere also. The time of occurrence is usually the late afternoon in the warmer months.

Whirlwinds

These are sometimes called 'dust devils' because they are made visible by the dust picked up by the spinning column of thermal convection. They are like small tornadoes, except that no cloud is involved. They are seen in dry countries with calm conditions and strong sunshine to heat the ground, mostly in the early afternoon in summer. There is a cyclonic wind (anticlockwise in the northern hemisphere), but only within some metres of the moving column.

Thunderstorms

Thunder accompanies lightning during the discharge of electrical voltages which travel between adjacent large cumulonimbus clouds and to the ground. The voltages result from uplift within each cloud during the release of instability in an atmosphere which is moist and unstable over a depth of at least 3 km. The vertical motions cause strong local winds at ground level during the storm. The motions are initially triggered either:

1 by strong insolation of the ground (causing 'air-mass thunderstorms' about 30 km across on summer afternoons, above warm parts of the ground, such as a dry area with low albedo);
2 by turbulence of the air (especially on windward coasts);
3 by the wedging action of cold fronts (causing rapidly moving 'squall-line thunderstorms' along the front, maybe inducing tornadoes);
4 by low-level convergence of winds (according to the shape of the terrain);
5 by hills.

There are also 'convective-complex thunderstorms', due to low-level warm, moist winds, forming a ring of storms several hundreds of kilometres across, moving slowly and lasting 12–16 hours. Typically, this last kind occur in the early hours, inland at mid-latitudes.

Thunderstorms are most numerous near the equator, e.g. in Java and central Africa. There are 136 thunderdays annually in Jakarta, and 240 in Kampala. In a drier climate like that of Sydney there are only about 26. In the USA, thunderstorms are most numerous in Florida and Colorado, for different reasons. Around the world, there are some 2000 storms in progress at any moment.

SURFACE WINDS

The routine measurement of wind speed was discussed in Chapter 2 (see pp. 47–51). Here we consider estimates by means of either the Beaufort Scale, the tattering of flags, or the distortion of trees.

The Beaufort Scale

It has been customary since 1838 to estimate the wind speed at sea from observations of the kind shown in Table 6.2. The wind speed is related to a power of the Beaufort Scale value (B), i.e. the speed equals $B^{1.4}$ m/s. This gives 9.5 m/s for Beaufort scale 5, which is close to the 10 m/s in Table 6.2.

The Table 6.2 estimate of wind speed at sea is equivalent to an hourly mean value, since waves result from prior wind conditions. The wave height is seen to depend on the square of the wind speed, but some inaccuracy in using the Beaufort Scale arises from the dependence of wave height on atmospheric instability. The sea's appearance depends also on the swell, tides, rain, water depth and currents. Despite this, experienced observers can estimate the wind within 1.5 m/s on 60 per cent of occasions.

Or, ashore, within 10 per cent of the true speed. In practice, Woodhead (1972: 237) reported typical errors of 15–30 per cent in estimating winds in Africa by this means. In any case, estimates using the Beaufort Scale may be in error by appreciably more than the desirable maximum error of 0.5 m/s quoted in Table 2.2 for a normal climate station.

Similar indications can be used to estimate air movements indoors – 0.1 m/s feels stuffy, 0.5 m/s is sufficient to make a candle flame flicker, 1 m/s lightly stirs loose papers and 1.5 m/s prevents deskwork.

Tattering of flags

Wind can also be estimated by observing the tattering of flags over a few weeks or more. Special cotton flags are used, about 25 cm high and 35 cm long, each swivelling freely on a 1.5 m mast. The flags fray along the outer edge, and the loss of area reflects the total air movement (the 'wind-run') during the flag's exposure. Typical rates of tatter are 2–14 cm^2/day. Rain also affects the rate, though this complication can be minimised by trimming the flags each week. The correlation between wind-run and tatter is typically 0.63–0.91, or 0.89 between tatter and the total period of gusts above 17 m/s.

Unfortunately, the scatter of results, even after averaging measurements on three adjacent flags, is too great to allow reliable calibration of the method. In two out of three tests, an empirical equation based on previous tattering predicted a wind-run only about half of that actually observed. One concludes that the method is better for rough comparisons between sites than for estimating wind speeds numerically.

Distortion of trees

An even more approximate estimate of winds over the past few years can be obtained from the sideways distortion of certain trees, especially persimmon, poplar, juniper and larch. Sitka spruce growing in places higher than 610 m from sea level in Wales are deformed by the prevailing south-west winds, and beech trees above 400 m are considerably distorted where the annual average wind speed exceeds 6 m/s.

One cause of deformation is the desiccating of trees on the side buffeted by strong winds. The tree's shape is affected also by the abrasion due to blown ice, the breaking of branches, the accumulation of glaze in storms, or wind pressure during the growing season. The result is a shortening of branches on the windward side, especially when strong winds carry snow or occur during the growing period. In the case of persimmon trees, the growing period is chiefly late spring and early summer, so deformation reflects the winds at that time especially.

Trees cannot survive where winds are consistently over 12 m/s.

Elsewhere, the deformation is determined by the direction of the 'dominant' winds (defined in terms of both frequency and strength) rather than by the 'prevailing' (i.e. the most common) wind direction. So maps have been made of the directions of tree deformation to show the patterns of strong winds about certain mountains in Japan and in the USA. Such a map for the Adirondack Mountains in New York State was corroborated by occasional direct observations of wind direction.

An attempt has been made to use observations of tree deformation in assessing the speed of winds, as well as their direction. For this, it is necessary to categorise tree shape and to link each category to an average wind speed. Trees tend to have a symmetrical form when winds average less than 3 m/s during the growing season, whereas there are no branches on the upwind side where the wind averages more than 6.5 m/s. However, the error can be 2 m/s on account of the variability of wind direction.

Wind data

Measurements from wind instruments must be examined before being stored, as there are about four errors per 1000 items of data – in Britain at least. The data can be checked by a computer, which tests that each hourly mean wind speed, gust speed or wind-direction value falls within a reasonable range, that it differs plausibly from the previous and subsequent values, and that zero wind speed is associated with zero direction. An occasional histogram of recent wind directions shows whether there has been a zero shift or a loosening of the mechanical coupling between the wind-direction vane and the recorder.

Wind data are commonly displayed in the form of a 'wind-rose' as in Figure 6.4. It is an improvement on a 'web-rose', in which lines connect the ends of the spokes. A wind-rose summarises the frequency with which winds fall into particular categories of direction and speed. There are eight categories of direction and four of speed in Figure 6.4, i.e. the rose is based on a table of data with eight rows and four columns, each box containing the number of occasions that the winds are within the relevant direction and speed categories. The length of each segment of a spoke of a wind-rose is proportional to the frequency of winds within the particular speed range. Data for winds from intermediate directions, such as the north–north-east, are shared between the adjacent main classes (e.g. north and north-east respectively) in proportion to the numbers already there.

Instead of the wind-speed categories shown in Figure 6.4, one might use classes corresponding to the Beaufort Scale, or the following: either (a) below 1.8 m/s, 1.8–5.9, 6–11 and over 11 m/s, or (b) below 5 m/s, 5–9 and above 9 m/s. The number of categories should be governed by the number of data available (see p. 111).

Table 6.2 The Beaufort Scale of wind speeds.

Beaufort scale no.	*Description*	*Nautical observation*	*Land-based observation*	*Equivalent wind speed at 10 m*		
				knots[a]	*km/h*	*m/s*[b]
0	*calm*	Sea like a mirror	Smoke vertical	0(1)	0	0
1	*Light air*	Ripples on sea	Smoke drifts, wind vanes immobile	2(4)	4	2
2	*Light breeze*	0.3 m wavelets with glassy crests	Wind felt on face, leaves rustle, reading a newspaper becomes difficult	5(7)	9	4
3	*Gentle breeze*	1 m breaking wavelets, scattered white caps[c]	Light flag is extended, hair disturbed, leaves in constant motion, dune sand disturbed	9(11)	16	6
4	*Moderate breeze*	White caps, 1.5m waves	Thin branches move, dust, dry soil and paper raised, sitting outdoors uncomfortable, hair blown about	14(15)	24	8
5	*Fresh breeze*	2.5 m waves	Small, leafy trees sway, wavelets on lakes, limit of agreeable wind	18(19)	34	10
6	*Strong wind*	White foam, spray, 4 m waves	Large branches sway, umbrella hard to use, whistling in wires, hair blown straight	24(24)	44	12
7	*Moderate gale*	Heaped sea, some foam blown from crests	Whole trees move, interferes with walking	30(29)	56	15
8	*Gale*[d]	8 m waves, long crests, much blown foam in streaks	Twigs break off, difficult to stand in gusts	37(35)	68	18
9	(T0)[e] *Strong gale*	10 m waves, reduced visibility	Removes tiles, tents and people are blown over in gusts, damage to TV aerials and shrubs.	44(41)	82	21
10	(F0)[e] *Storm*	Heavy rolling sea, overhanging crests, sea white with foam	Trees blown down, mobile homes damaged	51(47)	95	24
11	(T1) *Violent storm*	16 m waves, spray obstructs visibility, all crest edges blown into froth	Few trees uprooted, light garden furniture lifted, many roofs severely damaged	59(54)	110	28
12	*(F1) Hurricane*	Sea white, air filled with foam and spray	Caravans moved, moving cars blown off the road, small buildings overturned.			32–50
	(F2)		Large trees uprooted, objects blown as missiles			51–70
	(F3)		Forests flattened, trams overturned, cars lifted from ground, buildings demolished			71–92

Table 6.2 continued

Beaufort scale no.	*Description*	*Nautical observation*	*Land-based observation*	*Equivalent wind speed at 10 m*		
				knots[a]	*km/h*	*m/s*[b]
	(F4)		Trees debarked by flying debris, cars rolled to disintegration or thrown a distance			93–116
	(F5)		Strong frame houses carried some distance to disintegrate, reinforced concrete structures badly damaged, cars fly beyond 100 m	117–42		

Notes: [a] The bracketed values are speeds relating to the version of the Beaufort Scale published by Graham (1982: 327).

[b] The values in the last column, for Scale numbers 0–11, are the corrected values from Alcock and Morgan (1978: 273).

[c] The fraction of the sea covered by white caps is 0.01 per cent at 5 m/s, 0.1 per cent at 10 m/s, 0.3 per cent at 13 m/s (Katsaros *et al.* 1987: 472).

[d] A 'blizzard' has over 16 m/s and visibility below 150 m (Takahashi 1985: 165). In the USA, 'high-wind warnings' are issued when winds over 18 m/s are expected for at least an hour. A 'severe blizzard' involves winds above 23 m/s, zero visibility and a temperature below 12°C (Takahashi 1985: 165)

[e] The terms T0 and T1 in the first column refer to the Torro classification, and F0–F5 to the Fujita scale of tornado winds. Elsom and Meaden (1984: 322) quote the Torro Scale number as equal to [(B/2) − 4], where B is the Beaufort Scale number.

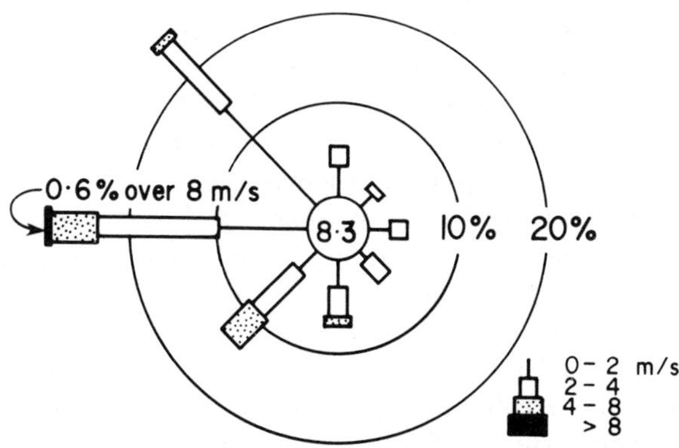

Figure 6.4 Wind-rose for Silverwater (a suburb of Sydney) during the cooler half of 1976, based on measurements every three hours. It can be seen that winds were mainly from the west, and on almost 10 per cent of days the winds were both from the west and less than 2 m/s, for instance. There was no measurable wind on 8.3 per cent of occasions

WIND PROFILE

Winds are usually stronger at a greater height above the ground, and the vertical pattern of the variations of speed and direction is called the 'wind profile'. It can be observed by means of the sideways drift of an ascending balloon or by wind measurements on a tall mast. A typical profile is that of 10-minute mean winds at different levels on the Post Office Tower in London, e.g. 11 m/s at 43 m, 13 m/s at 61 m and 20 m/s at 190 m. Alternatively, one can fly a kite and measure the angle and tension of the string to determine the wind on the kite at various elevations.

Two well-known formulae – the logarithmic and the power-law equations – describe the variation of wind speed with height above the ground. These are discussed below.

Logarithmic equation

An equation can be derived theoretically, as follows:

$$uz = us.\ln[(z-d)/zo]/\ln[(zs-d)/zo] \quad \text{m/s} \qquad (6.2)$$

where uz is the wind speed at height z above the ground, ln signifies the 'natural logarithm' of the term that follows, and us is the wind at a fixed standard height (zs, usually 10 m). The term d is a constant for a given surface, called the 'zero-plane displacement'. It is about two-thirds of the height of the elements of surface roughness, whether blades of grass or city buildings, etc.

Another surface-dependent term in Eqn (6.2) is the 'roughness' zo. It is governed by the height and density of the surface irregularities. Values are given in Table 6.3, which reconciles different values given by various authors. For many surfaces, the roughness is about an eighth of the height of the irregularity. In the case of a height h (m), each element with a sideways area s (m^2) and a plan area p (m^2), i.e. there are 1/p elements per square metre of ground, the roughness zo equals 50 h.s/p centimetres.

Eqn (6.2) assumes a 'neutral' atmosphere, which is neither stable nor unstable. It occurs when winds are strong and there is only modest heating or cooling of the ground by radiation. On this basis, it is possible to calculate the wind speed above a particular surface from nearby measurements where the roughness is different (see Note 1, p. 334). Also, Eqn (6.2) permits calculation of the wind speed at one height from measurements at another. This yields Figure 6.5, which enables measurements at any height to be standardised, i.e. replaced by the equivalent at the standard height of 10 m.

Roughness values in Table 6.3 are related to the forced-convective heat-transfer coefficient mentioned on p. 105. The coefficient equals 1300/r, where r is the diffusion resistance in Eqn (3.28) (Linacre 1964a: 687). On

Table 6.3 Typical, approximate values of surface roughness

Surface	*Height of surface element: metres*	*Surface roughness (zo): centimetres*
Ice, mud flats	–	0.001–0.005
Open water, wet soil and sand	–	0.03[a]
Snow	–	0.5[b]
Rough sea	–	0.5
Grass	0.01	0.3
	0.1	2
	0.5	10
Maize crop	3	100
Airfield, fallow farmland	–	3
Orchard	3	30
Suburbs	–	110–50
Pine forest	5	90
	30	200
City	–	300

Notes: [a] The value for water can vary widely. Munn (1966: 175) quoted values between 0.0001 cm and 50 cm.

[b] The roughness depends on the nature of the surface beneath the snow. Sevruk (1982: 15) reported 0.5 cm for a uniform layer of snow more than 10 cm deep, but 2 cm for patchy snow less than 10 cm thick on black fallow soil. Harding (1986: 11) measured values between 0.02–0.38 cm, the median of 14 values being 0.18 cm.

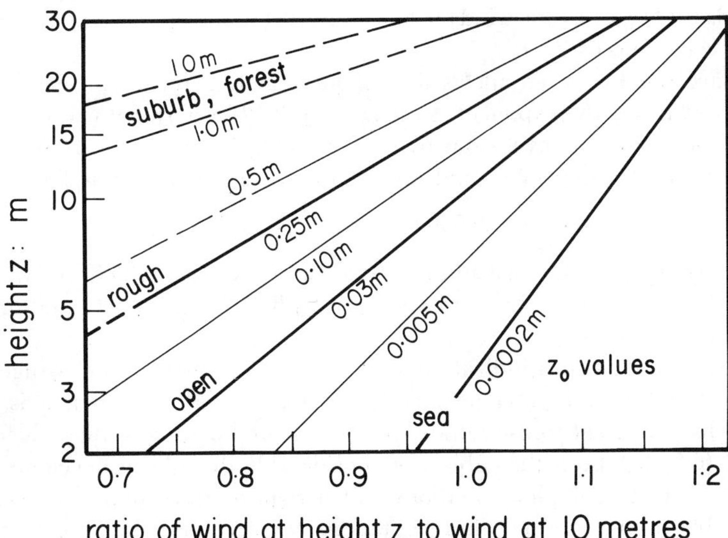

Figure 6.5 Diagram for converting a wind speed at one height over a particular surface to the equivalent speed at 10 metres, with a standard exposure (Wieringa 1977: 43, and 1980: 967, by permission of the World Meteorological Organization.)

the assumption of a neutral atmosphere and a nil zero-plane displacement, the resistance equals 6.3 $(\ln[z/zo])^2/u$ seconds per metre, where z is the height at which the wind speed is u. So the coefficient is given by 200 $u/[\ln(z/zo)]^2$, i.e. 2.6 u $W/m^2.C°$ for a lake with a roughness of 0.03 cm (see Table 6.3) and winds measured at 2 m. This estimate resembles the empirical value of 2.3 u units mentioned on p. 105.

Power-law equation

The following equation is free of the assumption of a neutral atmosphere, implicit in Eqn (6.2). It is an empirical expression involving an exponent (a) which depends on the degree of atmospheric stability (Deacon 1949):

$$uz = us\ (z/zs)^a \qquad m/s \tag{6.3}$$

The exponent for a neutral atmosphere is about 0.14 (i.e. one-seventh) and then Eqn (6.3) is known as Archibald's Law. A value of 0.14 is often assumed and is appropriate for strong winds and smooth landforms. However, stable conditions at night create a more rapid variation of wind speed with height near the ground (see Figure 4.6), which implies a higher value of the exponent. Conversely, thermal convection during the daytime stirs the lowest 200–300 metres of the atmosphere, which reduces any differences of wind speeds, so that the exponent becomes less than 0.14. Typical values are 0.1 for the instability of light winds in daytime, and 0.4 for calm and stable conditions at night. Strong winds bring the exponent towards 0.14, e.g. 0.4 for winds below 2 m/s, 0.17 for 4–10 m/s, and 0.14 for higher wind speeds.

Surface roughness too affects the exponent in Eqn (6.3). An increase of roughness raises the exponent, so it may be 0.12 over smooth water, 0.17 over an open site, 0.25 over suburbs and 0.5 over a city. An empirical relationship between the exponent, and the roughness (zo: *metres*) is as follows:

$$a = 0.24 + 0.096 \log (zo) + 0.016 (\log zo)^2 \tag{6.4}$$

where log signifies the logarithm to base 10 of the term which follows. So the exponent is 0.13 over a grass surface with a roughness of 0.02 m, for example.

These considerations indicate that effects on the exponent for changes of speed, stability, and roughness, are individually known, though it is not clear what exponent to use for any particular combination of these factors.

Eqn (6.3) has been the subject of considerable theoretical development, leading to more complex equations which require input data not readily obtainable from a normal climate station. Neither Eqn (6.2) nor (6.3) is likely to represent the actual profile beyond a few dozen metres from the ground because upper winds may be complicated by local irregularities of topography.

Different adjacent roughnesses

The effect of surface roughness in Eqn (6.2) causes some adjustment of the wind profile downwind of the boundary between surfaces of different roughness. Figure 6.6 illustrates wind which blows first over the sea and then over low grassland, for which the roughnesses are 0.03 cm and 2 cm, respectively (see Table 6.3). There arises a sloping 'transition zone', below which the profile is determined by the downwind surface and above which the upwind surface still determines the profile. Within the zone, there may be considerable 'wind shear', i.e. variation of wind with height.

The slope of the transition zone is shown in Figure 6.6 as about one in 13, but the value is uncertain. One in 50 has been quoted, but a slope as steep as one in 5 can be inferred from the following formula in some circumstances. It is for the transition zone height (ht), expressed in terms of the downwind surface roughness (zo in *metres*) and the distance (x) downwind of the boundary:

$$ht = 10\,(zo)^{0.4}\,.\,x^{0.6} \qquad \text{metres} \qquad (6.5)$$

Thus, the transition zone is 68 m high at 330 m downwind over grassland, for which the roughness is 0.02 m. A similar formula given by Elliott (1958: 1048) implies a slope of one in 9.

An alternative to Eqn (6.5) is shown in Table 6.4, where allowance is made for the atmospheric stability. There the slope (x/ht) is seen to range from one in 2, to one in 40. Elsewhere, Hanna (1987) has observed that the slope of the upper boundary of a sea-breeze is one in 10 for the first 2 km from the coastal edge, and thereafter one in 33. More recent work shows a fetch-height ratio of 20 to one, the 'fetch' being the distance downwind of the boundary.

Table 6.4 Height (ht) of the top of the transition zone in the atmosphere, at a distance x metres downwind of a change of surface roughness

	100 m		*1000 m*	
Distance downwind x	*ht*[a]	*x/ht*	*ht*	*x/ht*
Unstable atmosphere	45	2.2	250	4.0
Neutral	22	4.5	120	8.3
Stable	7	14	25	40

Source: Hogstrom and Hogstrom 1978: 945.

Note: [a] A roughness (zo) of 1 metre is assumed for the rougher downwind surface.

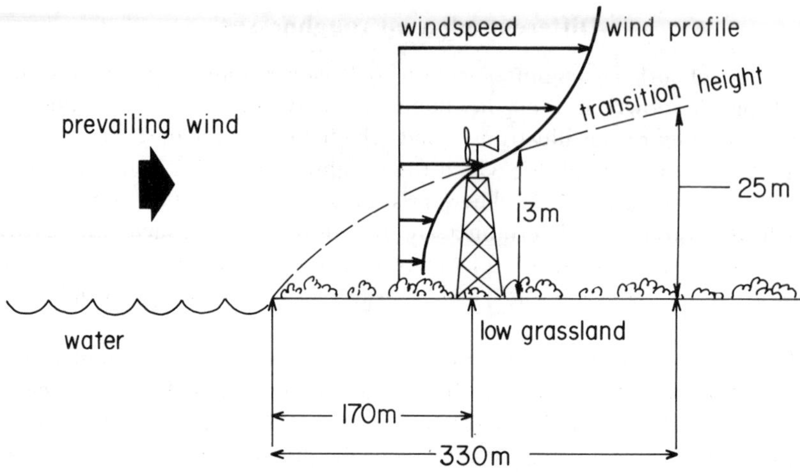

Figure 6.6 Effect of a change of surface roughness on the wind profile at places downwind. In this example the blades of the wind-energy collector would be affected by winds controlled by the sea's roughness upwind.

One concludes that a slope of about one in ten is representative. An implication of this is that measuring the wind at 10 m needs uniform ground for at least 100 m upwind (see 'Climate stations', p. 51). And the environment of a particular surface, the zone which affects the surface but is hardly affected by it, is above a height about equal to a tenth of the surface's diameter (p. 29).

TIME VARIATIONS

The strength and the direction of surface winds vary from year-to-year, season-to-season, day-to-day and hour-to-hour. The year-to-year fluctuation is the 'inter-annual change', the season-to-season variation the 'annual change' (i.e. the swing *within* a year), whilst hour-to-hour alterations are part of the daily change.

Winds may vary considerably from one year to the next. An example of inter-annual variation comes from Lerwick in Scotland during 1959–63 when the annual means were 6.9, 6.3, 7.1, 5.7 and 5.9 m/s, respectively. Annual mean wind speeds at Southport in England ranged between 5.2 and 7.3 m/s during 1897–1954 with the annual means rising fairly steadily from 5.7 m/s in 1897 to 6.7 m/s in 1915, then falling to 5.7 m/s again by 1945, before rising once more.

Annual and monthly variations

The changes of wind speed from season-to-season depend on location. In

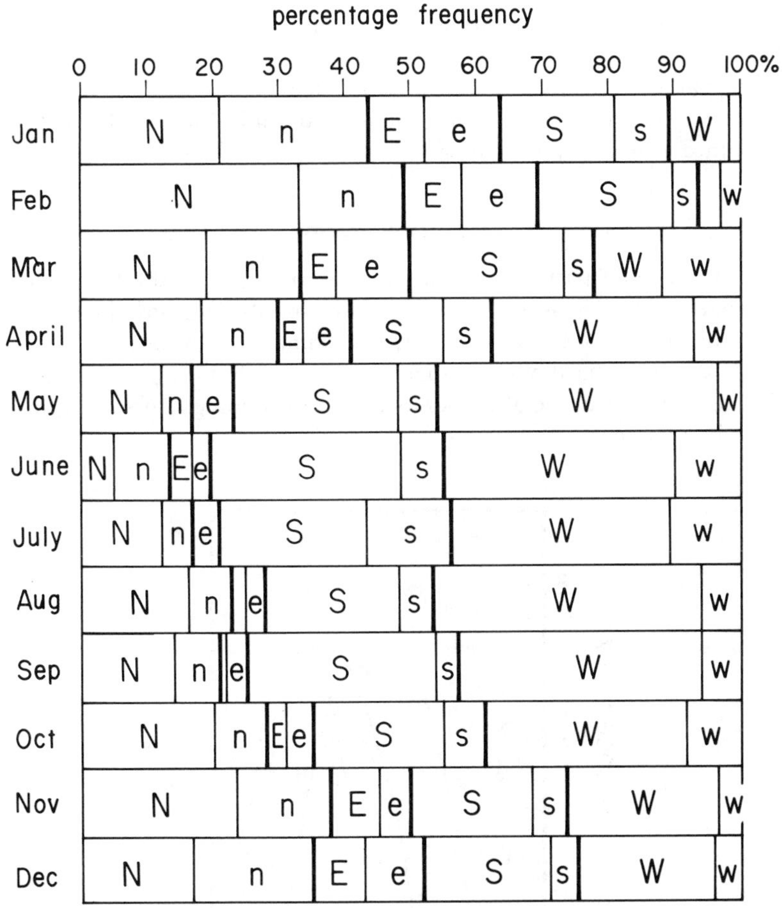

Figure 6.7 Seasonal variation of the directions of gradient winds at Sydney during 1968–70 (McGrath 1972). Capital letters denote winds over 5 m/s and lower-case letters lighter winds, from the indicated quadrants. N means strong winds from the north, for example

some places, monthly mean surface winds are stronger in summer than winter, as in parts of Australia. The ratio of July/January winds is 2.1 at Karachi and 3.4 at Seistan in Iran, for example. In New Zealand the corresponding summer/winter ratio is less at 1380 m where it is 1.1, than at 900 m, (where it is 2.2). On the other hand, strong winds occur mostly in winter in Canada, interrupted by long periods of light winds. Likewise, the ratio of January/July speeds is 1.28 at Ramallah in Jordan, and in London also the wind speeds are greater in winter than summer. In the USA, there is most wind in winter on the mountains, but most in spring on the Great Plains and in Illinois, and most in summer at Palmdale in

California. The annual variation is especially notable near mountains in hot climates.

The variation within a year may be demonstrated in a diagram like Figure 6.7, which shows a swing, from north and south winds in summer, to south and westerly in winter.

Daily variation

The typical daily pattern of winds discussed earlier in the first section of this chapter (p. 208) leads to variations of speeds like those in Figure 6.8. The daily range is greatest in hot, dry climates with clear skies. In ten American cities the daytime range in winter is from about 5 m/s in the morning to 6 m/s in the afternoon, with corresponding figures of 4 m/s and 6 m/s in summer.

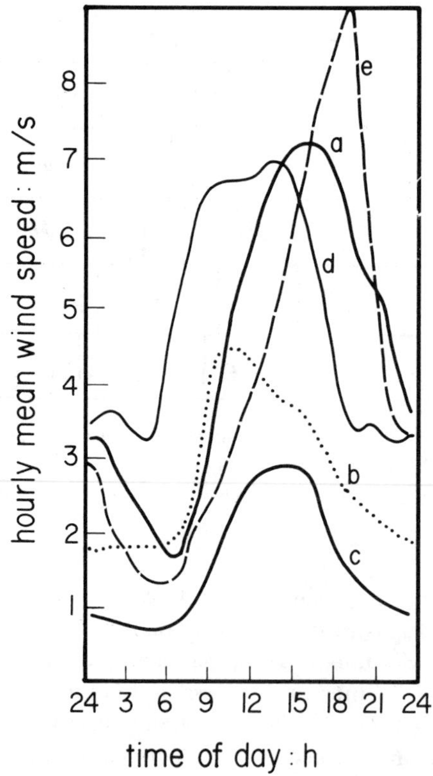

Figure 6.8 Daily variation of surface winds in October at (a) Victoria West (South Africa), (b) Belgaum (India), (c) Madras (India), (d) Hobsonville (New Zealand), (e) La Quiaca, Argentine (Tagg 1957: 44, by permission of ERA Technology Ltd, UK)

Most places have lower speeds at night. In Illinois, nocturnal wind speed is only about three-quarters of that of the daytime wind in winter, and half that of the daytime wind in summer. At East Malling in England, the daily range of winds at 10 m is from 2.5 m/s at about 2 a.m. to 6.5 m/s at 1 p.m. in summer, on average. The nocturnal reduction of surface wind speed is accompanied by an increase of speed at levels above 100 m or so (see Figure 4.6) due to the ground inversion unhitching the upper layer from the ground's friction.

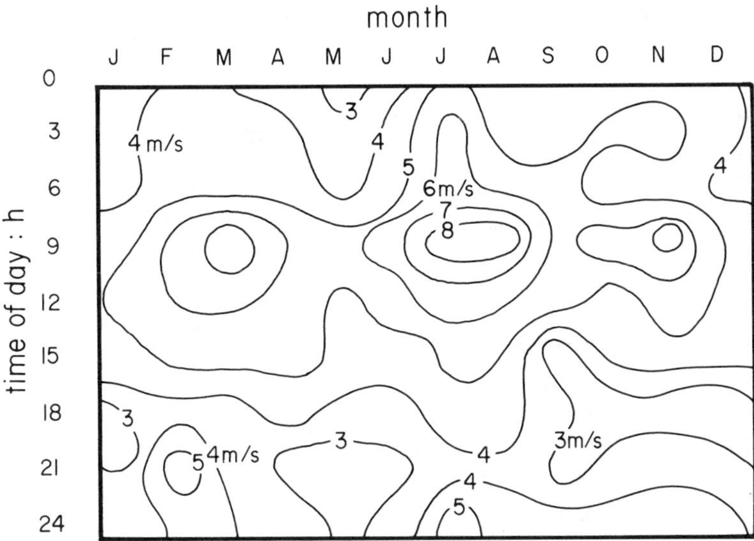

Figure 6.9 Effect of month and time of day on the wind speed at Aoulef in Algeria (Warne and Calnan 1977: 967, by permission of the Institute of Electrical Engineers, UK)

The daily variation at different times of the year is indicated by an isogram such as Figure 6.9. In the present example, the strongest winds occur at about 9am in summer, and the lightest in the early evening in autumn and winter.

A diagram like Figure 6.10 is sometimes used to show the normal daily fluctuation of winds in a particular month. It is called a 'hodograph'. An arrow from the origin to any of the apexes of the irregular loop is a vector whose direction is that from which the wind comes at the particular hour of day. The arrow points into the wind, like a wind-vane. Its length is proportional to the average speed.

Another presentation of measured wind data is Table 6.5. This shows the wind speed most likely to follow any given wind, at any particular time of the day, in Sydney. The table allows estimates of trends for several hours

Table 6.5 Estimates based on wind records during October 1963 at Observatory Hill (Sydney) of the most likely wind speed after the next three hours

Present wind: m/s	*Median wind 3 hours later: m/s*							
	3 a.m.	*6*	*9*	*Noon*	*3 p.m.*	*6*	*9*	*Midnight*
0–0.5	0.2	1.2	4.5	1.2	[a]	0.2	0.2	0.2[b]
1.0–1.5	0.2	3	4.5	3	2	0.2	[a]	[a]
2.0–2.5	1.2	3	1.2	3	[a]	0.2	0.2	[a]
3.0–3.5	1.2	3	2	2	1.2[c]	2	0.2	2
4.0–5.0	0.2	5	4.5	5	5	3	0.2	4.5
Over 5.0	4.5	5	5	5	5	3	4.5	5

Source: Hawke *et al*. 1975: 25.

Notes: [a] Insufficient data

[b] e.g. if the wind at midnight is between 0 and 0.5 m/s the most probable wind at 3 a.m. is 0.2 m/s.

[c] e.g. if the wind at 3 p.m. is between 2.5 and 3.5 m/s it is likely to be 1.2 m/s at 6 p.m.

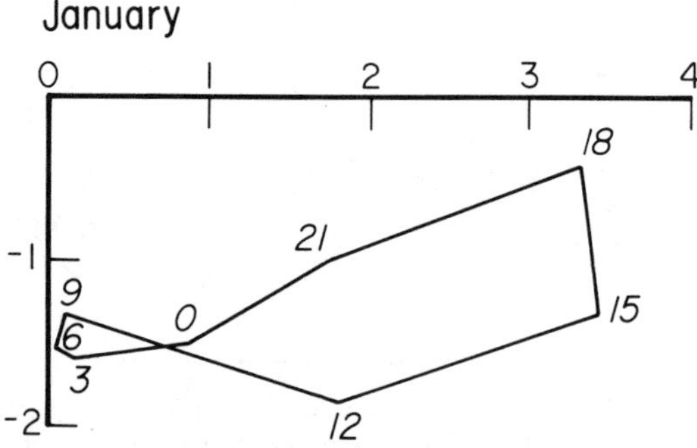

Figure 6.10 The hodograph for winds at Sydney in January during 1964–8 (McGrath 1972). For instance, at 1800 hours (i.e. 6 p.m.) the wind is about 3.5 m/s from just south of east, whilst from 3–9 a.m. the winds are about 1.5 m/s from the south, on average

ahead, e.g. a wind of 2 m/s at noon is most likely to be followed by one of 3 m/s at 3 p.m., and that is probably followed by 1.2 m/s at 6 p.m., and so on. Of course, the reliability of the predictions lessens with each additional increment of time, and a similar table would be needed for each season. The usefulness of Table 6.5 depends on the degree of correlation between winds three hours apart. Such a table could be constructed for wind direction also.

The wind speed is commonly measured only at 9 a.m. and 3 p.m., whereas one may want the whole-day mean at the standard height. In East Africa the speeds measured at 2 m height at those times (i.e. U_9 and U_3), give the daily mean at 10 m as $0.27\ (U_9 + U_3)$ m/s.

FREQUENCIES OF VARIOUS WIND SPEEDS

The frequency of various wind speeds at a particular place may be shown by the kind of wind-rose illustrated on p. 215. But Figure 6.4 indicates the frequency in only four ranges of speed, which is a rather crude summary of the measurements. The histogram in Figure 6.11 is more detailed. A smooth curve has been fitted to the histogram, on the assumption that deviations from the curve are due to random processes, and the curve represents the 'probability density function' p(V), analogous to that in Figure 4.10 or to the data in the third column of Table 4.10. However, a closer fit between data and curve is achieved by plotting the 'cumulative distribution function' C(V), like the fourth column in Table 4.10. The 50th percentile (i.e. the median) generally approximates the long-term mean and is about 30 per cent more than the modal wind.

Wind data may also be presented in a different but related form, where the vertical axis shows the 'exceedance' E(V), i.e. the likelihood that a value of wind speed *exceeds* the figure shown on the horizontal axis. The exceedance E(V) of a wind V equals [1 − C(V)]. In addition, the horizontal-axis values can be divided by the long-term mean wind speed, so the speeds are 'normalised', being replaced by ratios. This facilitates comparison between exceedance curves from both calm and windy places. Examples are shown in Figure 6.12, revealing variations of the exceedance curve according to the location in Sydney and to the season of the year.

Most commonly, wind data are presented in the form of a 'velocity-duration curve', as in Fig. 6.13. Sometimes the horizontal axis is labelled in terms of the hours per year, instead of the percentage of measurements.

GUSTS

There is particular interest in the left part of Figure 6.13, where the curve shows the frequency of the highest wind speeds. These winds occur only briefly and in bursts, called 'gusts'. Gusts are important as regards architecture, highway design, and outdoor structures such as wind-energy collectors; they also cause damage to agriculture. Storm gusts blew down Scotland's Tay Bridge in 1879.

A gust is an extreme case of the customary fluctuations of the wind, a pulse with a speed one or two Beaufort-Scale units higher than the average. It is defined as a positive departure from the mean over a specified time, where the departure lasts for not more than two minutes. A negative departure is a 'lull'. One particular kind of gust is called a 'squall', which is a wind that rises suddenly to at least 11 m/s, lasts for at least a minute and then dies away abruptly.

Gusts are due to 'chaotic vorticity' i.e. random circular stirring of the air. This sporadically mixes volumes of air from higher layers of the

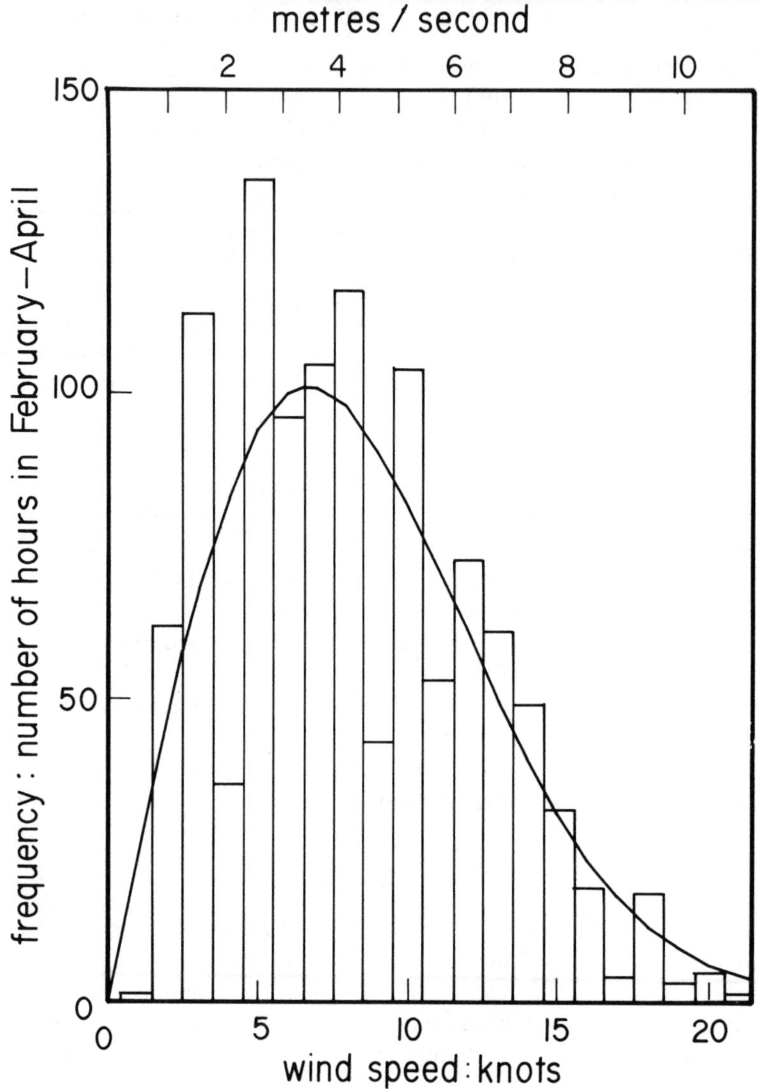

Figure 6.11 Frequency distribution of February–April winds at Hatyai, Thailand (Exell 1982, by permission of the Australian Institute of Physics.)

troposphere into the slower air near the ground, temporarily bringing the greater speed and different direction of the upper wind. The latter is generally more clockwise in the northern hemisphere (see 'Causes of wind', p. 203).

The strongest gusts are generated by the turbulence within storms. Less

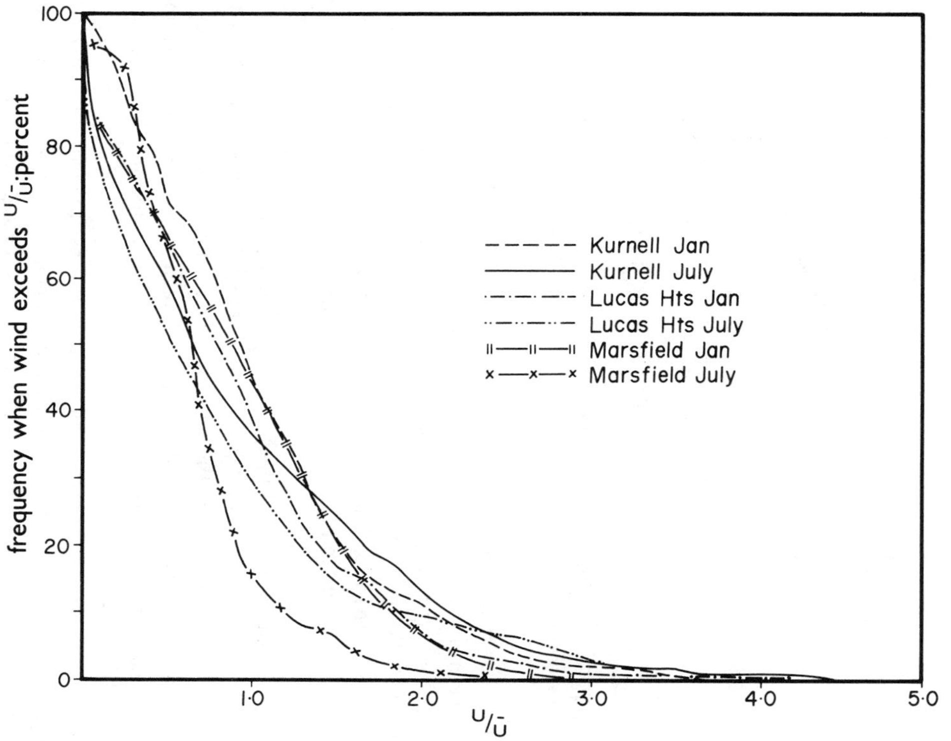

Figure 6.12 Exceedance curves in 1978 for Kurnell (on the coast at Sydney), Lucas Heights (15 km inland) and Marsfield (20 km inland). The horizontal axis shows the wind speed as a ratio of the mean for the particular place

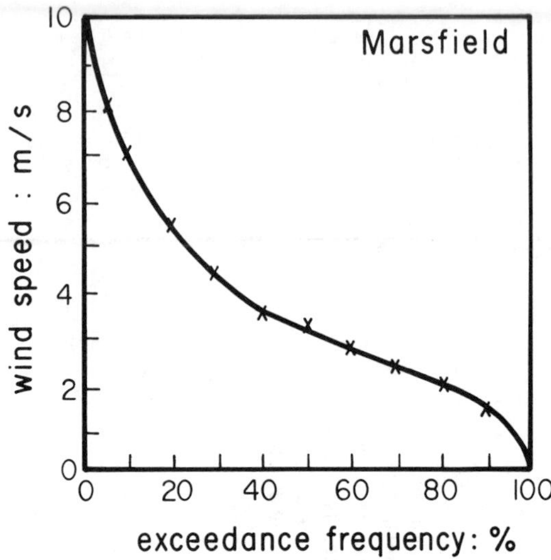

Figure 6.13 Velocity-duration curve, showing the percentage of measurements at Marsfield *greater* than any wind shown on the vertical axis.

violent gusts can occur if there is thermal convection due to intense irradiance from a clear sky when light winds prevail.

The eddies within gusts are of various sizes. Eddies with a diameter of the order of 100 m or more tend to last for only about two minutes, whereas eddies larger than a kilometre endure for ten minutes or so. (These distances and times imply rates around a metre per second, which exceed those in Table 1.3.)

Speeds of the winds within gusts are given here in terms of metres per second. On the other hand, American practice is sometimes to describe a gust in terms of the *time* for the wind to cover a mile. Canadian authorities record the annual extreme hourly mean, whilst the fastest 3-second mean each year is noted in Australia, and annual maximum 10-minute mean gusts in Britain.

There is a regrettable lack of uniformity as regards the period of averaging in deriving gust speed. This means that care is needed in comparing values of extreme winds, since averaging over longer periods leads to lower values. The fastest 10-second gust is typically 1.5 times the maximum 17-minute mean, for instance.

Indices of gustiness

Gust Factor

The highest 3-second average speed during a gust may be compared with the mean over a much longer period P, and the ratio is called the 'Gust

Factor' (G). It depends on P, on the height at which measurements are taken, and on the kind of terrain nearby. The comparison period P is often taken as an hour, and then a Factor of 2 applies in extensive (frontal) storms, but 3 in local (convective) thunderstorms. A Factor of 2 generally applies in towns, and 1.6 in open inland areas, 1.5 with onshore coastal winds and 1.35 over the sea. The higher values over rougher terrain are due to increased stirring and reduced surface wind speed.

The Gust Factor is smaller if measurements are made well above the ground. For instance, G at Cape Canaveral was found to be 1.65 for 10-second gusts at 10 metres height, but 1.38 at 50 m and 1.28 at 150 m. Factors given in the Australian Manual of Meteorology (1966) are 1.5 at 8 m and 1.3 at 50 m, which implies that gust speeds increase with elevation less rapidly than average winds do.

The Factor is altered if the period of measurement is different. The 'instantaneous' gust speed (over 2–3 seconds) is about twice the hourly mean at Hong Kong, whereas the 1-minute gust is only 1.28 times. G is 1.59 for 1-second averaging at a typical site in Britain, but 1.47 for 5 seconds, 1.43 for 10 seconds and 1.24 for 1 minute. For a mean wind of 10 m/s, the Factor is 1.50 for 5-second gusts and 1.10 for 10-minute gusts in Sydney. Comparable figures are given in Table 6.6.

Table 6.6 Effect of the periods of averaging on the value of the Gust Factor (G), which is the ratio Ug/**u**, where Ug is the peak wind speed (i.e. the average over a brief period d), and **u** is the mean over a much longer time (P)

Peak period: d	*Reference period: P*	*Gust Factor*
Instantaneous	10 minutes	1.4–1.6
Instantaneous	5 minutes	1.2–2.0
3–5 second	10 minutes	1.3–1.9
0.5 second–1 hour	1 hour	1.0–1.6
Instant–1 minute	1 hour	1.3–2.1
1 second	1 hour	1.6

Source: Davis and Newstein 1968: 373

G is much greater if the maximum gust is compared with the annual mean wind speed instead of with the mean of the surrounding hour. This is exemplified by values from Lerwick in Scotland where the long-term average annual maximum gust is 37 m/s and the annual maximum hourly mean is 24 m/s (implying a factor of about 1.5), whereas the annual mean is 6.4 m/s, implying 5.8. The ratio of peak gust to annual mean is about 10 for cities in Australia and 7 in the USA.

Another influence on G is the design of the anemometer. A cup anemometer gives a factor about 5 per cent more than a Dines instrument does because of its faster response to changes. Also, the factor is less with stronger hourly-mean winds, e.g. 1.76 for 10 m/s winds but 1.50 for 40 m/s winds.

Finally, the presence of obstacles upwind alters the Gust Factor, according to wind direction.

The magnitude of the Gust Factor may be used in conjunction with an extended version of Figure 6.5 to gauge the height from which the pulse of stronger wind has come, i.e. the size of the eddy. For instance, a factor of 1.5 measured at 10 m over open land has descended from about 210 m.

Gust Coefficient

An alternative to the Gust Factor is the 'Gust Coefficient', which is the standard deviation of instantaneous winds divided by the mean. The ratio is typically 0.09 at 10 metres height offshore and 0.15 over land.

Gustiness Ratio

A similar measure of turbulence is the 'Gustiness Ratio', which is the *range* of wind speeds divided by the mean. It may be about 0.3 over the open sea, so that a mean wind of 15 m/s would be accompanied by gusts of 17.3 m/s (i.e. the mean plus half the range, when the range is 0.3 times the mean). The ratio at an open coastal site may be 0.5, for example, whereas the greater roughness of an urban horizon leads to a ratio there as high as 2.0. So pulses of 30 m/s occur in a city wind of 15 m/s.

Gust speeds

The strongest winds measured have been 84 m/s, with gusts up to 103 m/s, during a hurricane on top of Mt Washington in New Hampshire. Gusts up to 88 m/s have been measured in Antarctica, whilst winds occurring once in 20 years in the USSR and the West Indies range from 20 to 54 m/s. Such maximum values in Britain are from 38 m/s near London, to 54 m/s on the west coast of Ireland. The strongest gust measured at Cape Reinga at the northern tip of New Zealand was 55 m/s. However, conventional cup anemometers respond inadequately to brief gusts ('Measurement of wind', p. 49), so all their measurements are uncertain.

Long-term modal values of annual maximum gusts in Australia range from 20 m/s in the north-east, to 31 m/s at the southern tip of Western Australia and Tasmania. The maximum gust velocities in Australian thunderstorms range from about 15 m/s at Cairns, 19 at Hobart, 23 at Darwin and Brisbane, 24 at Perth, 25 at Sydney and Melbourne, to 28 m/s at Alice Springs. Gust speeds occurring once in 25 years are higher in tropical cyclones, ranging from 38 m/s at Cairns to 45 m/s at Brisbane, and 56 m/s at Exmouth and Onslow in Western Australia

Data from seven places in New England (USA) show that the monthly fastest wind speed Umax (in terms of the shortest time for a one-mile wind-run) equals (2.88 Umon + 2.4) m/s, where Umon is the monthly mean speed. So a mean of 4 m/s, for instance, implies gusts of up to 14 m/s. In Australia, the hourly-mean gust speed near the ground equals the long-term mean wind speed plus a third of the hourly-mean *gradient* wind, because gusts result from the momentary intrusion of the upper air.

Terrain and height affect the extreme wind, so the range is 23–32 m/s for maximum winds at 40 m in five American cities, but 31–46 m/s at 16 m above airfields. A rougher terrain usually reduces the extreme wind but increases the variation with height

Effect on wind direction

Fluctuations of wind speed in a turbulent atmosphere are linked with variations of wind direction, though not consistently. Swings of wind direction recorded at 100 m at Brookhaven (NJ) were of five quite different types. With the first, the record shows swings of 90° or more occurring when winds are about 2 m/s and the lapse rate is 11 C°/km, i.e. slightly above the adiabatic rate (see 'Estimating screen temperature, pp. 73–4, 84), implying thermal instability. Second, swings of 45–90° occur with a similar lapse rate but winds of 7 m/s, i.e. neutral conditions. Third, 15–45° swings are associated with a lapse rate of 14 C°/km and 4 m/s, i.e. extreme thermal instability. Fourth, cloudy conditions with winds of 10 m/s and a stable lapse rate of about 5.7 C°/km result in a steady trace with fluctuations of only 15°. Fifth, a wavering smooth record of wind direction results from inversion conditions.

Similar measurements over open grassland in England also show that wind direction fluctuates according to the speed and atmospheric stability. In *stable* conditions the fluctuation decreases in faster winds, e.g. 50° with 2 m/s but 10° with 8 m/s. The opposite is shown in Figure 6.14, where winds are seen to be more erratic when faster, the reason being that the atmosphere is *unstable* on summer afternoons. The range of direction is about 35° in moderately unstable winds of 2–11 m/s, but increased to 90° in more unstable conditions.

ESTIMATING EXTREME WINDS

Structures in the open must withstand the strongest wind likely to be encountered within the structure's working life. A rule-of-thumb is that temporary structures erected for the construction phase of some large project must cope with the strongest wind likely to occur once within five years, most 'permanent' structures are reckoned to have a life of 50 years, whilst buildings like hospitals and communication centres – which are

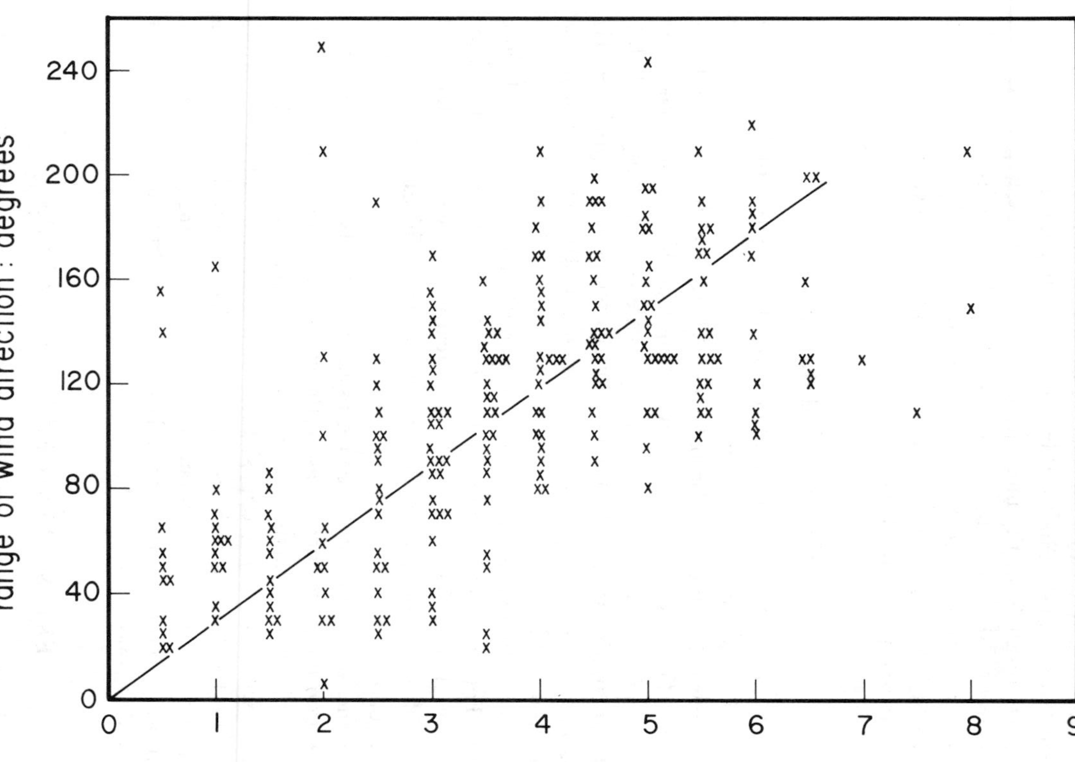

Figure 6.14 Effect of hourly mean wind speed on the hourly range of direction, at about 3 p.m. in summer at Marsfield (D. Merrikin 1982: private communication)

needed after any disaster – should withstand the strongest wind within a century. The relevant wind speeds at Sydney, for instance, are 34 m/s, 44 m/s and 47 m/s, respectively.

There is no absolute maximum wind speed at a place, since, if any particular value were assigned, it always remains possible that future fluctuations may eventually throw up an even stronger wind. Instead, one estimates what is called the '100-year wind', for instance, this being the speed likely to be exceeded only once in a century, on average. That means that the chance of its occurrence in any particular year is 1 per cent.

It is not easy to assess the extreme wind speed likely to occur once within 100 years when available measurements span a much shorter period. The estimation requires using those measurements to assess the likelihood of each value being exceeded, and then extrapolating to find the wind with a 1 per cent chance, as discussed below. The extrapolation may be either graphical or based on some equation representing the cumulative distribution function (see 'Frequencies of various wind speeds', p. 225). An equation is chosen which facilitates extrapolation by providing a straight-line graph of the likelihood against the wind speed, and by minimising the distance of extrapolation. Whichever equation is chosen, one assumes that the data set is homogeneous (see Chapter 4, 'Climate data', pp. 122–3).

It is usually stipulated that measurements over at least 20 years are needed to estimate long-term extreme values. If only three annual maximum 10-minute gust speeds from Prestwick (UK) are used to deduce the 50-year speed there, values are obtained which differ by up to 20 per cent from the more accurate estimate from 24 annual maxima. One way round the difficulty of too few data is to consider the maximum gust speeds in each month (instead of each year), so that we have 36 values instead of only 3 from which to extrapolate. Then we can deduce the 100-year value in terms of the wind exceeded once in each 1200 months. In this case, the

Table 6.7 Values of the annual maximum gust speed at Washington (DC) during 1908–46

Year	1908	1909	1910	1911	1912	1913	1914	1915
Gust: m/s	23.7	26.8	22.8	34.4	29.9	31.3	29.1	26.8
Year	1916	1917	1918	1919	1920	1921	1922	1923
Gust: m/s	25.9	23.2	24.6	24.6	22.8	24.6	25.5	20.6
Year	1924	1925	1926	1927	1928	1929	1930	1931
Gust: m/s	23.2	23.7	23.7	23.7	20.6	20.1	23.2	21.9
Year	1932	1933	1934	1935	1936	1937	1938	1939
Gust: m/s	25.4	25.0	26.3	20.6	30.8	25.0	25.0	25.0
Year	1940	1941	1942	1943	1944	1945	1946	
Gust: m/s	21.0	23.2	24.6	27.7	24.6	20.1	24.1	

Source: Landsberg 1958: 81.

Note: The mean value is 24.7 m/s, and the standard deviation 3.2 m/s

Table 6.8 Ranked wind speeds from Table 6.7

Annual maximum wind U: m/s	*ln U*	*Number of values less than or equal to U*[a]	*100 C (U) percentile*[b]	*Exceedance 100 E (U)*[c]	*Return period*[d]*: T years*	*Weibull reduced variate*[e]	*Gumbel reduced variate*[f]
20.1	3.00	2	5	95	1.05	−3.02	−1.10
20.6	3.03	5	13	87	1.15	−1.97	−0.71
22.8	3.13	9	23	77	1.30	−1.34	−0.39
23.2	3.14	13	33	67	1.49	−0.92	−0.10
24.1	3.18	18	45	55	1.82	−0.51	+0.23
24.6	3.20	23	57	43	2.33	−0.17	+0.58
25.0	3.22	27	67	33	3.03	+0.10	+0.92
25.9	3.25	30	75	25	4.00	+0.33	+1.25
27.7	3.32	34	85	15	6.67	+0.64	+1.82
29.9	3.40	36	90	10	10.0	+0.83	+2.25
30.8	3.43	37	93	7	14.3	+0.98	+2.62
31.3	3.44	38	95	5	20.0	+1.10	+2.97
34.4	3.54	39	97	3	33.3	+1.25	+3.49

Notes: [a] This is related to the term m in Eqn (4.1); [b] i.e. 100 m/(N+1), the number of values (N) being 39 (see Table 6.7); [c] i.e. 100 − 100 C(U) per cent; [d] i.e. 1/E(U), see pp. 109–10; [e] i.e. ln [ln T]; [f] i.e. − ln [−ln C(U)].

desired figure would be the 99.92 percentile (i.e. 100 × 1199/1200). However, an objection to this procedure is the non-homogeneity of monthly values due to the annual fluctuations of wind speeds (see 'Typical values', pp. 117–25; 'Time variations', p. 223).

We shall use the data in Table 6.7 to demonstrate three alternative procedures for estimating the wind exceeded once each 100 years, on average. The data are ranked in Table 6.8, which also refers to the cumulative distribution function C(U), the exceedance E(U) and the return period T, discussed in 'Climate data', p. 110.

The normal distribution

The simplest assumption to make is that the values in Table 6.7 vary from a central figure in a completely random fashion. This assumption can be tested (see 'Scatter of values', pp. 130–3). If the distribution is indeed normal, the probability density function is given by Eqn (4.6), for which the cumulative equivalent is as follows:

$$C(U) = 1 - \exp\,(-\,U^2/2s^2) \qquad (6.6)$$

where C(U) is the chance that the annual wind maximum is less than or equal to a stipulated value U, and s is the standard deviation of the whole set of values. In other words, C(U) is zero when U is zero, and unity when U is very large. The equation describes what is called a 'Rayleigh

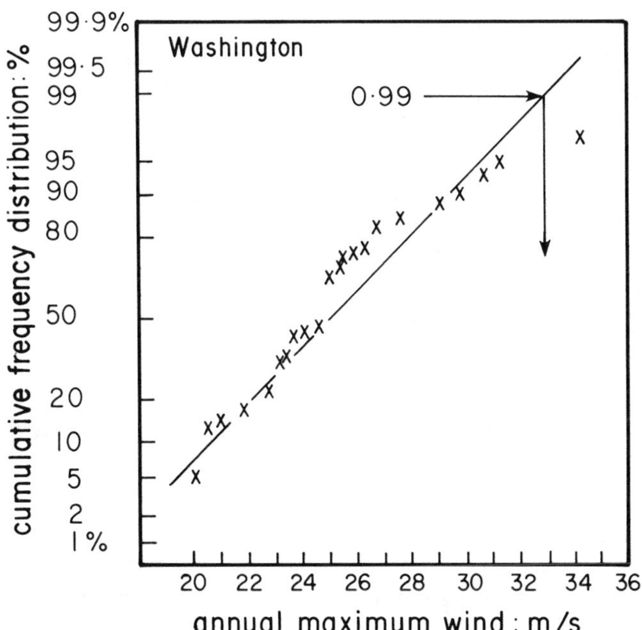

Figure 6.15 Plot of values of the annual maximum wind at Washington (see Table 6.7) against a vertical scale representing a normal distribution

distribution', characterised by the standard deviation. The equivalent 'probability-graph' is shown in Figure 6.15.

The data in Figure 6.15 are from Table 6.8. A line has been fitted and extended to intersect the 99 per cent ordinate at 33 m/s, so this is the estimate of the 100-year gust. In defining the straight line, the highest 10 per cent of values were ignored since they are the least reliable as they depend on only a few unusual measurements. Figure 6.15 shows that the Washington data do not lie close to a straight line, so there is little certainty about the inferred 100-year wind. The reason for the poor fit in this case is that the data in Table 6.7 do not consist of all the measurements over the 39-year period, but include only the extreme value for each year; therefore the data are neither regularly nor randomly selected. By contrast, complete sets of regular measurements often do fit Eqn (6.6) with useful accuracy.

An alternative to a Rayleigh distribution of wind speeds is one which assumes a normal distribution of the natural logarithms of the measured values. It has the merit of compressing the right-hand side of the horizontal scale, thus reducing the distance to the 99 per cent value. This gives a satisfactorily straight line of points representing the cumulative distribution of 3-hour mean wind speeds over 10 years, for instance. Unfortunately, inspection shows that data from the second column in Table 6.8 do not lie any closer to a straight line on log-normal graph paper than do points in Figure 6.15. A different transformation of the horizontal variable is to take the square-root of the wind speed, which again squeezes the right-hand end of the scale but still fails to linearise a plot of the Washington data.

The Weibull method

If the data do not fit Eqn (6.6), they may perhaps fit the more general Eqn (6.7) instead:

$$C(U) = 1 - \exp - (U/w)^k \tag{6.7}$$

This is variously called the Frechet equation, the Weibull equation, the Goodrich law or the Fisher-Tippett type II equation. It resembles Eqn (6.6) but is more flexible, allowing for different values of the Weibull 'scaling factor' w and of the 'shape factor' (or 'dispersion factor', k) to achieve a better fit of the equation to the measurements.

Eqn (6.7) is usually expressed in terms of the equivalent exceedance E(U), which equals [1 − C(U)]. It follows from Eqn (4.2) that the inverse of E(U) is the return period T, so that Eqn (6.7) gives the following:

$$\ln [\ln T] = k (\ln Ut - \ln w) \tag{6.8}$$

where the left-hand side is called the 'Weibull reduced variate' and is the same as ln [–ln {1−C(U)}], and Ut is the wind speed with a return period T. Plotting the reduced variate against (ln Ut) in Figure 6.16 yields a straight line whose slope is k, whilst (ln w) is given by the line's intersection with the horizontal axis. The reduced variate for 100 years is 1.53, and for the Washington data the corresponding value on the horizontal axis is 3.44, so the 100-year wind is exp 3.44, i.e. 31 m/s.

Eqn (6.8) shows that (ln w) equals (ln Ut) when the reduced variate is zero. So (ln w) is 3.275 in Figure 6.16, i.e. the scaling factor w is 26.4 m/s. This is just over the mean of the 39 values. In fact, w is commonly about 10 per cent more than the mean.

Figure 6.16 does not show any impressive conformity of the Washington data to a straight line. On the other hand, a satisfactory fit has been reported for other data from temperate latitudes. Agreement is better if there are only a few calm periods.

The shape factor k is 2.0 for the Rayleigh distribution (Eqns (4.6) and (6.6)). In practice, k is indeed commonly in the range 1.7–2.5, provided all measured values are included in the set, and not merely the annual fastest winds, for example. The factor depends on the location, being 1.1–1.8 for seven places in Canada, for instance, and 1.5–2.1 in Alaska. In other states of the USA, the following relationship obtains where **U** is the mean wind speed:

$$k = 0.9\, \mathbf{U}^{0.5} \tag{6.9}$$

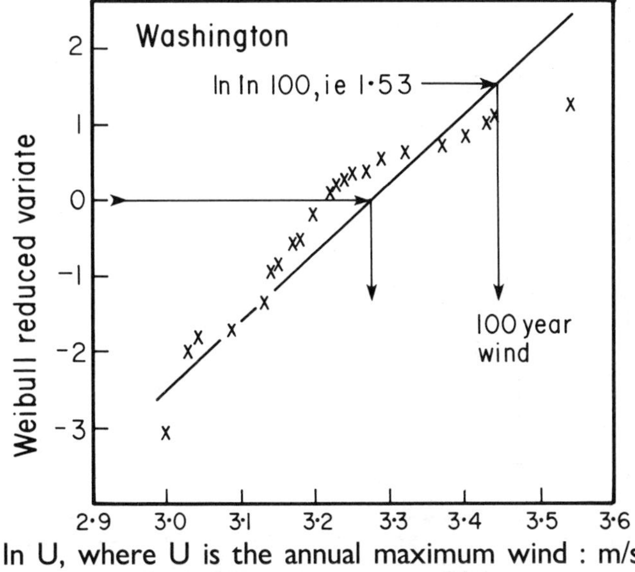

Figure 6.16 Plot of the Weibull reduced variate against the natural logarithm of values of the annual maximum wind speed at Washington (see Table 6.8)

A similar relationship, but with 0.73 in place of 0.9 has been reported by other authors. Thus, k is 2 if the mean wind speed is about 6 m/s.

Quite different values of k emerge from taking a set of annual *maximum* wind speeds, ignoring all other measurements, as in Table 6.7. Figure 6.16 gives k equal to 9. Elsewhere it has been found that 9 is the usual figure for extreme winds in extra-tropical storms, but 4.5 for tropical storms.

The Gumbel method

A procedure popularised by Gumbel (1954), is based on the Fisher-Tippett type I equation for the cumulative frequency distribution C(U), as follows:

$$C(U) = \exp\{-\exp[-g\,(U - Um)]\} \tag{6.10}$$

where 1/g is the Gumbel scaling factor and Um is the modal value of the set. The two terms play the same roles as the standard deviation and the mean value in Eqn (6.6), and as the Weibull scaling factor w and exponent k in Eqn (6.7). The chief difference between the Weibull and Gumbel distributions is the suitability of the former for complete data sets (such as all the 3 p.m. winds), whilst the Gumbel distribution is useful for extreme values, e.g. a set of only the monthly maximum daily average winds.

Eqn (6.10) can be developed into an analogue of Eqn (6.8), thus:

$$-\ln[-\ln C(U)] = g\,(U - Um) \tag{6.11}$$

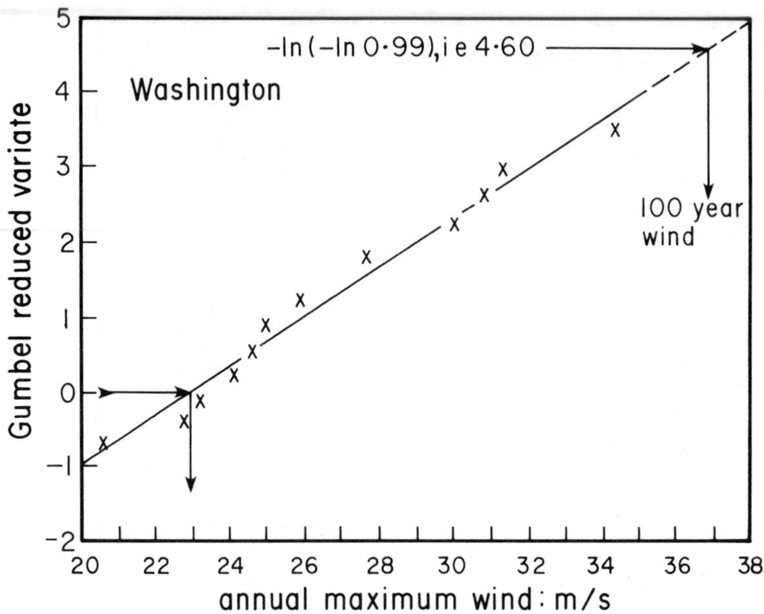

Figure 6.17 Plot of the Gumbel reduced variate against the annual maximum wind speed at Washington (see Table 6.8)

This leads to the kind of graph in Figure 6.17 and the Washington data yield the line which is shown. The velocity corresponding to a zero reduced variate on this line is shown by Eqn (6.11) to be the modal wind Um and, in this case, is 23 m/s. The slope of the line (g) is 0.33 s/m, i.e. the variate increases from zero to 3.0 when the wind value increases from 23 m/s to 32 m/s. Also, the 100-year wind, which corresponds to a probability C(U) of 0.99 (for which the Gumbel reduced variate is 4.6), is read off the straight line as 37 m/s. This may be compared with 32 m/s and 36 m/s obtained earlier. The Gumbel value is much more likely to be correct, because *all* the points in Figure 6.17 lie so close to the straight line that it is accurately defined. This illustrates the superiority of the Gumbel procedure to the Weibull method and confirms that the latter tends to slightly underestimate the 100-year value.

Knowledge of g and Um allows rapid estimation of the extreme wind for any specified return period, provided it is longer than about 10 years. The required equation is as follows (see Note 2, p. 335):

$$Ut = Um + (\ln T)/g \qquad \text{m/s} \tag{6.12}$$

The term (ln T) has the value of 3.9 for a 50-year value, 4.6 for 100 and 6.9 for a 1000-year value.

The literature contains several examples of the derivation of the 100-year extreme by the Gumbel method. Bridgman (1978: 14) obtained the following formula for winds at Newcastle (New South Wales), yielding 50 m/s for the 100-year wind:

$$Ut = 32 + 3.9 \ln T \qquad \text{m/s} \tag{6.13}$$

In this case, the Gumbel scaling factor (1/g) is 12 per cent of the modal wind. Table 6.8 gives a ratio of 13 per cent and similar data for Sydney give 12 per cent. Figures discussed by WMO (1966: XIII.10) yield a ratio of 21 per cent, and others from Nigeria give 36 per cent. The median of these ratios is about 15 per cent

Gomes and Vickery (1976: 35) used a set of the maximum wind speeds in each of 201 storms in Sydney during 1960–9. In this way, they derived a series large enough for Gumbel analysis, which would not have been possible from only ten annual maximum values. There were about 20.1 storms annually, so the 1005-storm value derived by Gumbel analysis is equivalent to the 50-year value (i.e. 50 × 20.1 = 1005), which proved to be 43 m/s.

Two other cases are dealt with in Note 3, pp. 335–6. The first is best handled by assuming a Gumbel distribution of extreme winds, whilst the other involves a Weibull distribution of all 3-hour mean wind values. More sophisticated procedures are necessary if neither Weibull nor Gumbel graphs of the measured data give a straight line to extrapolate.

Gumbel analysis is applied to other climatic elements as well as to extreme winds. For example, the method is used to estimate the maximum flood likely to affect a dam across a river within the dam's assumed lifetime. It has also been used for estimating extreme temperatures.

SPATIAL VARIATION OF WINDS

Wind speeds vary with latitude, proximity to the oceans, elevation on a mountain and with the shape of the local topography. This is important when seeking either to collect wind energy, to protect crops, to plan the ventilation of a city, or to design buildings, for example.

Winds are weak at the equator. The strongest winds in the southern hemisphere are at 40–70°S, where 90 per cent of the surface is unobstructed ocean. Gales over the oceans are most common at 35–55°S in July and at 40-70°S in January.

Coastal winds may be appreciable. On average they exceed 9 m/s at a coastal site in north-west Tasmania, for example. Winds are similarly strong north of Auckland in New Zealand, the mean being 9.2 m/s at 10 metres above the ground, with a peak gust of 27.7 m/s during six months of measurements. Notably high speeds are found at Mawson at 68°S on the

Antarctic coast where the annual average is about 22 m/s .

On the other hand, inland winds tend to be relatively modest. For instance, the daily wind in the centre of Australia is about 2.2 m/s on average, and at Canberra 1.7 m/s, compared with 4.4 m/s at Perth and 3.3 m/s at Sydney on the coast. Wind speeds in Sweden halve within 20 km of the sea. Similarly in Denmark, the surface wind at the coast is 38 per cent of the gradient wind, whereas it is only 21 per cent at 25 km inland.

Averages in the most populated parts of the USA are only 2–5 m/s, and merely 7 out of 348 climate stations in that country show a yearly mean above 7 m/s. That is unfortunate, since winds of at least 8 m/s are needed for economic collection of wind power. Such winds are found on coastal capes, and on the plains of northern Texas and of Kansas. Most wind power is found in Wyoming, and least in Florida.

The effect of hills

Winds tend to be stronger on hills, because of the wind profile discussed in 'Wind profile' pp. 216–20. For example, a survey in South Australia revealed five places with long-term mean winds more than 8 m/s, and all are on hills at least 100 m above the surroundings. The increase of speed on some Scottish hills is by 1 m/s for each 80 m elevation, which resembles the increase from 2.1 m/s to 3.7 m/s between 220 m and 350 m on a hill in Lancashire (England) or the increase by 1.8 m/s at places 200 m higher on slopes in Oregon.

Nocturnal inversions within valleys, and shelter by hills, greatly reduce the surface wind. As a consequence, the average wind 12 m above a hill at White Cliffs (New South Wales) proves to be as much as 71 per cent more than at 5 m above a site only 10 m lower. In other words, a minor change of location can appreciably alter the wind energy available.

Table 6.9 Effects of various factors on the ratio of the surface wind speed at a site, to that above open level ground nearby

Terrain	*Time*	*Values of the ratio* *3–5 m/s*	*6–10 m/s winds*
Summits above 50 m	Day	1.45	1.15
	Night	1.75	1.48
Lower summits	Day	1.35	1.10
	Night	1.65	1.35
Windward slopes:			
upper part	Day	1.25	1.05
	Night	1.50	1.25
lower part	Day	1.00	0.95
	Night	0.85	1.00

Source: Goltsberg 1969: 48

The ratio between winds at high and low elevations, respectively, depends on the shape of the land, time of day, and wind direction and speed. Typical ratios are given in Table 6.9, which shows that they are greater at night-time on the upper parts of windward slopes at the summits of high hills when light winds are dominant.

It may be possible to infer the wind speeds on hilltops from measurements of the sideways displacement of balloons rising from flat land nearby. Support for this comes from Davidson *et al.* (1964: 30), who reported similarities between the change of wind speed up the side of a mountain and the change vertically in free air above adjacent plains. Likewise, seasonal mean winds over 22 years, at 14 m above an isolated 839 m hill in Alaska, were within 7 per cent of those at the same altitude but above low-lying land 27 km away. Barry (1981: 53) reported the following relationship between hilltop wind speed (Uh) in Europe and the speed in free air at the same height above sea-level (Uf):

$$Uh = 2.1 + 0.5\ Uf \qquad m/s \tag{6.14}$$

We infer that strong winds are slowed by the mountain's friction, but light winds are accelerated by the vertical convergence.

Peculiar winds may be induced by the shape of a hill, as in the confusion at the brink of a scarp. There may be strong wind-shear at the summit, with winds at 20 m above the ground which are twice those at 2 m. In addition, wind may accelerate on a ridge (defined as high land whose length is at least ten times its height) if the wind is at right angles and the atmosphere is stable. Winds 8 m above the top of an isolated round hill in Scotland 100 m high are 70 per cent stronger than they are upwind. Druyan (1985: 98) reported a speed-up equal to 2H/L, where H is the hill's height and L the distance from top to halfway down, e.g. a 220 m hill, where L is 275 m, causes acceleration by 60 per cent.

Strong winds across a mountain range lead to bands of windiness, parallel and downwind of the range. These result from vertical oscillations of winds which strike the mountains and are deflected upwards and then bounce down again if there is an inversion layer above. The distance from the range to the first band of strong surface winds is around 20 km, though proportional to the wind speed. Lacy (1977: 70) reported an instance of surface winds of 15 m/s at 13 km upwind of a range, 26 m/s on the ridge, 40 m/s 5 km downwind, 10 m/s at 20 km, 48 m/s at 30 km, 6 m/s at 42 km and 15 m/s at 47 km. So the wavelength of the oscillations was about 20 km again.

The case of separate mountains is different. Winds tend to flow round a mountain when the atmosphere is stable but over the mountain when there is neutral stability, i.e. when winds are stronger. There tends to be a horizontal flow around the hills below a critical height, which again depends on the atmosphere's stability.

Winds in a valley depend on whether the valley is open (heading to the sea or extensive plains) or closed. As an example, the Parramatta River Valley in Sydney broadens out to the ocean, and a shallow layer of cold air flows steadily down the valley on most winter mornings. On the other hand, in the Hawkesbury River Valley nearby, the river flows through a narrow gorge that throttles the escape of air at night so that cold air draining from the surrounding hills is impounded, creating a deepening inversion (Figure 6.18).

A different situation exists in the San Gorgioni pass in southern California, which is an open valley that funnels the wind, accelerating it, so that the annual mean velocity is 8 m/s. This has led to the installation of numerous windmills to capture the wind energy. Such gaps between mountains create strong winds only when the gradient wind is parallel to the valley's length. Wide valleys are best – open at both ends, with smooth walls and flat floors.

A narrow gorge induces undesirable turbulence, and valleys with steep sides are usually only poorly ventilated. So the surface wind is only 22 per cent of the gradient wind at a place in Scotland where the horizon is 5 degrees above the level of the valley floor, but 42 per cent at a spot only 2 degrees below the horizon.

Toposcale effect

There can be large differences of wind speed within a particular region, on account of hills or the coast. For example, there are considerable differences of direction and wind strength within only 8 km amongst certain mountains in the USA, shown by the various directions of leaning and deformation of the trees there. So a distant anemometer is unsuitable as an indicator of the windiness of a place, especially in mountainous terrain or with an atmosphere whose stability is far from neutral.

The standard method of estimating the wind at a point between widely spaced anemometers involves taking a weighted average of the anemometer measurements, the weighting depending on the distance (di) between the point and the respective anemometer. The weighting is chosen to make the nearest measurement most influential in the calculation of the estimate. Commonly the weighting factor is $[1/di^2]$, but Johnson and Linacre (1978) used $[\exp(-0.1di^2)]$. The latter has the advantage of not becoming near-infinite when di is small, and any error of a very close instrument is better moderated by readings from other instruments. Thus the estimated wind speed Ue at a particular site is given as follows:

$$Ue = \Sigma\,[Ui \exp(-0.1\,di^2)]/\Sigma\,[\exp(-0.1\,di^2)] \qquad \text{m/s} \qquad (6.15)$$

where Σ signifies summation of the bracketed term that follows, and Ui is the wind measured at anemometer i, which is at a distance di (in kilometres) from the site. Verrall and Williams (1982) used [exp $(-0.04\ di^2)$] as the weighting term, which effectively limits the radius of the influence of any particular anemometer to about 10 km.

The pattern of winds within a region at a given moment is known as the 'wind-field'. One needs to know this to calculate the movement of air pollution within a city, for instance. A worthwhile refinement for determining the wind-field at a particular time near a coast involves separating the area controlled by a sea-breeze from the rest of the region. This requires a preliminary assessment of the wind-field at the given time to find the line of greatest convergence, i.e. the sea-breeze front. Then the interpolation procedure is repeated, but separately on each side of the line.

A different method of estimating the wind speed in each square of a map grid is to use Linacre's concept of the 'windiness ratio' (Linacre and Barrero 1974). This is defined as the average ratio of a wind speed in a grid square to the speed measured simultaneously by a permanent, continuous recorder, placed centrally in the region. The ratio may vary slightly between the morning and afternoon because of different regimes of cold-air drainage and sea-breezes (see 'Causes of wind', pp. 203–9), but it is usefully constant for any particular place, determined principally by the geography. Places with equal values of windiness ratio can be linked by isopleths, as in Figure 6.18. Low values occur in sheltered valleys.

The nature of the windiness ratio allows the value for each grid square to be steadily improved by adding ratios from new measurements obtained with a mobile anemometer. A square without measurements can be allotted a ratio by interpolation between squares with values. Thereafter, the wind at any time in a certain grid square can be estimated by multiplying the square's windiness ratio into the wind measured at that time by the central reference anemometer. Better still, the wind in the square can be deduced from simultaneous measurements by two or three continuous recorders in nearby squares, each weighted according to the anemometer's distance and also according to the windiness ratio of the square containing an anemometer. In other words, the wind speed (Ue) in a square whose windiness ratio is Re is estimated thus:

$$Ue = Re.\Sigma\ [Ui\{\exp(-0.1\ di^2)\}/Ri]/\Sigma\ [\{\exp(-0.1\ di^2)\}/Ri] \quad \text{m/s} \qquad (6.16)$$

where Ui is the wind measured at station i, which is at a distance di from the certain square, and for which the windiness ratio is Ri. The equation may be tested by using it to estimate the wind for a square where a measured value is already available. In this way, Johnson (1979: 437; 1982) showed that it is more accurate than other methods of interpolation which ignore the climatological information implicit in the windiness ratio.

An alternative to mapping windiness ratios is to map the ventilation in terms of the percentage of time when wind speeds exceed some specified

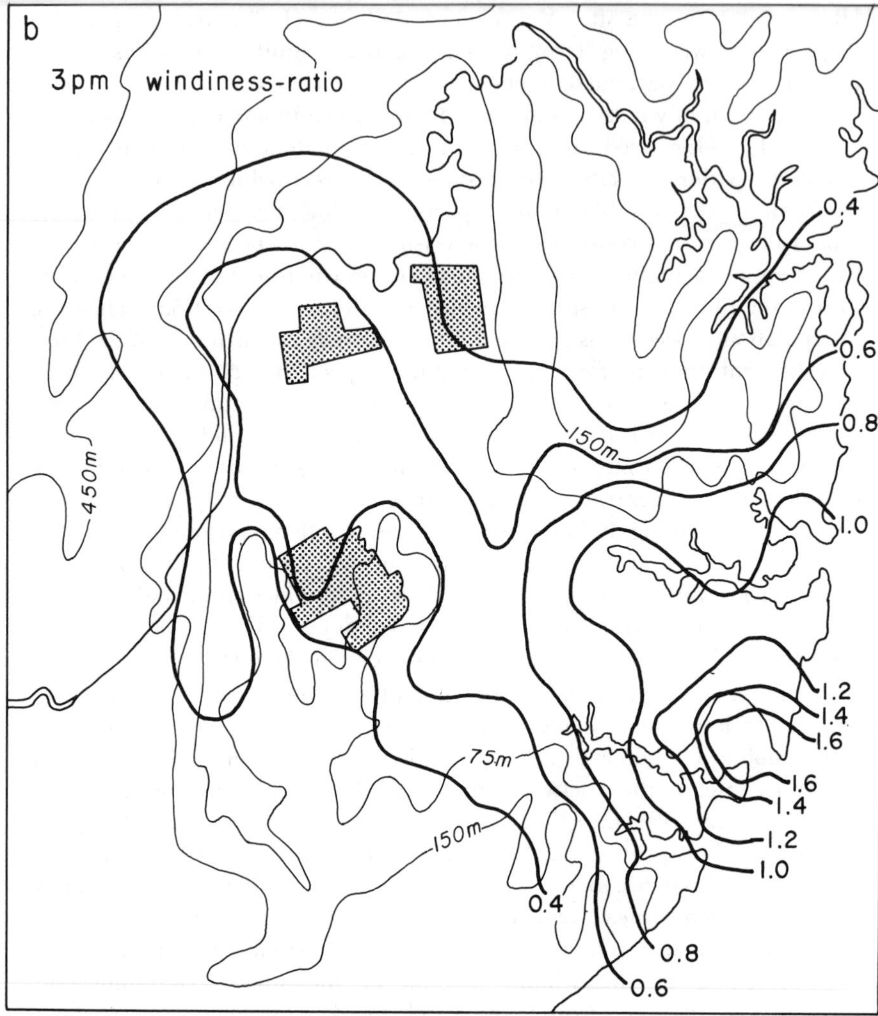

Figure 6.18 Isopleths of windiness ratio in the Sydney region (Keene 1981). The diagram shows an area of 35 × 40 km^2. The Hawkesbury River flows from the north (top centre) and then eastwards, whilst the Parramatta River flows eastwards to the sea (centre right). Shaded areas indicate alternative sites for a proposed new airport; the central one is seen to be slightly better ventilated than the other two

value. This is useful in assessing which localities are either susceptible to high concentrations of air pollution or, on the other hand, can provide wind power reliably. Wendland (1982: 423) took 3.6 m/s as the threshold value in surveying part of Illinois.

Extreme winds (see 'Estimating extreme winds', pp. 231–9) also vary with terrain. A rough rule for winds in a valley is that the extreme is 0.85

times that at a standard, open site, whilst hilltop winds are 1.25 times as fast. To estimate the extreme wind at a site from a frequency distribution of measurements, it is better to use a few measurements at that place and then fit a Weibull equation than to use a long record of measurements from another place nearby.

Selecting a site to collect wind energy

Obviously, any wind-energy collector should be located at the windiest place available. In general, one looks for places where there are frequently steep atmospheric-pressure gradients (e.g. in the path of intense low-pressure systems), in long sloping valleys parallel to prevailing gradient winds, and on high plains, exposed ridges and coastal sites. A ridge should preferably be either straight or concave athwart the prevailing strong winds. An isolated hill has the advantage that it benefits from the air's acceleration at the top from whichever direction the wind comes, whereas a mountain range or ridge offers this only when the wind is more than 60° from the axis of the range.

The strongest winds are offshore, but location of a wind-energy collector at sea would necessitate a construction capable of withstanding the worst possible storm. There would also be problems of corrosion, the danger to shipping, the need to transmit the power to shore, and so on.

A procedure for selecting a site is outlined in Note 4, pp. 336–7. The final choice would be made in the light of the following data: the frequency distribution of hourly mean wind speeds and directions; the frequency and speed of turbulence and gusts; the persistence of calms and high winds; the roughness and slope of the ground nearby; wind shear within the lowest few dozen metres of the atmosphere; the frequency of storms, lightning, hail, tornadoes and earthquakes; extremes of irradiance and temperature; precipitation (including snow and ice); and the risk of salt spray and blown dust.

WIND POWER

There is considerable energy in the wind at about 25 m height where most wind-energy is collected. The wind over the USA, for example, has enough power to generate 75 per cent of the electricity consumed. Over most of the American mid-west and in northern Scotland the wind-power available exceeds the solar energy. Already, wind-energy collectors on a single Hawaiian island produce about 2 MW of electricity, and in Tanzania there are studies of the use of wind power for electric light, cooling, grinding flour and pumping water. A single valley in Crete has thousands of wind pumps. So there is appreciable small-scale use of wind energy, especially in remote areas, along with great interest in the possibilities of large installations for

Table 6.10 Hypothetical example of errors involved in calculating the average wind-power

Wind speed range: m/s	*Mean of range: m/s*	*Fraction of time: %*	*Power: W/m²*
0	0	25	0
0.5–5	3	40	7[a]
5.5–10.5	8	30	98
11–17	14	5	88
	Total	100	193
Central value	8.5[b]	100	393[c]
Mean	4.3[d]	100	51[e]

Notes: [a]i.e. $0.40 \times 0.64 \times 3^3$; [b]i.e. $(0 + 17)/2$; [c]i.e. 0.64×8.5^3; [d]i.e. $0.25 \times 0 + 0.40 \times 3 + 0.30 \times 8 + 0.05 \times 14$; [e]i.e. 0.64×4.3^3.

generating grid electricity. Wind offers some alternative to fossil fuel as a source of power, and hence a reduction of the greenhouse gases which cause global warming.

The rate of collecting the kinetic energy of a wind of speed U by a collector of area A, is $d.A.U^3/2$, where d is the air's density (about 1.29 kg/m^3 at 15°C and 1000 hPa). So a wind of 10 m/s on to 10 m^2 contains about 6.5 kW. If the wind speed U is expressed in mph, the power is given by 0.058 U^3 W/m^2, if in knots then 0.088 U^3 W/m^2, if in km/h then 0.014 U^3 W/m^2, and if in metres/second then 0.64 U^3 W/m^2.

Estimates of the wind's power are always inexact on account of the fluctuations of wind speed and the cube dependence of power on the speed. An increase from 8 to 10 m/s almost doubles the available power. One consequence is that the power of a fluctuating wind cannot be calculated accurately from the average wind speed. This is shown in Table 6.10. The central value between 0–17 m/s provides an estimate of available power (393 W/m^2) which is much more than the estimate derived from grouping the winds into four ranges (193 W/m^2). On the other hand, the mean wind speed (4.3 m/s) yields only 51 W/m^2, which is far too low. The discrepancies are due to the fact that the average of cubes (the true answer) is greater than the cube of an average. This is another example of the principle discussed in Note 3 of Chapter 3.

The argument against using an average wind might be applied against taking average values within each of the 0.5–5 m/s, 5.5–10.5 m/s, 11–17 m/s ranges in Table 6.10. If there were an equal distribution of wind speeds in a category, the mean would yield an underestimate of the power. Fortunately, this is offset by the greater number of slower winds in each category, with an average less than halfway between the category limits. As a result, the procedure shown in the first part of the table is commonly accepted for calculating the total wind energy at a place.

The ratio of the true power (e.g. 193 W/m^2 in Table 6.10) to the value based on the mean velocity (e.g. 51 W/m^2), is a useful figure, allowing easy calculation of the true power from the mean wind speed. The ratio is 3.8 in the hypothetical case above. However, the ratio is not a constant. It depends partly on the general windiness. In Northern Ireland it was found to be about 8 for light winds, but 1–3 in winds of useful strength. Measurements from eight American sites indicate that the actual wind power is about five times the value calculated from the annual mean wind speed where the latter is about 1.5 m/s, but the ratio is only two where the speed is 8 m/s.

The ratio depends also on the period of the mean velocity. When the *daily* mean wind is 5 m/s, the ratio at ten places in India is within 0.9–1.4. The ratio for *monthly* mean winds of 5 m/s is within 1.0–3.4, i.e. somewhat more. Ratios are 1.7–4.2 for *annual* average winds of about 7.5 m/s.

Measured wind power

A wind fast enough to be useful as a source of power exceeds about 8 m/s, equivalent to over 300 W/m^2. There are spots with average winds above 8 m/s in most regions of New Zealand, and in the USA from Texas to Nebraska, in the Appalachian mountains, on the high plains of Wyoming, and on the north-west coast.

Wind speeds in Australia are generally much lower. Average power ranges from only 42 W/m^2 at Brisbane to 190 W/m^2 at Robe on the South Australian coast, and 450 W/m^2 at an exposed site on the north-west coast of Tasmania. The last figure is like the 460 W/m^2 measured at 10 m elevation near Auckland in New Zealand.

A happy circumstance is that solar energy and wind energy may be complementary in mid-latitudes. In many places, solar energy is dominant in summer and wind energy in winter.

Available power

Unfortunately, only a fraction of the wind's energy can be collected. A theory due to Betz in 1927 shows that one could collect only 59 per cent, even with a perfect machine. In fact, a modern wind-energy collector (WEC) extracts less than 45 per cent of the energy, and with older machines this 'power coefficient' is below 20 per cent. A wind pump in Delhi used only 9 per cent of the energy of the wind when it blew at 7 m/s. The difference between what is theoretically possible and what is practicable is partly due to the inability of a WEC to accelerate fast enough to capture gust energy.

The energy that can be captured is shown in Figure 6.19. This is a kind of cumulative wind-speed distribution curve (Figure 6.13). In this case, the

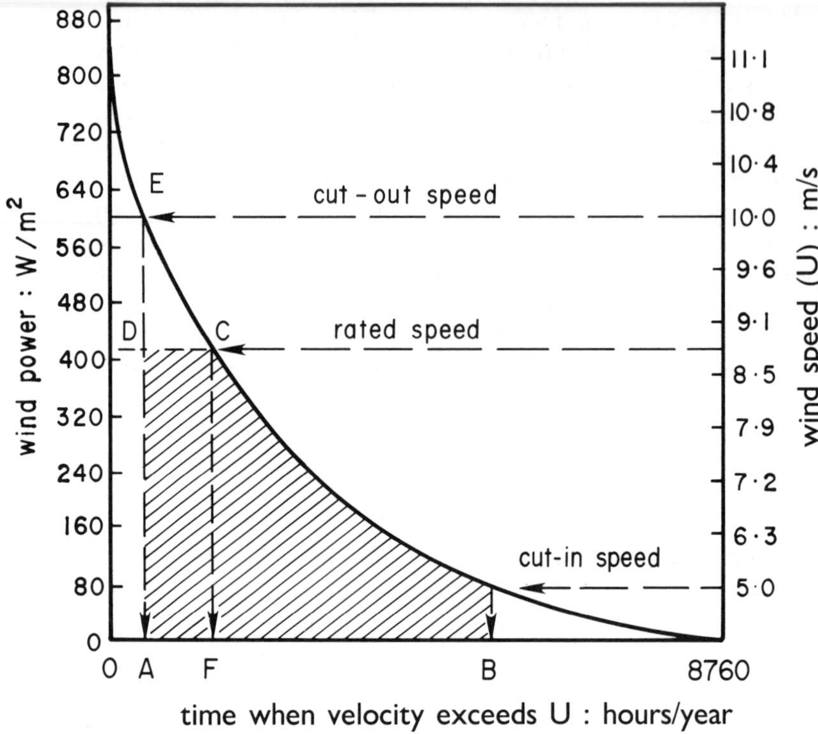

Figure 6.19 Hypothetical power-duration curve for a particular site and elevation. For instance, the distance OA represents the number of hours annually when the power of the wind exceeds an amount represented by the distance AE. (The cut-out speed is unrealistically low in this diagram, to save space). The speeds on the right vertical axis correspond to the powers shown on the left

linear vertical scale on the left of the diagram is expressed in units of 0.64 U^3, where U is the wind speed shown on the right. The shaded area represents the useful power, of which 59 per cent is potentially available for collection. It is expressed in units of kilowatt-hours each year, from each square metre swept by the WEC blades. The area is defined partly by the power-duration curve and partly by the characteristics of the wind-energy collector. Winds are too low below the 'cut-in speed', and the WEC operates fully only at the 'rated speed', which is the optimum speed for the particular design. Above that speed, the WEC sheds some of the available power to maintain a steady output. The WEC has to be shut down completely when winds exceed the 'cut-out speed' or 'furling point', to protect it from damage.

The cut-in wind speed depends on the WEC design, but typically is 3–5 m/s. The rated speed is usually about twice the local mean wind speed, and the cut-out speed is about 50 per cent or more above the rated speed, e.g. around 18 m/s. Houghton (1985: 935) gave an example of a cut-in speed of 4 m/s, a rated speed of 9 m/s and a cut-out speed of 20 m/s. Median values for 14 other WECs rated at 100 kW or more are 36 m for the blade diameter, rotating at 37 rev/min, with a rated wind speed of 13 m/s.

Golding (1955) compared the calculated power output with that actually obtained from a 13 kW WEC with an 8-m diameter rotor. In only one month out of eight was the difference less than 10 per cent, and for two months the calculation had an error exceeding 50 per cent. Part of the explanation lies in the following practical difficulties of determining the amount of power in the wind:

1 Incorrect specification of the anemometer height (e.g. the power at 10 m is overestimated by about 40 per cent if the anemometer is actually at 20 m) and the difference between the top and bottom of the circle swept by the WEC blades.
2 Insufficient number of ranges of wind speed in the equivalent of Table 6.5: too few ranges produce an overestimate, which may be 20 per cent with four of them, or 10 per cent with ten.
3 Variation of atmospheric density due to temperature or elevation
4 Insufficient duration of record, yielding an atypical sample.
5 Unsatisfactory anemometer accuracy, exposure, maintenance and use.

It is clear that any calculation of the available wind energy is likely to be only approximate.

7

PRECIPITATION

Water is essential to life. We need two or three litres each day simply to replace what is lost in breathing, sweating and body wastes. So its availability is a prerequisite for habitation, and people in an arid country like Australia live mainly where precipitation roughly matches the rate of lake evaporation for most months of the year. But the kind, intensity and frequency of rainfall affect many other aspects of modern life too – crop yields, traffic accidents, house design, hydro-electricity generation, the incidence of acid rain, tourism and so on. In this chapter, we will consider the formation of rain and its distribution in space and time, as well as features of dew and snow.

INTRODUCTION

The amount of water consumed in a country is related to its population's size and affluence. Someone of a poor nation manages on 15–40 litres daily (L/d), whereas people of developed countries use very much more. Consumption at Addis Ababa is 30 litres per head daily (L/hd.d), but it is over thirty times as much in the USA. The figure in Saudi Arabia is 130 L/hd.d, within Sydney about 400, in Cairo 540 and in tropical Darwin 900 L/hd.d.

The amount of domestic water consumed per *household* was about 3000 L/d in California in 1964, 850 L/d in the UK and 1440 L/d in Melbourne. The figure for Australia as a whole was 946 L/d in 1983.

There has been an increase of consumption due to more people and greater wealth. The amount per head in Adelaide and the total consumption in the USA have risen by 20 per cent per decade. The average water consumption in Sydney was 520 L/hd.d in 1977 (45 per cent being for domestic use and 25 per cent for industrial) but is expected to be 740 L/hd.d by the year 2000. The demand has risen more than tenfold in the USA between 1900 and 1980, and almost fourfold in the USSR between 1965 and 1985.

The figures become much larger if one includes each person's share of the water used in irrigation agriculture, in countries such as Israel, Australia and Japan. There, about 75 per cent of all water used is consumed in irrigation. The national average in Australia in 1977 was about 3000 L/hd.d, largely due to what is used for crops. In the USA, the amount used in irrigation has increased from 2700 L/hd.d in 1950 to 4500 L/hd.d in 1980. In different units, the annual total amount of water per head in the UK in 1979 was 225 m^3, but 456 in France, 707 in Japan, 939 in Sweden, 1242 in Australia and 2720 m^3 in the USA.

The global average consumption of water expected in the year 2030 will be 2900 L/hd.d (instead of 1600 L/hd.d in 1975), of which 83 per cent will be for irrigation. Such a global average divided into the amount of fresh water available, assumed constant, implies a maximum world population of 10–25 billion if the water were used equably. This is only a few times the present population. However, global warming may increase the amount of river-flow available.

Increasing demands on the available water supplies have led to bigger dams and a larger fraction of river-flow being impounded. So the difference between normal flows and rates of consumption is declining, reducing the safety margin in times of drought. Also, there is a climatological problem resulting from the common practice of using dams to generate electricity as well as to store water for irrigation. Farmers want the water released in summer, whereas electricity consumers want the water saved for power in winter. The management problem is to compare the risk of low rainfall in summer with that of cold spells in winter.

Another example of the value of precipitation data concerns the use of rain-water in a city like Jakarta. Much of the domestic water there is pumped from wells and needs to be sterilised by boiling. This consumes considerable kerosene, whose combustion creates air pollution. An alternative would be to drink pure rain-water, collected from rooftops into tanks, provided there are no problems of acid rain or disease. If the average household contains five people and needs, say, 200 L/d of potable water, a population density of 5000 per km^2 would require a monthly rainfall of only 6 mm, if all the rain could be collected and used. Assuming 10 per cent of the area is covered by roof, the minimum rainfall would need to be 60 mm/mo, which is exceeded in Jakarta in 10 months of the average year. So a substantial contribution to reducing air pollution might be made in this way.

Many other problems can be tackled if we have sound measurements and proper analysis of the data. Examples are the design of water storages, the planning of drainage channels for flood mitigation, the estimation of crop growth, the assessment of legal claims for damage due to wet weather, and so on. In what follows next, we deal chiefly with rain formation and patterns of precipitation. The measurement of rainfall was discussed in Chapter 2, pp. 51–61.

PRECIPITATION PROCESSES

The processes of rain formation govern the variations of precipitation in space and time, discussed later (see pp. 264–96). Aspects to consider here include the atmospheric moisture, the formation of clouds and the creation of rain.

Atmospheric moisture

The amount of water vapour in the atmosphere is a measure of what is available to become cloud, with the chance of subsequently becoming rain. The amount turns out to be strongly correlated with the surface atmosphere's dewpoint temperature Td, as follows:

$$\ln W = a.Td + b \tag{7.1}$$

where a and b are constants and ln W is the natural logarithm of the 'precipitable water'. The latter is the depth of water formed if all the vapour above were condensed at the surface. In practice it is determined by calculations based on radiosonde measurements of the water content of the air at each height above the ground. Values of the constants in Eqn (7.1) are given in Table 7.1, whose median values in the bottom row yield the following:

$$\ln W = 0.06\ Td - 0.02 \tag{7.2}$$

Such an expression works best in winter away from the coast, for monthly rather than daily values, and at high latitudes. The effect of latitude is shown by the correlation coefficient between estimates from the equation, and actual measurements. It is 0.89 at New Delhi (at 29°N), but less than 0.25 at Madras, at 13°N. One reason for a low correlation at low latitudes and near the coast is that there the surface air does not represent the whole depth of the atmosphere, but only the sea-breeze.

It follows from Eqn (7.2), and from the roughly exponential dependence of saturation vapour pressure on dewpoint (see Note 6, Chapter 2), that the precipitable water W (cm) is given approximately by [0.15 e], where e is the water-vapour pressure (hPa) at screen height. However, such an expression can involve errors of 30 per cent in the USA, unless the constant is adjusted to 0.16 in summer and 0.19 in winter.

Measurements of the precipitable water around the globe are shown in Figure 7.1. This implies a roughly similar pattern of surface atmospheric water-vapour pressure and hence of the dewpoint. For instance, 45 mm in Figure 7.1 (i.e. 4.5 cm) means a water-vapour pressure of 30 hPa (i.e. 4.5/0.15), and hence a dewpoint of 24.1°C (see Note 7, Chapter 2).

Above sea level, there is a notable reduction in the amount of water vapour available for forming clouds and then rain (see Figure 3.11).

Table 7.1. Published expressions like that in Eqn (7.1) for the natural logarithm of W (the precipitable water vapour in the atmosphere, in centimetres of water equivalent), in terms of the dewpoint temperature Td

Author	*Place*	*Period* [a]	*Expression*	*Correlation* [b]
Reitan 1963: 778	USA	Month	0.061 Td + 0.11	0.98
Bolsenga 1965: 431	N. Hampshire	Day	0.077 Td + 0.12	0.85
Smith 1966: 726	North hemi.	Year	0.07 Td + 0.03	–
Ojo 1970[b]: 226	Nigeria	Month	0.051 Td + 0.27	0.93
Atwater and Ball 1976	USA	Month	0.071 Td – 0.02	–
Tuller 1977: 209	Auckland	Year	0.07 Td – 0.24	0.88
Rao *et al.* 1980: 1080	0°S	Year	0.024 Td + 0.91	0.49
	20°S		0.030 Td + 0.39	0.56
	40°S		0.063 Td – 0.18	0.92
	60°S		0.069 Td – 0.09	0.96
	80°S		0.037 Td – 0.40	0.88
Viswanadham 1981: 8	0°S	Month	0.054 Td + 0.22	0.62
	25°S		0.044 Td + 0.17	0.75
	40°S		0.061 Td – 0.16	0.72
	60°S		0.057 Td – 0.11	0.94
Iqbal 1983: 93	Canada	Season	0.055 Td – 0.05	–
McGee 1984: 283	S. Africa	Month	0.079 Td – 0.08	0.93
	Medians		0.06 Td – 0.02	

Notes: [a] Period of averaging the measurements.
[b] Coefficient of the correlation between observed values and estimates from the given expression.

Approximately, the water-vapour pressure halves for each 7 km extra elevation. This compares with the halving of total atmospheric pressure for about each 5 km.

The connection between surface dewpoint and rainfall is exemplified by observations at Harare (Zimbabwe) that rain occurs in December–March on 75 per cent of days when the 9 a.m. dewpoint exceeds 15°C, but on only 31 per cent of days with a lower dewpoint. The rainfall is caused by the high dewpoint, since the increase of surface humidity is *followed* by rain.

Formation of cloud

The link between the surface dewpoint and the water content of the atmosphere suggests a similar relationship between surface conditions and the occurrence of cloud. Indeed, a relationship like this was discussed in 'Relationships', pp. 138–43. However, the correlation coefficient in Figure 4.13 is low, implying that the occurrence of cloud depends on several other factors in addition. This is confirmed by the differences between the patterns in Figures 5.16 and 7.1.

Cloud is the result of cooling air below its dewpoint temperature. This

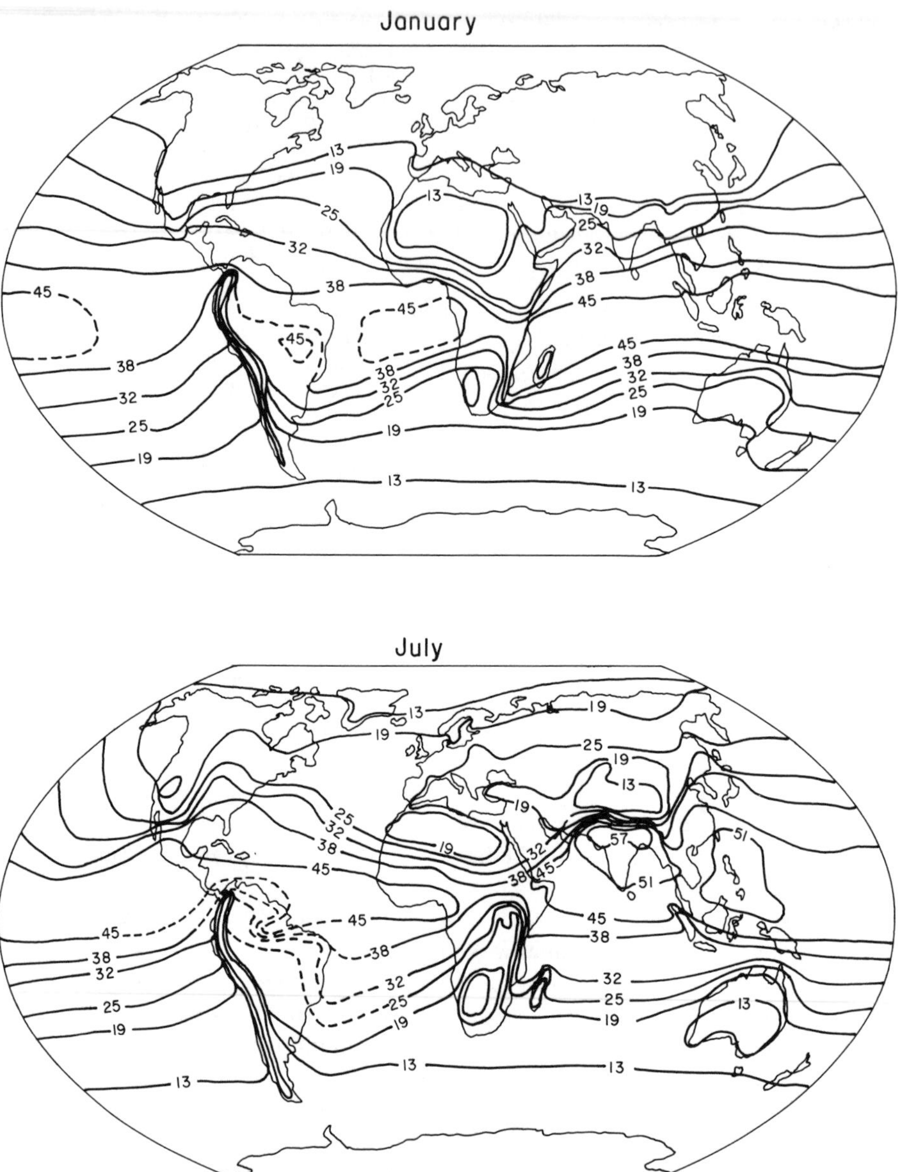

Figure 7.1. Global patterns of precipitable water (mm) in January and July (by permission of Dr. S.E. Tuller)

happens either by contact with a cold surface (e.g. a cold ocean), by admixture with an almost saturated cold air-mass, or by uplift. The last, uplift, is the chief cause of atmospheric cooling and is itself the consequence of either (i) convergence of low-level winds, (ii) the wedging action of an advancing front between different air-masses, (iii) convection, or (iv) hills. Each will now be considered.

Low-level convergence

Convergence near the ground can be either on a large scale, as at the Inter-Tropical Convergence Zone near the equator, or on a small scale. One example of the latter is the uplift over small islands caused by the confluence of sea-breezes from opposite sides. Similarly, colliding sea-breezes from opposite sides of the narrow strip of land north of Auckland in New Zealand lead to a line of clouds and showers. There is a convergence of katabatic winds at night in valleys of the Andes, and of nocturnal land-breezes from either side of the Straits of Malacca. Also, much of the cloud in West Africa is due either to large-scale convergence induced by the Easterly Tropical Jet above, or to coastal convergence resulting from deceleration of onshore winds on encountering the land's friction.

Frontal uplift

The frontal zone between different air-masses is typically slanted, with a wedge of cold air advancing beneath the warmer air (if it is a cold front) raising the retreating air-mass. The result is a complex sequence of clouds, including an extensive band of stratus cloud above and along the frontal zone (Linacre and Hobbs 1977:130).

There is less frontal uplift at the lowest latitudes, where adjacent air-masses are equally hot and often equally moist, with little contrast of conditions. Frontal cloud is formed chiefly in mid-latitudes.

Convective uplift

Convection due to buoyancy of the lower layers of the atmosphere is due to their being either warmed or moistened, or both. The density of air at 20°C is 3.4 per cent less than at 10°C. If 5 per cent of the dry air is replaced by water vapour (moist air is not dry air *plus* water vapour) the respective molecular weights (i.e. 18 units for water vapour and 28 for air, which is mostly nitrogen) mean a reduction of density by 1.7 per cent.

On the topo-scale or meso-scale, convection is due to solar heating of the ground in a spatially irregular fashion, due to differences of albedo and wetness. This leads to distinct columns of rising air ('thermals'), surmounted by separate cumulus clouds. It takes place especially in fair weather after the ground has been moistened by previous rain, and is seen most commonly at about noon at low latitudes. However, the process is self-limiting (i.e. there is negative feedback), since the cloud which is created cuts off further solar heating of the ground and so lessens uplift and the formation of more cloud.

Widespread cloud inland may be caused by the blowing of polar winds in mid-troposphere over warm moist air beneath, from lower latitudes. This occurs in the American mid-West, for instance. Again it leads to instability and hence extensive, deep cloud, with subsequent tornadoes and thunderstorms (see 'Causes of wind', pp. 210–1).

Convective ascent also occurs above a warm sea. There is a high correlation between cloudiness and sea-surface temperature. The convection may lead to a tropical cyclone (see 'Causes of wind'), where cloud extends as spiral bands on a synoptic scale. This is associated with particular latitudes and parts of the oceans.

Another process leading to maritime cloud at low latitudes is the nocturnal cooling of the top of a moist, mid-troposphere layer by the loss of long-wave radiation to space. The resulting super-adiabatic lapse rate creates instability and hence convective uplift.

Hills

Hills cause convergence by obstructing the lower winds. Air may be forced upwards, causing cooling to dewpoint perhaps, so that 'orographic' cloud results. This is commonly stratus. However, if there is a conditionally or convectively unstable atmospheric layer, orographic uplift may increase atmospheric instability to such a degree that cumulo-nimbus cloud forms.

Data from Japan show that there is an increasing number of days of fog (i.e. cloud at the ground) at higher elevations up a mountain, as far as about 1500 m above sea level. At yet higher elevations there is a *decrease* of cloud.

Formation of rain

The link between cloudiness and rainfall is exemplified by the data in Table 7.2 from Darwin. The climate there is monsoonal, i.e. there is a switch around of the prevailing wind twice a year. Moist north-westerly winds from August–December come from equatorial seas, whilst east or south-east winds at other times come from Australia's dry interior. Increasing extra-terrestrial radiation from July–December (see Note 12, Chapter 3) promotes convection, which leads to cloudiness and rain.

The association between cloudiness and rainfall is indicated by the similarities between Figure 5.16 and a global map of rainfall, and between the curves in Figure 7.2. There are similarities between the global

Table 7.2. Effect of solar position (shown by the monthly mean extra-terrestrial irradiance, Qa) on the atmospheric pressure p, and hence on the direction of the prevailing wind, and thus on the dewpoint of the incident air-mass which affects the cloudiness and precipitation at Darwin, 12°S

Month	*Jan*	*Feb*	*Mar*	*Apr*	*May*	*Jun*	*Jly*	*Aug*	*Sep*	*Oct*	*Nov*	*Dec*
Qa[a]	459	451	429	387	347	324	330	366	408	441	454	458
p[b]	4.4	4.5	5.5	7.7	9.4	10.8	11.5	11.2	10.6	9.3	7.0	5.4
Wind[c]	W	W	W	E	E	E	E	NW	NW	NW	NW	NW
Dewpoint °C	18	16	16	6	1	1	0	0	2	4	10	14
Cloud[c,d]	5.1	5.2	4.7	3.8	2.8	2.3	1.9	1.8	2.2	2.6	3.9	4.5
Rain: mm/mo	411	314	284	78	8	2	0	1	15	49	110	218
Raindays	18	16	16	6	1	1	0	0	2	4	10	14

Source: Gentilli 1971: 306.

Notes: [a] In units of W/m^2, from Table 8.3.
[b] In units of hPa, after subtracting 1000 hPa.
[c] Winds and cloudiness values refer to 3 p.m. measurements when the surface is linked by convection to the gradient wind conditions.
[d] In units of oktas.

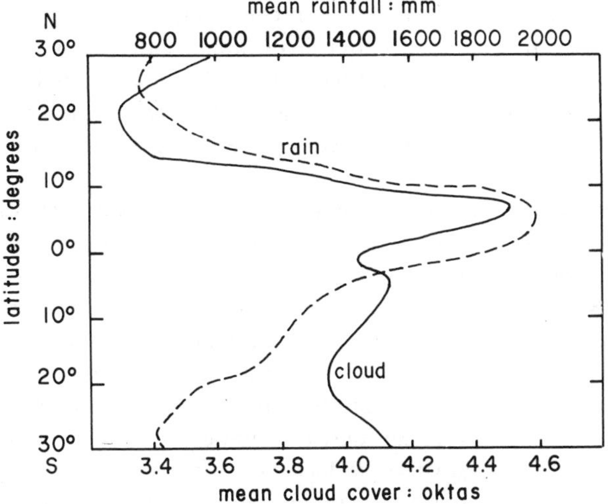

Figure 7.2. Variations of cloud and rain with latitude (Atkinson 1971: 6–7).

latitudinal variations of mean cloudiness and annual rainfall, a maximum of each occurs near the equator, and minima near 25°N and 25°S. However, cloudiness increases with latitude above 50°N and 50°S, whereas rainfall decreases. So the presence of cloud is no guarantee of rain.

Cloud is a necessary but insufficient condition. Precipitation results only when several criteria are all met – there must be enough cloud droplets of suitable size, sufficient 'condensation nuclei', ample cloud depth and appropriate temperature conditions. A deficiency of nuclei might perhaps be remedied by the artificial seeding of suitable clouds, e.g. stable orographic clouds with cloud-top temperatures between −15°C and −21°C.

There have been many attempts at cloud-seeding to enhance rainfall since the pioneering work of Schaefer in 1946. This involves putting powdered chemicals into clouds to trigger the formation of raindrops from cloud droplets. Vonnegut successfully used silver iodide for this purpose in 1947. Langmuir carried out Project Cirrus in New Mexico during 1949–50, seeding clouds each Tuesday to Thursday, and apparently stimulating a weekly cycle of rain. Experiments started in Australia in 1947 and there was subsequently much study of the conditions which lead to worthwhile increases of rainfall. However, research work was terminated in Japan by 1974 and in Australia in 1982, apart from a continuing experimental program in Tasmania, Australia's wettest state. The difficulties in Australia are the infrequency of suitable clouds, the natural variability of rainfall and the high costs of airplane operation.

The position in the USA was summarised by the American Meteorological Society (AMS 1981: 87) as abbreviated here:

> Precipitation amounts from certain cold-orographic cloud systems apparently can be increased under favourable conditions with existing technology in the western United States. However, possibilities of decreases in orographic precipitation also exist under certain conditions, and indiscriminate seeding of all orographic clouds is not recommended.
>
> Strong statistical evidence bolstered by cloud measurements has led to general acceptance of the results of 12 years of experiments with winter cumulus clouds in Israel, which indicate rain increases of about 15% in the target area. More guarded success is ascribed to statistically determined increases in precipitation from convective rainbands in winter cyclonic storms near Santa Barbara, California.
>
> Attempts to increase precipitation from warm-season convective clouds have indicated local increases under certain circumstances and, under other circumstances, local decreases.
>
> There are indications that precipitation changes, increases or decreases, may occur at considerable distances from the primary seeded areas.
>
> In all cases where indications of precipitation increases have been suggested, confirmatory experiments are required before any of the technologies can be considered scientifically proven.
>
> (American Meteorological Society 1981: 87)

Note that cloud-seeding has sometimes led to an actual reduction of rainfall. Only in Israel has an encouraging 13–18 per cent increase of rainfall been clearly demonstrated. There the biggest effect on rainfall is found 35–50 km downwind of the treated area. The increase is probably due to an extension of the duration and area of rainfall, rather than to higher intensity.

More recently, K. Bigg has analysed past research in Australia which seemed to show a steadily declining effectiveness of cloud-seeding above a catchment, compared with rainfalls in a nearby unseeded control valley. He proposes that the decline resulted from an eventual effect in the *unseeded* valley – bacteria created by cloud-seeding in the target valley act as raindrop nuclei and persist for months, propagate widely, and later reach the control area. So the relative decline may be due to unexpected effectiveness in the control valley, rather than ineffectiveness in the target valley.

However, cloud seeding seems less important in controlling rainfall than the cloud's liquid water content per unit volume, and that depends on the extent of the prior uplift. In particular, rainfall depends on the existence of numerous cloud droplets larger than 0.3 mm.

The modal raindrop size is 1.8 mm when precipitation is at a rate of 12 mm/h, 2.1 mm when 50 mm/h, 2.4 mm when 90 mm/h, 2.0 mm when 146 mm/h, and 1.2 mm when above 170 mm/h. The sizes are remarkably uniform, because smaller drops either aggregate or hardly descend and larger ones disintegrate as they fall.

Categories of rainfall

Rainfalls differ with regard to their time of occurrence, extent, intensity and duration. The various kinds of precipitation are usually classified according to the process chiefly responsible for the preceding *cloud*. So rainfalls are labelled with reference to the prime importance of either low-level convergence, cyclonic activity (e.g. frontal movement), convection or hills. Such a classification has two disadvantages. First, it is rather indirect to label rain according to the process of forming something else, i.e. cloud. The formation of cloud is not synonymous with forming rain, as already explained. Second, the procedure obscures the fact that cloud (and hence rain) often results from a combination of convergence, orography and/or convection, not a single factor alone. However, for convenience, we will abide by the usual convention, considering in turn the kinds of rainfall consequent on convergence, hills, fronts, convection and tropical cyclones, respectively.

Convergence rainfall

Sea-breezes, and nocturnal winds which are due to horizontal temperature-differences (see 'Causes of wind', pp. 203–7), cause convergence and clouds on the topo-scale or meso-scale, and then rain, which tends to be brief. Large-scale convergence, as over Africa, leads to prolonged rainfall with occasional dry spells of a few hours or days.

Orographic rainfall

An interesting mechanism that explains the enhanced rainfall on hills up to about 1000 m is the 'seeder-feeder' process, first proposed by Tor Bergeron in 1968. The process involves a high layer of cloud that provides 'seed' rain even in the absence of the hills, and then that rain falls through a lower layer of orographic cloud which 'feeds' the precipitation, multiplying it. The same mechanism probably operates when snow falls on high ground.

Frontal rainfall

Frontal rains tend to occur at mid-latitudes and continue as the frontal zone is passing overhead. Thus, if the slanting zone extends 300 km horizontally and happens to advance at 40 km/h, there is rain for some 8 hours at any point in the path of the front. Such rainfall is usually only light since the rate of uplift in the example would be only 0.2 m/s, i.e. about 5 km in 8 hours. If the cloud's liquid water content is 0.5 g/m^3, for instance, and half becomes rain, uplift by 0.2 m/s implies a rainfall intensity of merely 0.4 mm/h.

Convective rainfall

Heavier rainfalls occur when fronts trigger convective uplift. An example is the stimulation of convective rainfall by the lake-breeze front from Lake Victoria in East Africa. Convective rainfall occurs also either *behind* a cold front, on account of the movement of cold, moist air over a warm surface, or *ahead* of a warm front.

Convective clouds may involve rapid uplift, e.g. as fast as 30 m/s. They tend to arise where the air is warm and holds appreciable moisture, near the equator especially. The rainfall may consist of a third of the water vapour that flows into the cloud, so that over 25 mm usually falls at the centre of a tropical storm.

The rain falls as heavy showers. Typically, 30 of the convective showers, out of about 314 which occur in Jakarta each year, involve at least five minutes of rainfall at a rate exceeding 60 mm/h. Such storms are usually associated with thunder and lightning, resulting again from the rapidity of the uplift caused by intense thermal convection. A convective storm like that lasts only half an hour or thereabouts, and covers about 20–50 km^2. The discharge of the instability over that area (i.e. within that atmospheric 'convection cell') triggers similar disturbances in adjacent cells. This all

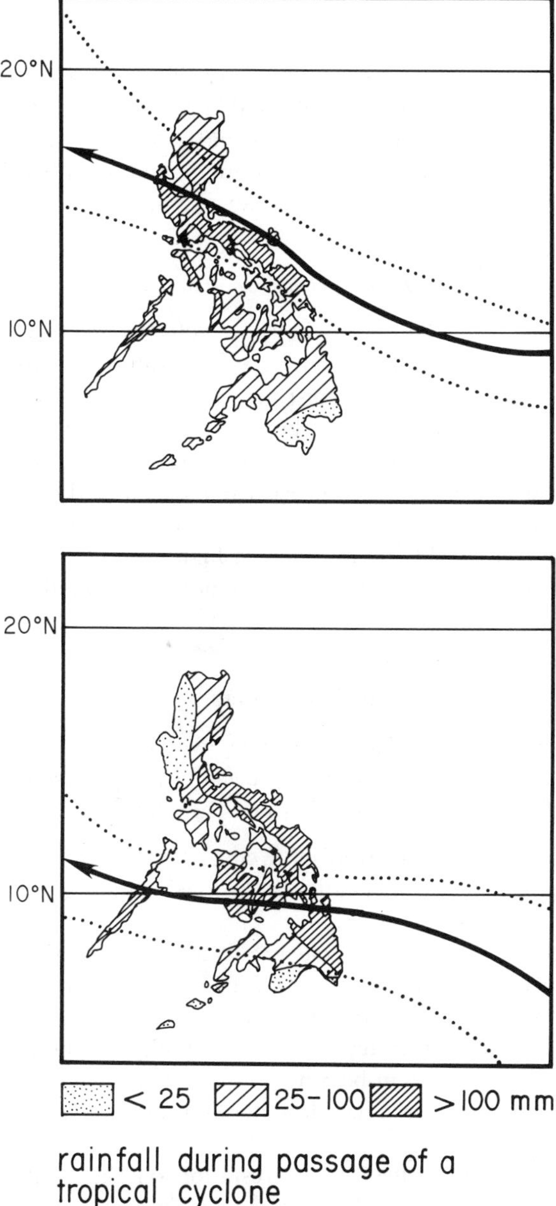

Figure 7.3. Average rainfalls per cyclone over the Philippines, associated with particular paths of the tropical cyclones (Philip. Meteor. Agency, by permission of Dr Aida Jose of the Philippines Atmospheric Geophysical and Astronomical Services Administration)

leads to a pattern of rainfall which is spotty in both time and space.

Fair-weather cumuli, especially at temperate latitudes, cause brief and isolated showers of a more modest intensity. They tend to occur in the afternoon after there has been prolonged solar heating of the ground, followed by evaporation, convection and cloud growth. As a result, there is a strong correlation between monthly rainfall and monthly mean daily maximum temperature in Mozambique, for example.

Tropical-cyclone rainfall

A tropical cyclone (see 'Causes of wind', pp. 209–10) involves both intense thermal convection and low-level convergence, resulting in extremely rapid uplift of a continuous supply of moist air from the sea beneath. Rainfalls may be at a rate of 30 mm/d at 400 km from the eye, above 300 mm/d at 100 km, with even higher intensities at 25 km radius. The precipitation is greater over land than over the sea, because of the added convection from the hot ground in summer and the extra uplift caused by hills. Australian tropical cyclones may yield 700 mm/d onto high ground. The passage of a single tropical cyclone nearby brings rain to most of the Philippines (Figure 7.3).

Surface effects

It seems plausible that surface conditions which influence evaporation thereby affect the subsequent rainfall, especially in the case of meso-scale convective precipitation. The following evidence can be offerred:

1 There is more rainfall downwind of a city.
2 There is a slight increase of cloudiness observed over large irrigated areas in Khartoum and Texas.
3 There is a relationship between sea-surface temperature near Sydney and the rainfall in various suburbs (Figure 7.4), the dependence being weaker inland. However, the connection may be indirect – rather than the sea-surface temperature governing the rainfall, both may be influenced by the onset of onshore, rain-bearing winds. This would explain why changes of sea temperature take place about a month *after* changes of the inland rainfall.
4 In addition, a possible effect of deforestation on the local rainfall has been discussed at least since the time of Gay-Lussac (1778–1850). It has been suggested that forests would increase rain because of more evaporation from the dark and rough surface, or that an increase of summer precipitation by 10–20 per cent would result from belts of trees planned for the eastern USSR. Likewise, it was widely held in the USA in the last century, that 'rain follows the plough', i.e. that planted crops

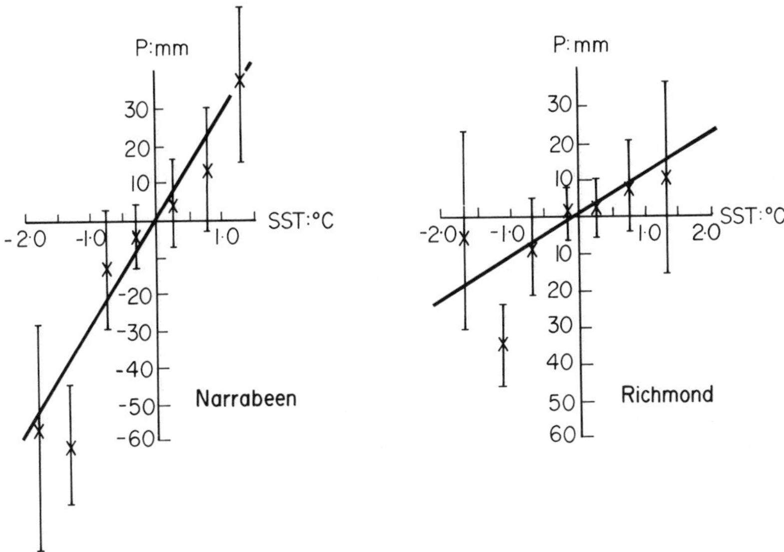

Figure 7.4. Effect of the sea-surface temperature at 3 km offshore from Sydney on rainfalls at 0.3 km and 50 km inland, i.e. Narrabeen and Richmond, respectively (Hirst and Linacre 1978: 326, by permission of the Australian and New Zealand Association for the Advancement of Science)

and trees increase precipitation, which increases vegetation, and so on in a virtuous spiral. However, there is no evidence to support this. A study of an area near Uppsala did show higher rainfalls over forested areas, but that may simply reflect the tendency for forests to be at higher elevations. More recently, from 1949 to 1980, a study in China involved comparing the changes at three climate stations within about 150 km of each other whilst the area around the lowest station had the forest half removed. The rainfall apparently decreased by 4 per cent, i.e. by about 55 mm/a, compared with that at the control stations . But such a small reduction due to deforestation might be fortuitous, or caused by stronger winds over the rain gauge once trees were cleared away. Furthermore, studies in Oklahoma have shown that fair-weather cumulus is *decreased* over dense tree cover, indicating *less* rainfall.

Despite such arguments, the likelihood is that surface conditions have little effect on rainfall, except in special circumstances, for at least four reasons:

1 Variations of surface conditions are usually not felt beyond a few hundred metres from the ground, which is much below the level at which rain is formed.

2 Trends of rainfall in 29 catchments in India during the decade after the removal of forest showed a decrease of rainfall in 10 cases and an *increase* in the other 19.
3 The sequence of evaporation, advection, condensation and precipitation normally takes several days. By that time the wind has separated the place of rainfall by some hundreds or thousands of kilometres from the place of the original evaporation. So 90 per cent of the rain on to the Mississippi basin, for example, comes from ocean surfaces far upwind.
4 Occasional flooding of either Lake Eyre (whose area is 9,300 km^2 in a desert region of South Australia) or of the Salton Sea in California does not cause heavier rain than usual.

Only in circumstances like those of the huge Amazon Valley, where high temperatures occur with little wind and heavy rainfall, is precipitation linked to local evaporation. And perhaps bands of vegetation about 100 km wide across semi-arid areas would stimulate some convective rainfall.

SPATIAL VARIATION OF RAINFALL

The annual amount of precipitation depends chiefly on the geography, i.e. the latitude, distance from the ocean, elevation and shape of the terrain. As regards latitude, the global annual precipitation at 10-degree intervals from 80°N to 50°S is 350 mm/a at 80°N, 350, 480, 590, 530, 600, 820, 1920, 1950 mm/a at the equator, 1710, 750, 660, 940, and 1160 mm/a at 50°S. Figure 7.2 shows that there are maxima at about 5°N (near the thermal and radiation equators, see 'Estimating screen temperature', pp. 72–3 and Linacre 1969a: 4), as well as at 45 degrees, corresponding roughly to the Inter-Tropical Convergence Zone and to the polar fronts of the global circulation, respectively. Less precipitation occurs near the two Tropics, where anticyclones are most common, and at the highest latitudes, where the precipitable water in the atmosphere is least, on account of low temperatures (Figure 7.1).

The effect of the distance from the sea is seen in Figure. 7.5 and is a common feature of isohyet patterns, especially where a mountain range shelters the inland. However, the effect depends on the direction of the prevailing wind. Even a place near the coast receives rainfall as though well inland if the wind comes from the interior.

Effect of elevation

Increased orographic uplift on windward slopes enhances the likelihood of clouds dense enough for rain to form. Examples of the increase of annual rainfall with elevation are given in Table 7.3. The increase appears to be greater at low latitudes, presumably because the warmer air can hold more

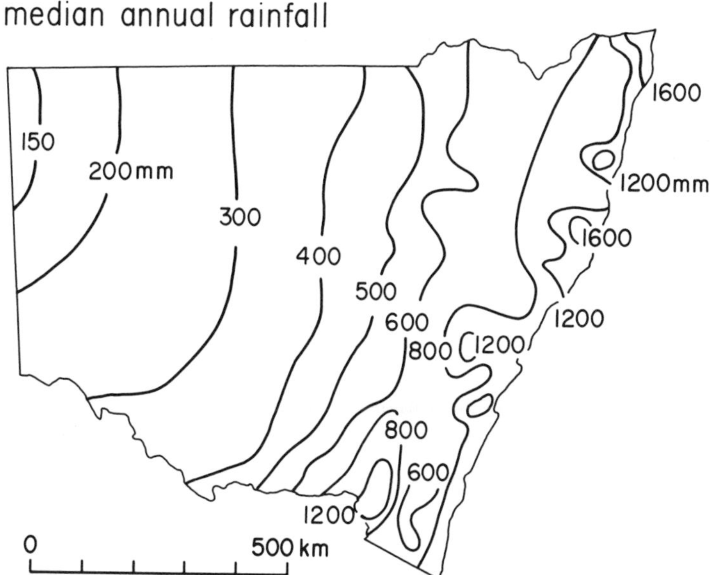

Figure 7.5. Median annual rainfalls in New South Wales during 1941–70 (Edwards 1979: 1)

moisture. Also, the increase with elevation is more consistently expressed as a percentage per 100 m than as an absolute increase, since values of the former have the lower coefficient of variation in Table 7.3. One infers that rainfall is not increased additively but *multiplied* by hills, in accord with the seeder-feeder mechanism (see previous section, 'Precipitation processes').

The absolute increase is more in summer than in winter, at least in Georgia (USSR), i.e. 0.4 mm per annum.metre instead of 0.2 mm/a.m. The reason is that hills particularly enhance the heavy showers of summer.

The gradient of rainfall with elevation depends not only on the latitude and season, but also on the orientation and on the elevation itself. The gradient increases with height, especially in winter at mid-latitudes, where cloud is not due to thermal convection. But there is not an increase of rainfall at the highest levels, because of the reduced water content of cold air. For example, there is 4000 mm/a at 1000 m on the windward side of the equatorial Andes but less than 600 mm/a near the peaks. Also, the gradient on Mt Kilimanjaro is 0.6 mm/a.m between 9 and 911 m and 1.9 mm/a.m between 813 and 1479 m, but only 0.7 mm/a.m between 1463 and 1829 m. Therefore, it can be seen that there is always an elevation on any high mountain, at which rainfall is at a maximum (Figure 7.6).

The height of maximum rainfall tends to be greater at high latitudes (see Table 7.4) and in dry regions. The maximum is at 1000 metres where rainfall is 6000 mm/a in the tropics, but at 2500 m if 1500 mm/a. And it is lower in winter.

Table 7.3. Effect of additional elevation on the annual rainfall

Place	*Approx. latitude: degrees*	*Elevation: metres from*	*to*	*Rainfall: mm/a from*	*to*	*Gradient mm/a per m*	*% per 100 m*	*Source*
Sweden	65	–	–	–	–	0.1	8	Smith 1979: 170
Scotland	55	9	333	–	–	–	7	Smithson 1969: 372
UK	53	0	500	714	1924	2.4	18	Ballantyne 1983: 379
and six other cases								
Germany	50	150	850	580	1000	0.6	7.6	Hann 1897: 303
		50–70 latitude		medians		0.6	9	
Washington	47	60	900	880	2450	1.9	11	Henry 1919: 38
Alps	47	600	2600	900	1800	0.5	3.3	Jones 1983: 248
and four other cases								
Wyoming	43	–	–	–	–	0.5	3	Schumaker *et al.* 1984: 154
and five other cases								
Arizona	34	0	2100	127	508	0.2	5.7	Osborn 1984: 1860
		30–50 latitude		medians		0.6	5	
S. Africa	29	10	925	69	229	0.2	12	Henry 1919: 34
Assam	28	–	–	–	–	8	11	Beckinsale 1957: 82
and ten other cases								
Mt Kilimanjaro	3	274	1107	1309	1959	0.8	4.8	Nieuwolt 1974: 18
		0–30 latitude		medians		3	11	
		0–70 *Coefficient of Variation*				1.3	0.69	

Effect of landform

The pattern of rainfall depends partly on the shape of the terrain. As an example, 88 per cent of the spatial variation of precipitation in western Colorado can be explained statistically in terms of the shape, defined by the elevation above sea level, the difference between the elevations of the highest and lowest places within 8 km, the number of degrees azimuth with land less than 300 m above the selected location, and the direction of the greatest openness to the sky. The effect of landform in the Sydney region can be seen in Figure 7.7, where the isohyets closely resemble the height

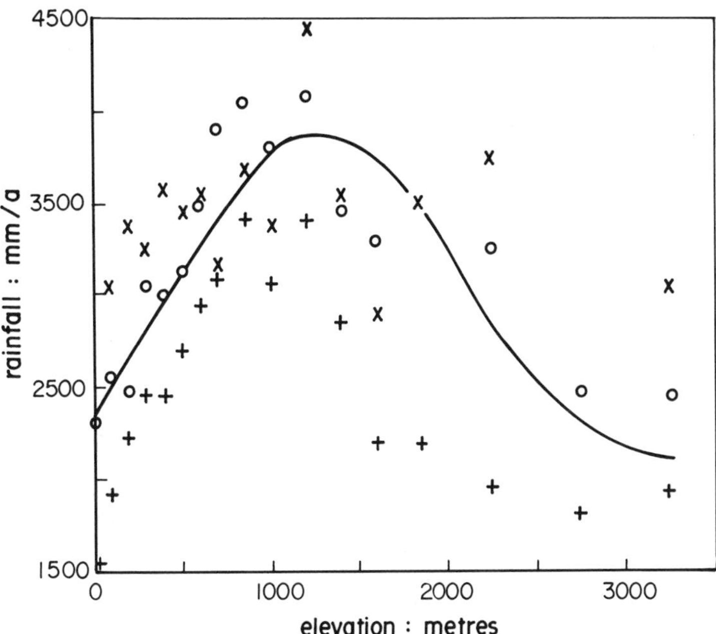

Figure 7.6. Effect of elevation on rainfall in Java (de Boer 1950: 425). The symbol × relates to data from west Java, o to central Java, and + to east Java.

contours and reveal considerable local variation. For instance, the annual rainfall is 750 mm/a and 1500 mm/a, respectively, at two places only 22 km apart. Parallelism of isohyets and contours has been found in British Columbia also.

Sometimes an increased rainfall is measured just in the lee of the peak, partly on account of the downflow of wind into any rain gauge there. Also, there may be bands of increased rainfall on the lee side, due to eddies induced by the landform. There is often greater calm in the lee, allowing the rain to settle into a gauge and be measured, whereas the stronger wind on the approach to the hill reduces the catch (see 'Measurement of rainfall', pp, 56–7).

More spatial variability is found with local convective storms at tropical latitudes, than with mid-latitude frontal rains which affect large areas. Convective storms in Israeli deserts, western Africa and Tanzania are only about 2–5 km across. A related feature is that the correlation between measurements exceeds 0.25 if the separation is between either 0–20 km, 40–70 km or 105–60 km, which implies fairly regular spacing of the storms.

Orographic rainfall may lead to large differences over small distances amongst hills. Two rain gauges only 6 km apart in Scotland collected rain

Table 7.4. Effect of latitude on the elevation of maximum rainfall

Place	*Latitude:* *Degrees*	*Median*	*Elevation:* *Metres*	*Median*
Andes	0		1200	
Equator	0		1200	
Mt Kilimanjaro	3		1500	
East Africa	5		1500	
Java			1250	
Central Java	8	5	1000	1225
"			1500	
Indonesia			1200	
Guatemala	15		900	
West Ghats	15		1370	
'Tropics'	20		1000	
"			1500	
Hawaii	20		1100	
Himalayas	28	24	2200	1325
Japan	36		1200	
California	38		1450	
New Zealand	43		1600	
"	45		2280	
Western USA	45	45	2500	
European Alps	45		3250	2250
Washington State	48		2000	
Carpathian Mtns	50		1850	

differing by 29 per cent. Andersson (1963: 325) quoted differences exceeding 100 mm/a within a kilometre's distance.

The distribution of snowfall also is affected by low-level convergence caused by hills or by an increase of surface roughness. Consequently, there is particularly heavy snowfall downwind of Great Salt Lake in Utah, for example.

Correlation of rainfalls

The correlation between rainfalls measured at two places depends on their distance apart, on the kinds of terrain and rainfall, and on whether one considers the daily, monthly or annual total precipitation. The correlation is highest for places which are close to each other, in flat country away from the coast where light rain affects a large area and when precipitation is considered in terms of annual amounts. Here are examples of these effects.

Effect of distance apart

As you would expect, there is a decrease of the correlation of rainfalls at places of increasing separation. In the lower Jordan Valley the coefficient of

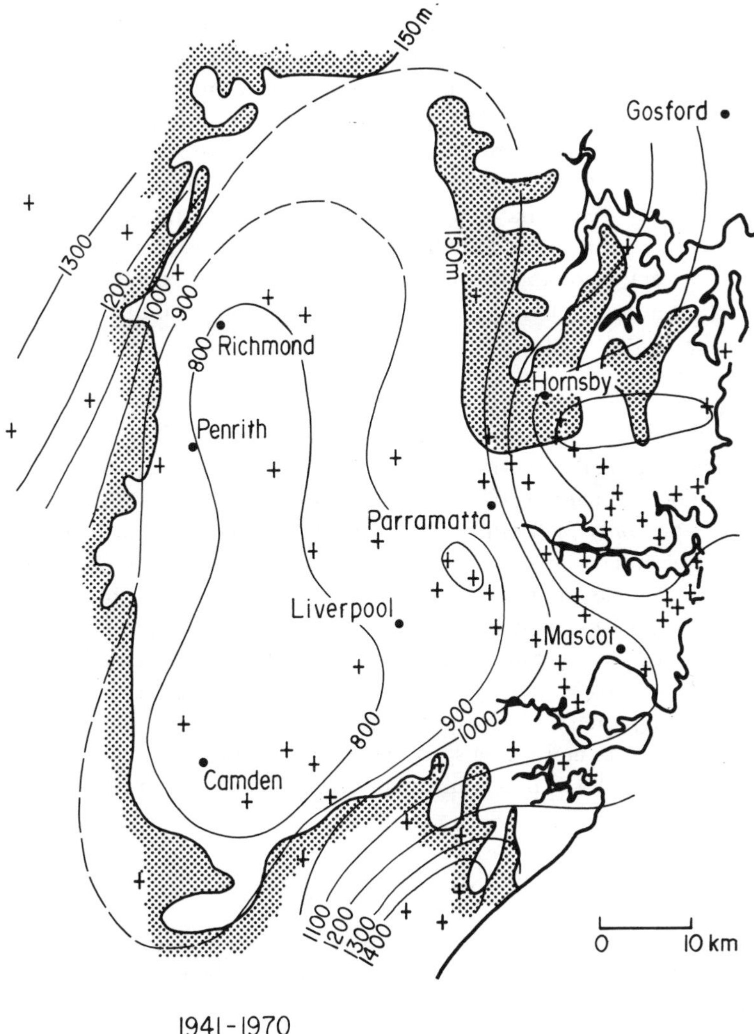

Figure 7.7. Average annual rainfalls (mm) in the region of Sydney during 1941–1970 (D. White 1976, priv. comm.). The shaded areas are more than 150 m above sea level. Crosses show the positions of the rain gauges

the correlation of daily rainfalls is 0.90 for places 10 km apart, 0.66 for 30 km and 0.40 for 50 km. At Otago in New Zealand, the coefficient for monthly rainfalls is 0.9 across 16 km and 0.6 across 80 km. Rainfalls during a minute in Germany, Florida and Hawaii are totally uncorrelated when the separation is 30 km, but the coefficient is 0.3 for 10 km, 0.5 for 5 km and 0.6 for 2 km. Summer storm rainfalls in Illinois correlate with a coefficient of 0.67 across 32 km, 0.78 across 16 km and 0.89 across 6 km.

The dependence of the correlation coefficient on the separation distance has been examined by considering *monthly* rainfalls during 1900–64 at 42 places in eastern Australia, compared with rainfalls at the same time at a central climate station. Around Charleville, for instance, the coefficient falls to 0.5 at a distance of about 430 km in an east–west direction, but 740 km in a north–south direction. Similar expressions apply to annual rainfalls in New Zealand and Australia.

Effect of rainfall intensity

The effect of rainfall intensity on spatial correlation appears to vary with location. The coefficient is 0.85 for light daily rains in Vermont at places 6 km apart, but only 0.50 when the intensity exceeds 25 mm/d. Synoptic-scale daily rainfalls in Illinois yield a coefficient of 0.97 for 15 km separation, whereas the intense and local convective rainfalls of Israel give only 0.28 for the same distance. The spacing for a coefficient of 0.6 is 3 km in summer in Arizona, but about 24 km for light rains in Israel.

There is a related seasonal effect. A coefficient of 0.8 applies across 80 km in Holland in January when rainfall is frontal, but across only 17 km in July when there is more intense convective rain. Likewise, the coefficient 0.6 applies across 8 km in a Tanzanian valley in July, but across 34 km in October.

Effect of duration

Annual rainfalls at places 24 km apart at Otago have a correlation coefficient of 0.9, whereas the distance for that coefficient for monthly totals is only 13 km.

ESTIMATING THE TOTAL RAINFALL WITHIN A CATCHMENT

Hydrologists may want to know how much rain is gathered within the catchment draining into a river, dam or reservoir, because that determines the adequacy of water supplies and the risk of flooding downstream. The quantity has to be estimated from measurements with rain gauges scattered throughout the catchment, so there is a sampling problem – how representative of the catchment's total receipt of rain are measurements by these few gauges? The error depends on the size of the catchment, the size, intensity and duration of the storms within the catchment, and the terrain.

The section 'Precipitation processes', pp. 252–64 showed that various processes lead to rain, and determine the area of each rainfall. For instance, low-latitude convection causes local, intense showers, whereas frontal rains are more gentle and more evenly distributed on a synoptic scale. As a consequence, the sampling error tends to be greatest where convection is dominant. The size of the error is illustrated by a study of rainfalls in Panama, using 36 gauges in an area of 300 km^2. One set of five gauges gave an average of 23 mm for a day's rainfall, whereas other sets of five gauges gave 12, 16 and 26 mm, respectively.

Rainfall intensity and the sampling error

Another factor affecting the sampling error is the amount of precipitation in a given period. For instance, the difference between a gauge's reading and the mean across a catchment in Illinois, varies as the -0.43 power of the rainfall, i.e. the gauge's readings are a better estimate when there is more rain. Therefore, totalling the rainfall over a longer time reduces the sampling error, as shown in Table 7.5.

Table 7.5. Effect of rainfall amount and duration on the error in assuming that a single central rainfall measurement equals the average over 260 km^2

	Mean error: mm		
Rainfall: mm	*Storm*	*Weekly total*	*Monthly total*
6	6.9	1.5	1.5
25	3.3	2.5	2.5
51	5.6	4.3	3.6

Table 7.6 confirms that the difference between a rain gauge's measurements and the spatial-average precipitation is less when the area for each gauge is small. Thus, the standard error of the measurements in a Japanese river basin of 840 km^2 is 25 per cent if there are three gauges, 10 per cent if 12, and only 3 per cent if 24. For 20,000 km^2 in Ohio, the standard error is 14 per cent if there are 16 gauges and 6 per cent if 80. In a catchment of 1300 km^2 the error is 15 per cent if there are five gauges and 32 per cent if only one.

About half the variance of daily values from gauges, each measuring the rainfall on to 180 km^2 in Montana, is due to the sampling error. The variation over 27 hectares may necessitate 553 gauges to reduce the standard error of the areal average monthly rainfalls to 2.5 per cent, or 31

Table 7.6. The percentage error in estimating a catchment mean rainfall of a storm from the average of measurements by separated rain gauges

Area per gauge: km²	*Median*	*Error: %*	*Median*	*Place*
13		4–10		Michigan
35		3		Japan[a]
70	53	10	9	
78		15		Michigan
140		18		Japan[a]
180		51		Montana
300	300	7	15	Ohio
1000		12[b]		
		25[c]		
12,000		12		India

Notes: [a] For storms of more than 30 mm
[b] In a catchment of 15,000 km²
[c] In a catchment of 2,000 km²

gauges for a similar accuracy in estimating the annual rainfall. Such large numbers of gauges are hardly practicable, and therefore any estimate of catchment rainfall is bound to be inexact.

The error is reduced if the gauges are arranged in a regular pattern. For instance, 20 gauges at random in 2200 km^2 in an English catchment incur an error of 1.6 mm/month, whereas a regular pattern of the same gauges have an error of only 1.2 mm/mo.

The effects of both the area per gauge and the overall size of the watershed are shown in Table 7.7. Thus, the average of daily measurements with five gauges in a catchment of 500 km^2 (i.e. 100 km^2 per gauge) may be 17 per cent different from the true catchment average.

The effects of rainfall intensity and of the area per gauge on the possible error are most concisely expressed by equations. They can be compared by calculating the error for the case of a 20-mm storm lasting an hour, with 100 km^2 per gauge in an area of 500 km^2, for example. The equations in Note 1, pp. 337–8 give 1.5 mm, 3.4 mm, 2.2 mm, 1.8 mm and 0.8 mm respectively. These may be compared with the 3.4 mm given by Table 7.7 (i.e. 17 per cent of 20 mm). The range of these figures illustrates the difficulty of determining the error precisely.

Appropriate spacing of rain gauges

Table 2.1 shows recommended spacings of 5–30 km between rain gauges, depending on the terrain. This is equivalent to 25–900 km^2/gauge, which accords with recommendations by other authors. Linsley *et al.* (1982: 63) proposed 600–900 km^2 per gauge for flat areas in general, but 100–250 km^2 in most mountainous regions, 25 km^2 on small mountainous islands and

Table 7.7. The effects of the area represented by each gauge, and of the number of gauges in a watershed, on the root-mean-square error of measuring the overall average daily rainfall in the USSR

	Area per gauge: km²						
	1	*10*	*50*	*100*	*500*	*1000*	*5000*
Number of gauges	Error: per cent of mean						
1	26	28	34	38[a]	47	54	66%
2	18	20	23	27	35	38	46
5	11	12	15	17	22	24	29
10	8	8	10	14	16	17	20
20	6	6	8	8	10	12	15
50	4	4	4	6	7	8	10
100	2	2	3	4	5	6	7

Source: Rusin 1970: 283.
Note: [a] i.e. the RMS error is 38 per cent of the mean of the measurements from two gauges each representing 100 km^2.

1500–10,000 km^2 in arid and polar regions. Forests may need more gauges. But even one rain gauge each 700 metres is barely adequate for estimating the average in a catchment on the Tanzanian coast, for instance, because of the spottiness of convective rainfalls.

In fact, the average spacing in the USA is equivalent to a single gauge per 576 km^2, in Australia to one per 256 km^2, and in Israel one per 25 km^2. One of the world's most dense networks (in England and Wales) has 4400 gauges, i.e. about 34 km^2/gauge, which results in a standard error of 3–4 per cent in estimating an area's overall precipitation. Errors are presumably larger in the many countries where even 1600 km^2/gauge is not feasible because of the cost.

Climatological problems require data which are as *complete* as possible, both spatially and temporally. These criteria differ from the needs of weather forecasting, which requires data which above all are *fresh*. Thus, in Kenya, for instance, there are about 2000 rain gauges, representing about one each 17 km on average, all valuable in providing climatological data. However, the only gauges useful as regards the short-term weather are the 50 where the measurements are reported promptly each day.

In the few cases where there are too many gauges, their readings merely duplicate those from adjacent instruments (see 'Measurement of rainfall', p.60). Instead, the spacing should be enough to bring the correlation coefficient between the daily readings at adjacent gauges to less than 0.9. For rainfalls above 13 mm/d in Vermont that would require spacing at more than 2 km, but 7 km for lighter rains. A spacing of 400 km^2 per gauge in Illinois corresponds to a coefficient of 0.8.

Theoretical study of the case of Uruguay shows that reducing the average spacing from the actual 42 km to the recommended 30 km (which implies almost doubling the number of rain gauges) would lead to a negligible improvement in the accuracy of determining overall rainfall.

An alternative criterion for spacing gauges is that there should be one for each distinct topoclimate. That might require determining a region's topoclimates by means of 'cluster analysis', a sophisticated statistical technique.

A refreshing simplification of the problem comes from a consideration of rain-gauge spacing from the point of view of accurately assessing the *run-off* from a catchment, rather than the rainfall itself. This focuses on the point of concern as regards flooding or rainfall impoundment. Theory reveals that only two or three well-spaced gauges are needed in a catchment to estimate the annual overall rainfall sufficiently accurately for this purpose.

Estimation of rainfall

One may want to estimate the rainfall at a place between gauges. The crudest method is to assume that the precipitation is the same as that measured somewhere nearby with similar vegetation. Another proposal is that short-term rainfalls at one site are the same as those at another, provided the two places differ in elevation by less than 200 m, have similar terrain within a radius of 5 km, have annual rainfalls within 10 per cent of each other, are similarly distant from the coast if within 50 km of it (the difference of distances being less than 20 km), and are separated by less than 150 km.

A further method of estimating precipitation is to use the average of measurements nearby, after each measurement has been weighted by the distance of the respective gauge from the particular place. Or the weighting can be in terms of the ratio of long-term measurements by the gauge, compared with measurements at a reference place, as in interpolating wind speeds ('Spatial variation of winds', pp. 242–4.).

Better is the development of a statistical relationship between (monthly) rainfalls and such controlling factors of geography as elevation, slope, orientation of the ground, exposure, latitude and distance from the sea. This requires data from many climate stations, as in the case of an interpolation formula based on data from 344 rainfall stations in south-east Australia. The root-mean-square-error of estimating annual rainfalls proved to be 57 mm/a, i.e. 6 per cent. There has to be a formula for each month, as in interpolating temperatures (see 'Estimating screen temperature', p. 79).

Graphical estimation of catchment rainfall

Instead of simply averaging the readings of the gauges in a catchment, it is customary to use either a weighted average or to calculate the pattern of isohyets for each storm in order to derive the catchment's total rainfall. The

Thiessen method of weighting the gauge recordings is illustrated in Figure 7.8. It involves forming triangles of lines joining measuring stations on the map, then the perpendicular bisector of each line is drawn so that the bisectors form a polygon about each gauge. The area of each polygon Ai is used to weight the relevant gauge measurement Pi, and the catchment average rainfall is the sum of the products Pi.Ai, divided by the catchment's whole area. The Thiessen method is convenient, since the areas Ai have to be calculated only once, until the pattern of gauges is altered. A disadvantage is that no allowance is made for topography.

The topography is taken into account in the alternative 'isohyet method'. The landform affects the long-term pattern of rainfall for any particular wind direction, and isohyets for any storm are drawn by hand in the light of these patterns and the gauge readings. It is then straightforward to measure the areas between the isohyets, multiply by the mean rainfall in each strip, and add the products together to obtain the volume of the precipitation on to the whole catchment. Unfortunately, the entire procedure has to be repeated for each storm.

The Thiessen and isohyet methods may give the same annual totals. Usually, the isohyet method is reckoned the more accurate, though it depends on the skill of the analyst and is more tedious than the Thiessen method.

In addition, there is the method described by Goel and Aldabagh (1979). This involves joining the positions of the gauges on the map by lines, none of which may cross another. Lines to gauges outside the catchment are included if the lines lie mostly within the catchment. The length of each line is measured and multiplied by the average of the rainfalls at the gauges at the ends of the line. Then these products are added and divided by the total

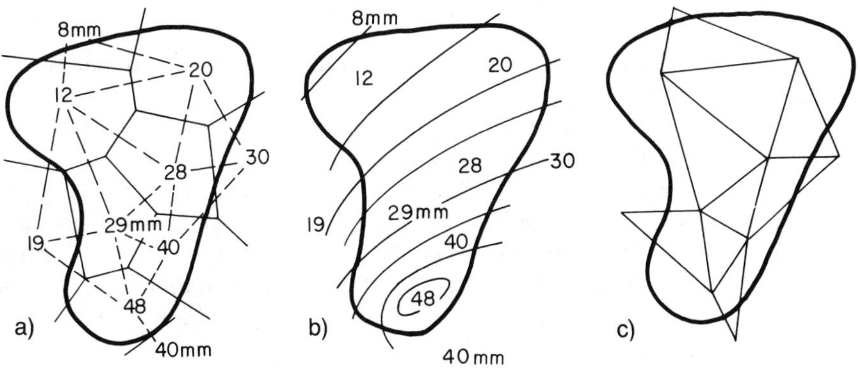

Figure 7.8 Procedures for finding the respective areas for weighting the measurements at individual rain gauges in a catchment: (a) the Thiessen polygon method, (b) the isohyet method with isohyets at 5-mm spacing, and (c) the method of Goel and Aldabagh (1979)

length of the lines, to obtain the catchment's average precipitation.

When applied to the example in Figure 7.8, the Goel and Aldabagh method gives 27 mm, while the unweighted average is also 27 mm, the Thiessen method gives 26 mm and the isohyet method about 24 mm. In this case, the scatter of the estimates is comparable with the measurement errors of individual gauges.

A different procedure is to impose a grid on the map of the catchment and to estimate the rainfall at each grid intersection, by interpolation between the measured values, perhaps by means of a regression equation expressing rainfall as a function of geographical factors at the gauges and the grid points. This could be done by computer. The intersection values can then be averaged.

LONG-TERM, ANNUAL AND SEASONAL RAINFALLS

The long-term average annual total rainfall is the usual index of the wetness of a climate. The calendar year is used as the unit, though it would be better to use July–June intervals in southern monsoonal climates and in northern Mediterranean climates, where the dry season occurs mid-year. This would avoid dividing a wet season across consecutive years.

Precipitation varies from year to year. There are several ways of describing the variability – the range, the standard deviation, the 'relative variability' and others. For instance, the 'relative variability' is the average difference from the long-term average as a percentage of the latter. The 'coefficient of variation' (Cov, see Eqn (4.5)) happens to be given as follows, at least over the Near East and northern Africa (Jones *et al.* 1981: 14):

$$\mathrm{Cov} = 100\, s/P = 950/P^{0.6}, \text{ approx.} \qquad \text{per cent} \tag{7.3}$$

Hence

$$s = 9.5\, P^{0.4} \qquad \mathrm{mm/a} \tag{7.4}$$

So that the standard deviation is 104 mm/a, if the mean rainfall is 400 mm/a.

The Cov is greatest at places at low latitudes, with arid climates, and is affected by ENSO events which are discussed later in this section.

The variation of annual rainfalls is often considered in terms of percentiles (Eqn (4.1)), such as the inter-quartile range or the variability index (see p. 126). The index exceeds 1.7 around Birdsville in central Australia, where it is dry, but is less than 0.5 at Zeehan on the west coast of Tasmania, which is wet. The figure is 0.62 for Sydney (see Figure 4.5) and 0.53 for Okavango (see Figure 7.9).

A rainfall of particular interest to farmers is the 25th percentile – called the 'dependable rainfall' Pd. There is at least this much in 75 per cent of years. Figure 4.5 shows that Pd at Sydney is 930 mm/a, whilst Figure 7.9 indicates 830 mm/a at Okavango; in this case a vertical line is drawn up

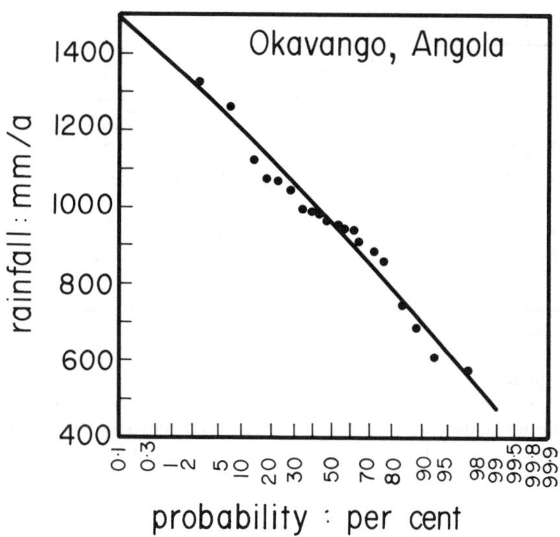

Figure 7.9. Distribution of annual rainfalls at Okavango (Angola) during 1950–70 (Gois 1974: 250). The horizontal axis shows the probability of the rainfall exceeding a value on the vertical axis. (By permission of the International Association of Hydrological Sciences)

from 75 per cent on the base, to the graphed line, and then across to the left axis. Table 4.8 shows that the 25th percentile is given by the mean rainfall minus 0.67 times the standard deviation, when the rainfall distribution is Gaussian, i.e. 'normal'. It might be thought this could be combined with Eqn. (7.4) to obtain the dependable rainfall anywhere, but rainfalls are Gaussian only in wet areas, whereas Eqn. (7.4) applies only in dry areas.

The wettest year in temperate climates commonly has about twice the mean rainfall, and the driest has about half the mean. In other words, the highest rainfall is four times the lowest. Obviously, this rule is only indicative. However, the ratio of extremes was 2023/504, i.e. 4.0, at Buenos Aires during (see Table 7.8), 3.7 at Sydney, and 3.4 at Katherine, at 14°S in the Northern Territory of Australia.

No rule like this applies in arid areas. There had been no rain for four years at Calama in the Chilean Atacama desert, and then there was a downpour of 64 mm. There are arid areas around the Gulf of Mexico where 200 mm/a fell one year and 1620 mm/a in another. At Dharan in Saudi Arabia there was 5 mm in 1946 and 187 mm in 1974.

Also, the ratio of extreme annual rainfalls increases as the years pass, allowing time for increasingly eccentric values to occur. Roughly speaking, the range increases in proportion to the logarithm of time.

Annual rainfalls over 700 mm/a may be almost Gaussian, as discussed (see 'Scatter of values' pp. 127–9 and illustrated in Figure 4.10. This is also shown by the linearity of data in Figure 7.9 and by measurements of

Table 7.8. Percentile rainfalls (mm) in each month, 1861–1968 at Buenos Aires[a]

Percentile	*Jan*	*Feb*	*Mar*	*Apr*	*May*	*June*	*July*	*Aug*	*Sep*	*Oct*	*Nov*	*Dec*	*Year*
100	348	249	545	405	245	174	212	278	349	367	284	318	2023
90	215	152	217	165	167	128	113	140	148	185	159	164	1284
70	108	95	149	111	97	81	71	80	102	104	102	113	1069
50	76	68	101	82	61	56	49	52	63	77	81	82	940[b]
30	50	49	59	54	34	34	29	33	40	54	53	58	848
10	19	17	31	21	11	13	9	14	19	24	28	30	701
0	3	0	2	5	0	1	0	0	3	4	12	5	504

Source: Hounam *et al.* 1975.

Notes: [a]i.e. January in 30 per cent of years has less than 50 mm rainfall.
[b]Note that the sum of a year's median monthly rainfalls does not equal the median annual rainfall, but is appreciably less, e.g. 848 mm/a instead of 940 mm/a. This is connected with the variability of monthly rainfalls being greater than that of annual totals.

rainfall at Kuala Kangsar in Malaysia, for instance. About 80 per cent of 503 places in Europe have a 'normal' distribution. The fraction is 76 per cent in East Africa, but 90 per cent if the wettest year is excluded in each case. A transformation of the annual rainfall data, taking their logarithms instead, gives a normal distribution in the case of Katherine in Australia, and such a transformation is needed in a significant minority of places in Australia and elsewhere.

There can be a great scatter of the annual rainfalls in a series, even when 30-year averages remain almost constant. Thus, the figure for England and Wales was 1290 mm/a in 1871 but only 615 mm/a in 1788, though the averages over 30-year periods (e.g. 1791–1820, 1821–1850, etc. to 1970) were all within 43 mm/a of each other.

Year-by-year variations of rainfall show influences apart from the random component, including periodicities (discussed in 'Periodic variations', pp. 134–5), the Quasi-Biennial Oscillation, sunspots and El Niño–Southern Oscillation events. These topics receive much attention in the climatological literature and will now be reviewed as regards precipitation.

Rainfall periodicities

An early example of the interest in regularities of annual rainfall was the claim by H.C. Russell in 1897 of a 19-year rhythm in Sydney's precipitation. Long beforehand, in 1626, Francis Bacon had noticed a rhythm of 35 years for rainfalls in Europe, rediscovered in 1889 by C.

Egeson and in 1891 by E. Brueckner, with maxima in 1703, 1745, 1776, 1815, 1848 and 1878. However, data were available for only 194 years, i.e. for no more than about five 'Brueckner cycles', which is hardly sufficient to substantiate any permanent regularity. Some other evidence does tend to confirm the periodicity, but several writers cast doubt upon it (Fairbridge 1967).

One rhythm which is well substantiated is the Quasi-Biennial Oscillation (QBO), a tendency for wetter times each 26 months or so, i.e. approximately two years. An example is the bias of monsoonal rainfalls in India during 1865–1970 towards alternately dry and wet years. In East Africa also, annual rainfalls for over 44 years show a rhythm of 26 months. 'The QBO is the only quasi-periodic oscillation whose statistical significance is clearly demonstrated' (Burroughs 1980: 156). The amplitude of the variation is up to 38 per cent of the average at places in New Zealand. The cause of the QBO is not yet established, but appears related to variations of the stratospheric winds at low latitudes, and of stratospheric ozone. Or the explanation may involve the 2–3 years taken for a rotation of the vast oceanic gyres.

Sunspots and rainfall

Less certain is the effect of the regular variation of the number of spots seen on the sun each year. These were mentioned in '*Gulliver's Travels*', published in 1726. A connection between their annual number and the weather in England was suggested by William Herschel in 1801, and the cyclic nature of the number was demonstrated by Heinrich Schwabe in 1843, with a periodicity of either about 11 years or else twice that, i.e. roughly 23 years. The latter is known as the 'Hales double sunspot period'.

There have been hundreds of scientific papers on the existence or absence of relations between annual sunspot number and weather. Information from a few of those written since 1980 is given in Table 7.9. Other evidence includes the following observations. Relatively dry and cold conditions during 1644–1713 (the Little Ice Age) coincided with the disappearance of sunspots (called the 'Maunder Minimum'), and low sunspot numbers during 1450–1550 (the Spoerer Minimum) and 1280–1350 (the Wolf Minimum) also coincided with advance of the world's glaciers. Wet years in Ethiopia over several centuries have preceded the year of most sunspots, whilst Ethiopian droughts have clustered around years with fewest sunspots. There has been a coincidence of major erosion of the New South Wales coast (i.e. strong and frequent storms at sea) in the years 1857, 1876, 1889, 1911, 1933, 1944, 1966 and 1975, all of which were years with least sunspot activity. Hines and Halevy (1977: 382) confessed having 'sought several means of undermining the claim [of a sun-weather correlation in their data] but have been obliged to admit its reliability at the 95%

Table 7.9 Aspects of climate reported since 1980 to be varying (approximately) in time with the annual number of sunspots

Aspect	*Place*	*Approximate periodicity: years*
Forest fires	Canada	10–11
Tree-rings[a]	Taiwan	11
Rain	S. Africa	10–11
Drought area	West USA	11
Rain	Rome	11
Avalanche deaths	Norway	12–13
Rain	India	10–11
Temperature, pressure	Europe etc.	11.1
Rain	NE USA	10–11
Rain	India	11
Drought	West USA	22
Drought	S. Africa etc.	20
Floods	India	22
Temperature	N. hemisphere	22
Drought	West USA	20–25
Drought	West USA	18.2 + 22
Drought	Iowa	22
Floods	India	22
Rain	India	22
Tree-rings[a]	West USA	22
Rain	Switzerland	20–22
Rain	India	20
Summer temperature	30° lat. inland	22.2

Note: [a] Trees in dry climates show extra growth when there is an especially wet season, i.e. the light-coloured band added to the cross-section of the trunk is wider than usual. Likewise for an unusually warm season in a cold climate.

confidence level'. They suggested that the sun triggers weather events that were about to happen anyway, bringing them slightly into rhythm with the sunspots.

Despite even reluctant evidence in favour, there has been trenchant criticism of reported relationships, notably by Pittock (1978, 1983). The periodicities in Table 7.9 are not all exactly the same, and the statistical significance of any relationship tends to be low, partly because of persistence in the data. Sometimes the series of available figures is too short for the minimum of five cycles needed to show the reality of any rhythm. Also, some of the reported correlations seem to have been achieved by suitable selection of the data. Moreover, there have been numerous studies showing an *absence* of any sunspot–weather correlation.

Even in cases of apparent regularities they may prove only temporary, i.e. 'non-stationary' or 'fugitive', as in the cases of Adelaide autumn rainfall, water levels in Lake Victoria in Africa, rainfalls in South Africa, the latitudes of anticyclones over eastern Australia and the May–October rainfall in Hong Kong. Perhaps the breakdown of the relationships is due to occasional switchings between alternative, quasi-stable patterns of global weather. Or there may be an interaction with a lunar cycle of 18.6 years, as in the records of annual rainfall at 126 places out of 136 in the north-east USA.

There is no connection between the QBO and any sunspot periodicity, at least as regards Indian rainfall. A study of over 70 years of data from 48 places showed a significant 2-year periodicity at 9 places and an 11-year rhythm at 7, but both periodicities were evident together at only 3 places.

Overall, the situation seems confused. Matters have not changed much since Tucker (1964) referred to the evidence as inconclusive and contradictory. In 1972 the Russian scientist Monin referred to the claims of sun–weather relationships as 'experiments in auto-suggestion', whereas Mitchell *et al.* (1979: 140) reported: 'As hitherto avowed agnostics with regard to many previous claims of sun/climate relationships, we find ourselves somewhat unnerved by our own data' [which pointed to significant connections]. One of the difficulties has been the absence of any convincing explanation of the physics relating sunspots to rainfall. In view of the contradictory evidence, it is premature to decide on the issue.

El Niño and the Southern Oscillation

We can be more definite about the influences of the El Niño and the Southern Oscillation. The El Niño (pronounced ell-nin-yo) is a warming of the sea surface off Peru by several degrees over several months. This appears to be linked with fluctuations of atmospheric pressure, winds and rainfall, right round the low latitudes, involving a see-saw of atmospheric pressures at opposite ends of the equatorial Pacific ocean, called the Southern Oscillation. Variations of seasonal-mean pressures at Darwin correlate inversely with those for Tahiti, with a period of around 38 months.

Now that oceanographers and meteorologists work more closely, the entire complex of El Niño and Southern Oscillation is known as the ENSO phenomenon. This affects the 'Walker circulation' of the atmosphere over the equatorial Pacific ocean, i.e. the atmospheric uplift above the ocean off Peru in a circulation involving subsidence and high atmospheric pressures around northern Australia. During an ENSO event, the temperatures off Peru are warmer than usual, whilst the ocean near northern Australia is less warm, reducing rainfall over eastern parts of the continent. At the same time there are droughts in Amazonia and India, but heavy rains west of the Andes.

It is not possible to foretell the onset of an ENSO event, but we can predict the later stages once it has begun. ENSO events occurred in the following years, especially in those shown in italics, 1726, *1728*, *1747*, *1763*,

1770, *1791*, *1804*, *1814*, 1817, 1819, 1821, 1824, *1828*, 1832, 1837, *1845*/6, *1864*, 1868, 1871, *1877/8*, 1880, *1884*/5, 1887/8, *1891*, 1896, *1899/1900*, 1902, 1905, 1911/12, 1914, 1918/19, *1925/6*, 1929/30, 1939, *1941*, 1953, *1957/8*, 1965, *1972/3*, 1976, *1982/3*, 1987/8. These are not regularly spaced, but occur about once each half-dozen years, on average. Fortunately, one ENSO event chanced to occur during the International Geophysical Year of 1957/8, when numerous meteorological and oceanographic measurements were made at the same time around the world, revealing the extent of the phenomenon and the clear connection between sea-surface temperatures and rainfall at distant places. ENSO events in the east Pacific ocean relate to monsoonal rains in India.

What are called 'strong' ENSO events lead to fewer tropical cyclones near eastern Australia. Also, the associated fluctuations of the Walker circulation cause abnormalities of winter mean pressures at Darwin which foreshadow lighter rainfalls in south-east Australia in the following spring. Unfortunately, the relationship may not be enduring; the correlation for 1954–74 was different from that for 1932–53.

Seasonal rainfalls

The relative wetness of summer or winter depends on the latitude. For instance, summer is wet at low latitudes in Australia because of tropical cyclones and monsoonal winds from equatorial seas, whereas the winter is wetter at latitudes above about 30° on account of rains associated with cold fronts. Actually, the boundary is 35°S on the east coast and 25°S on the west, the difference of latitude being due to warm seas in the east and generally cooler currents off Western Australia. So Perth (at 32°S on the west coast) has a highly 'Mediterranean' climate, i.e. a particularly dry summer and wet winter. The latter occurs because the oceans remain relatively warm in winter, making the atmosphere unstable, which leads to more rain. On the east coast, the fluctuation of rainfall depends on the seasonal shift of the latitude at which high-pressure systems move eastwards. South of their path, winds spin westerly from the anticyclones, from the arid interior of the continent, whereas easterly winds north of the anticlockwise highs bring moisture from the Pacific ocean.

At latitudes of about 35°S on east coasts, the rainfall may be fairly even through the year, as in Sydney and Buenos Aires (see Table 7.8) and Sydney. At the poles, there is most precipitation in summer when the warmer air can contain more water vapour. At the other extreme, at about 13°N in West Africa, up to half the season's rain may fall within the five wettest days.

A good measure of the tendency for rain to fall in one season, rather than uniformly through the year, is provided by the 'seasonality index' of Markham (1970). To calculate this, consider the rainfall in each month as a vector on a clock, so that, the rainfall in March (the third month) for instance, is represented by a horizontal line pointing to position 3, i.e. to the right on the page. The length of the line is proportional to the month's rainfall. Then the vectors are detached from the clock's centre and connected end-to-end, retaining their original directions. The rainfall represented by the straight-line distance from beginning to end of the 12-part line is the seasonality index.

It is useful to farmers, in predicting the rainfall during the remainder of a crop's growth, if significant persistence links rainfalls in various seasons so that present heavy rains foreshadow coming wet months ahead, for instance. This was tested by Dennett *et al.* (1983), who examined precipitation records from arid parts of West Africa. Unfortunately, there proved to be no seasonal persistence, which shows that processes governing the rainfall there take much less than two months to complete.

Thermal convection as a cause of rain near the equator (see Table 7.2) leads to two distinct rainy seasons, because the sun is overhead twice in the year between the Tropics. Thus, at 1°N in Uganda there is one wet season from March–May and another from August–November. Likewise, at Singapore (1°N) the sun passes overhead in late March and in late September, causing relatively wet periods in April–May and October–January, respectively.

MONTHLY AND SHORT-TERM RAINFALLS

The annual variation of rainfall (i.e. *within* a year) leads to a distribution of monthly totals which is like a truncated Gaussian distribution. In Sydney, for instance, there are far more months drier than the mean than months which are wetter. The average monthly rainfall in Sydney is calculated from curve N in Figure 7.10 to be 100 mm/mo, whereas the curve C shows the median to be only 70 mm/mo. In the same way, the September rain at Los Angeles is less than the mean in 27 out of 30 years, and two-thirds of August rainfalls at Gao in the Sahel are less than the mean. We say that the distribution is 'skewed', in contrast to the symmetrical distribution of annual rainfalls shown in Figure 4.10. However, a skewed distribution can be converted to a more symmetrical one by transforming the variable. The distribution of Australian monthly rainfalls can be transformed to normality by considering the rainfall to a power equal to some value between 0.25–0.67.

The scatter of monthly rainfalls in Java means that half the months have rainfalls within the range 130–260 mm/mo, when the mean is 200 mm/mo, for instance. The 25th percentile monthly rainfall (Pmq), the monthly equivalent of the dependable rainfall discussed earlier in this chapter (see pp. 267–71) is as follows:

$$\mathrm{Pmq} = 0.8\ \mathrm{Pm} - 30 \qquad \mathrm{mm/mo} \tag{7.5}$$

where Pm is the long-term mean total rainfall for that month. Similar

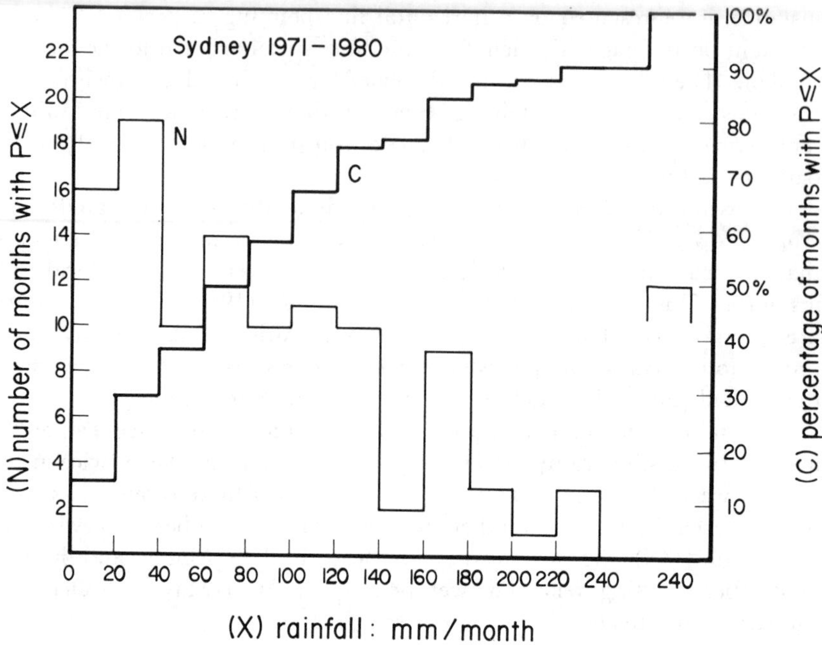

Figure 7.10. A frequency distribution (N) and cumulative distribution (C) of monthly rainfalls at Sydney during 1971–80

equations have been derived for 19 places in Thailand, Malaysia, 23 eastern states of the USA, and in Central America.

An empirical observation of Hargreaves (1981: 282) is that the 30-year maximum rainfall for a particular month of the year is around twice the average of that month's precipitation. This resembles the earlier approximate rule about annual rainfalls.

The year-to-year variation of a particular month's rainfall is often considerable, especially for months of the dry season. For instance, the coefficient of variation in India is around 40 per cent for months with about 350 mm/mo, but 90 per cent if only 100 mm/mo and 140 per cent if 50 mm/mo. Similar variability at a place in Java is shown in Table 7.10. Such instability of the rainfall pattern reduces the validity of characterising a place's rainfall by a single annual total.

There is frequently evidence of persistence of rain between consecutive periods of a month or less. For example, the chance of the second month being drier than usual in Australia after one dry month would be 50 per cent if there were no persistence, but in fact it is about 55 per cent. Monthly persistence has been reported from Malawi, according to the month of the year and to the precise location. In north Africa there is a correlation of

Table 7.10 Variety of rainfall patterns at Rembang (Java) during 54 years

Number of months annually		
Dry[a]	*Wet*[b]	*Percentage of years*
5–6	0–2	32
2–4	0–2	26
2–4	3–4	15
5–6	3–4	13
over 6	3–4	4
7	5	2
5–6	5–6	2
2–4	5–6	2

Source: Oldeman and Suardi 1976.

Notes: [a] A 'dry' month is one with less than 100 mm of rain.
[b] A 'wet' month has over 200 mm.

+0.63 between June rainfalls and the total subsequent precipitation until October, i.e. the wet season. On the other hand, there is no apparent month-to-month persistence in England and Wales. The consequent improbability of a long run of dry months there implies less likelihood of drought.

A missing monthly rainfall value can be estimated in any of the four ways discussed in 'Estimating in general', pp. 64–7. Possible surrogates for rainfall values are cloud amount or mean temperature. The latter depends on cooling – firstly, by the clouds which bring rain and, secondly, by the subsequent evaporation from the ground. Such a relationship between rainfall and temperature has been found at Khartoum and for Januaries at Alice Springs; both places are hot and dry (see pp. 79–80). More generally, we would estimate the missing monthly rainfall at place A by multiplying the simultaneous measurement nearby at B by the ratio of parallel long-term measurements at A and B. However, it may not be the nearest measurements which give the most accurate estimate. The error may be reduced by averaging the respective estimates obtained by using data from other places, in addition to those from B.

Monthly rainfall data can be used to design a tank for the storage of rain-water from a roof. First, get figures for average monthly rainfalls in the area. The annual total (P mm), divided by 12, times the roof area (A m^2), gives the maximum water supply available (W litres per month). The tank loses water in every month in which the rainfall equivalent to W L/mo (i.e. W/A or P/12 mm/mo) is more than the actual rainfall (P_{mo} mm/mo). Add up the differences ($W/A - P_{mo}$) for the longest sequence of such months, making a total T mm, equivalent to A.T litres to be stored. That's the proper tank size. A larger size allows for a drier year than usual.

Decadal and weekly rainfalls

The variation of rainfall week-by-week or from one 10-day period to the next is of particular relevance to agriculture, since such units of time are those used in planning field operations and for measuring crop development. The decadal unit is the sort of time between irrigations, at the rates of evaporation common in warm climates. Also, ten-day periods can be conveniently matched to the early, middle and late parts of each month.

The frequency distribution of weekly rainfalls at Grenfell was discussed in Chapter 4 (see Note 1, especially Table 8.4). Similar data from South Africa can be interpreted in terms of the return period for any particular value to be exceeded (Eqn (4.2)), yielding the graph in Figure 7.11. The remarkably straight line facilitates extrapolation to estimate the highest weekly rainfall likely within a century, for instance.

Rainfalls over a few days

Dry or wet spells are roughly defined as periods of a few consecutive days with the same weather. Precise definitions differ, so that a 'dry spell' in Britain means 15 days without at least 1 mm precipitation on any day, whilst any period with less than 2.5 mm on each day is a dry spell in Malawi.

The chance of a dry spell changes during the year, especially with a

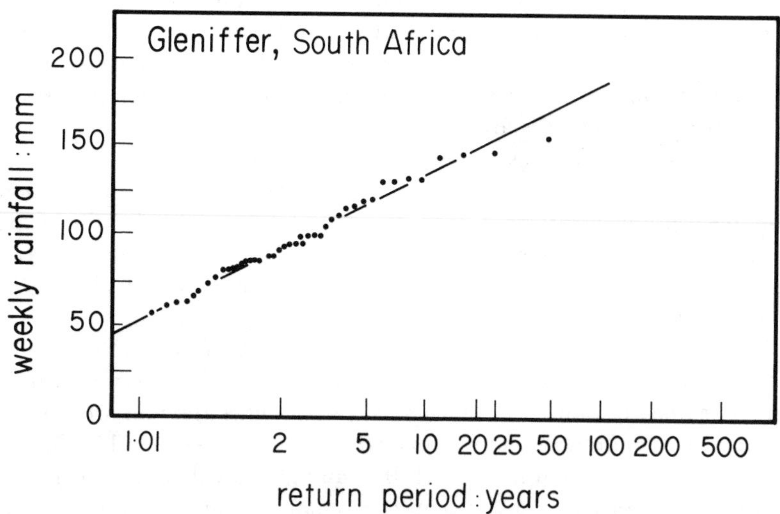

Figure 7.11. The weekly rainfall exceeded in each return period at Gleniffer in South Africa, at 29°S, 30°E and 1494 m elevation (Schulze 1980: 7, by permission of Dr R.E. Schulze)

monsoonal climate, so that the timing of the alternate wet and dry seasons varies greatly from year to year. For example, the dry seasons during 1959–68 near Manila in the Philippines began as follows: December 22, January 3, December 30, January 22, January 27, January 15, February 4, December 25 and December 17. There are 21 days between the earliest and latest of these dates. The durations of the preceding wet seasons were similarly variable, i.e. 292, 244, 236, 227, 240, 252, 260, 274, 207 and 211 days, respectively. The range is 85 days.

In 1831 G. Schubler proposed that the amount of rainfall is associated with phases of the moon. However, there have always been sceptics: the idea was opposed by Klauprecht in 1840 and by Dove in 1869. Since then, the evidence has remained confusing. On the one hand, it appears that there is least rain just before a new moon, and most near the time of full moon in New Zealand, at Forster and Mudgee in New South Wales, and in the USA. Lethbridge (1970) reported that thunderstorms in the eastern and central USA are most common after full moon. On the other hand, no significant correlation was found either in eastern China or at Sydney.

A study by Hanson *et al.* (1987) used American data over 80 years and showed statistically significant variations of daily rainfall in harmony with the lunar cycle. As an example, the chance of a day being wet in South Dakota in spring is 44 per cent in the first quarter of the lunar cycle but 55 per cent in the third. However, the phase relationship depends on the region, so that most rain falls during the moon's bulging phase in the north-west states, but at new moon in the east. The difference between the times of enhanced rain in various parts of the USA may perhaps be due to a tidal effect of the Moon's gravity on the Earth's atmosphere, inducing, say, more easterly winds. These would bring rain from the oceans to the eastern states, but dry air from the continental interior to the west coast.

Daily rainfall

The definition of a 'rainday' is a 24-hour period from 8 a.m. or thereabouts, with a rainfall exceeding some threshold amount. The threshold is 0.25 mm in the USA and most British countries. Elsewhere it may be 0.1 mm (which is the smallest amount measurable), 0.5 mm (the least that can be measured reliably), or 2 mm. A value of 2 mm is the smallest quantity significant in agriculture, since rain up to that amount may be held on foliage and not reach the ground. Snijders (1986: 525) reported that the criterion was 0.1 mm at most West African stations whose records he examined, but 0.5 mm at some and 1 mm at others.

The number of raindays each year in the USA varies inversely with the coefficient of variation of annual rainfalls. For instance, the Cov is 15 per cent where there are 150 raindays annually, 23 per cent where 100, and 37 per cent where 50.

The average precipitation on each day that rain occurs is a comparatively constant quantity, varying less than the rainfall totals. At Alice Springs, for instance, the rain per rainday varies between months of the year from 3.0 to 9.8 mm (i.e. by a ratio of 3.3), whilst the monthly totals vary from 6 to 39 mm/mo, i.e. by a ratio of 6.5. Wet months tend to have both more raindays and also heavier rainfalls.

The average amount of rain per rainday varies with latitude and season. Seven places below 1000 m elevation and around 4 degrees latitude receive about 11 mm per rainday, whereas seven at about 23° have 10 mm/d, seven around 40° have 6.2 mm/d, and seven around 59° have 3.4 mm/d. In Australia, there is a crudely linear relationship between rain/rainday and latitude, implying about 20 mm/d at the equator and 6 mm/d at 40°S.

In the tropics, the mean rainfall **Pr** during a rainday is given by the following:

$$\mathbf{Pr} = 7 + 0.029\ \mathrm{Pm} \qquad \mathrm{mm/d} \tag{7.6}$$

where Pm is the monthly precipitation. However, rather higher amounts and fewer raindays occur in northern Australia because of the occasional intrusion of tropical cyclones.

There is a seasonal effect on the rain/rainday. The amount varies from 6 mm/d in October to 30 mm/d in May at Dawa at 9°N in Ethiopia. At Edmonton at 53°N in Canada, the variation is from 2.3 mm/d in December to 6.2 mm/d in June. The figures are explained by the different kinds of rainfall dominant at various latitudes and seasons (see 'Precipitation processes', pp. 259–62).

There is a clear relationship between the average rainfall per rainday in the USA and the daily rainfall likely to be exceeded only once in two years. The two-year/day value is 36 mm/d for an average of 5 mm per rainday, 80 mm/d for 10 mm/rainday, and 110 mm/d for 13 mm/rainday. Such information is useful in considering soil erosion by heavy rain, for example.

Daily rainfall data generally show persistence, because weather systems producing rain last for more than one day (see 'Persistence', pp. 135–8). Another reason is that a single storm may extend from before 9 a.m., when the rainfall for yesterday is routinely measured and continue till later in the day, making both days count as raindays.

Records for January–July 1981 at Marsfield show an overall chance of rain P(W) equal to 58 per cent. But the chance increases to 75 per cent for a day following a rainday, indicating some first-order persistence. The first-order contingent probability P(W/W) approximately equals [0.8 P(W) + 0.2] in the Philippines (see Table 4.12), Australia and the USA. Correspondingly, the chance of a dry day after a dry day, i.e. P(D/D), equals [0.8 P(D) + 0.2].

Persistence varies during the year. It is more marked in Hong Kong during February–April, when large, slow frontal rains are likely, than when there are sporadic convective rains. In Kano, P(W/W) is 0.2 in June, 0.33 in July, 0.6 in August, 0.4 in September, and so on. In southern England there is positive persistence of dry days in summer for up to 25 days, and thereafter negative persistence, showing that anticyclonic weather tends to last 3–4 weeks.

The amount of rainfall on any particular day can be considerable, much more than the average per rainday. The record is the 1880 mm/d which fell in Reunion (at 21°S, in the Indian Ocean) in 1952. But there are other places with exceptional *average* amounts of rain/rainday such as Cherrapunji in Assam with 61 mm/d. Extremes of daily rainfall can be estimated with either the Weibull or the Gumbel equations, discussed in 'Estimating extreme winds, see pp. 236–7.

A study of winter rainfalls over 26 years in Melbourne has shown that the average rainfall on Sundays is 1.7 mm/d, and then during the week there is a steady increase to an average of 2.2 mm/d on Thursdays. So weekends are significantly drier than weekdays, presumably on account of weekday air pollution. Similar results have been found in Britain and elsewhere.

Variation within a day

The various processes discussed in the section 'Precipitation processes', pp. 252–64 lead to concentrations of rainfall at particular times of the day. The timing is important in agriculture and human society since rain at night can soak into the ground before being evaporated away, whilst rain during the day affects the outdoor activities of people, the trafficability of unpaved roads and so on.

In general, continental climates promote thermal convection, which leads to most rain in the afternoon, whilst maritime climates at the coast cause nocturnal maxima. Convection from equatorial continents leads to the sort of pattern shown in Figure 7.12. The afternoon maximum is common at higher latitudes too, in summertime. A similar maximum has been found at Uppsala, in tropical highlands, in Illinois in summer, on the south-east coast of the USA, at Kuala Lumpur, in the Negev desert, Fiji and Nigeria, and on mountain summits in China.

A nocturnal maximum of rainfall is observed at many low-latitude places, on account of the interaction of local winds, synoptic winds and the shape of hills nearby. For example, nocturnal land breezes from Sumatra and Malaysia collide above the strait between them, causing ascent and a tendency to rain at night. The same happens over tropical and steep valleys. Other cases of nocturnal maximum rainfall are found in the mid-west USA in summer, and in north-east Argentina and lowland areas of Papua New Guinea. A nocturnal maximum occurs during the rainy season in hot climates, in winter in temperate conditions, and on tropical coasts. A

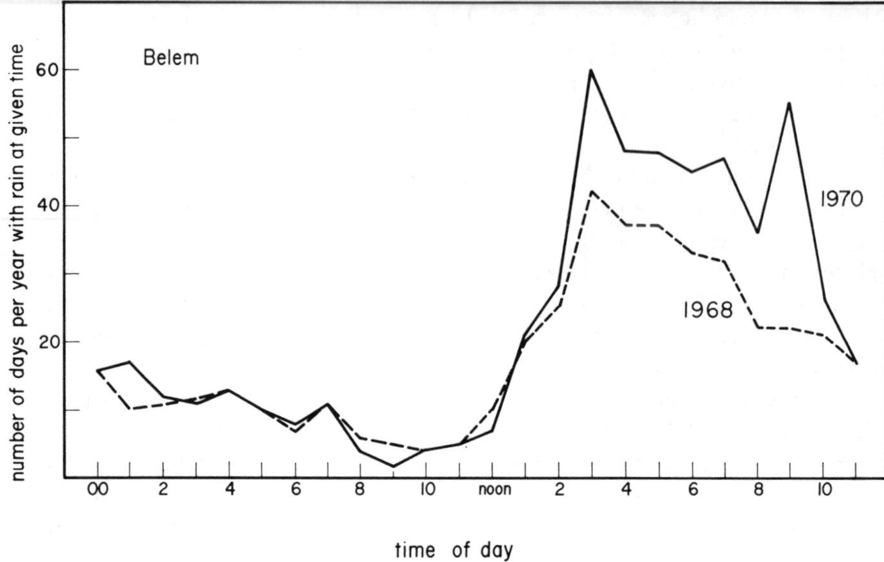

Figure 7.12. Diurnal variation of the rain at Belem (Brazil)

maximum near dawn is to be expected over the oceans since the surface is then comparatively warm.

Hourly rainfalls

The precipitation intensity is often described in terms of the total rainfall within an hour. This gives a quantity which is easily measured and conveniently expressed in millimetres depth. Figure 7.13 shows intensities in these units, and indicates that maximum rates are higher at low latitudes because of the greater prevalence of convective cumulo-nimbus clouds. These typically yield 20–30 mm/h, which is around five times what comes from high-latitude nimbo-stratus.

The fraction F of the rainfall in Idaho with an intensity *exceeding* **I** mm/h, is given as follows:

$$F = 13 \exp(-0.026\,\mathbf{I}) \qquad \text{per cent} \tag{7.7}$$

In this case the fraction is 8 per cent if the threshold value is 20 mm/h, for example.

Where it is hotter, there tends to be more convective rainfall which is relatively intense. So the fraction (f) of rainfall with an intensity of *less* than 1.8 mm/h is reduced:

$$f = 95 - 3.4\,T \qquad \text{per cent} \tag{7.8}$$

where T is the monthly mean temperature.

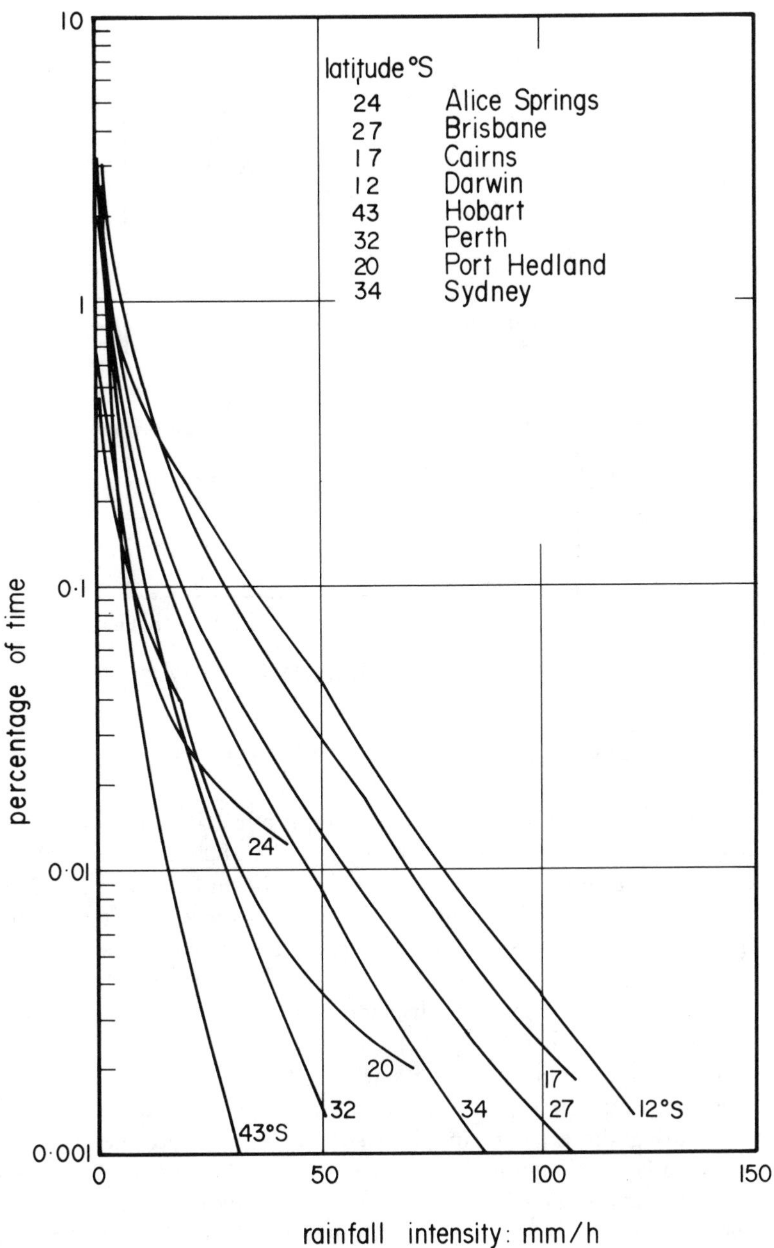

Figure 7.13 Fractions of a year with rainfall intensities exceeding a particular value at eight places in Australia (Flavin 1981). For example, the intensity at Darwin exceeds 100 mm/h for 21 minutes in an average year

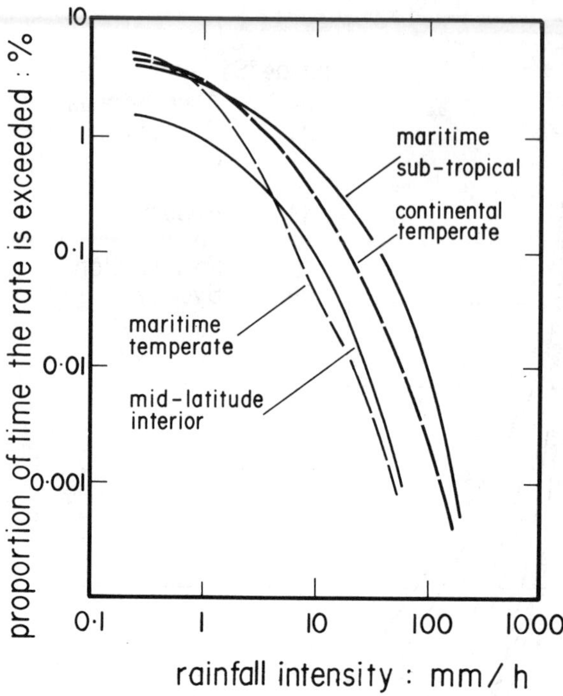

Figure 7.14 Effect of the type of air-mass on the frequency with which particular rainfall intensities are exceeded (Jones and Sims 1978: 1139).

The fraction of time when it is raining is affected by latitude. Even in the tropics, the average time is normally less than 800 hours annually, i.e. less than 9 per cent of the time despite the high annual rainfall and the large number of raindays. This surprisingly small percentage of time is due to the high frequency of brief, intense showers. There is milder rainfall at London (52°N), where 594 mm/a falls over 113 days each year, during only 5 per cent of the whole time. The fraction is 17 per cent in the wetter parts of Scotland.

The annual number of hours with a rainfall intensity above a certain value depends on the annual precipitation and on the kind of air-mass dominating the area. Figure 7.14 shows that a subtropical maritime air-mass leads to an intensity exceeding 20 mm/h for about 0.2 per cent of the time, i.e. about 17 hours each year. This roughly matches the 11 hours (plus or minus 5 hours) given by Table 7.11. Figure 7.14 also indicates that the intensities exceeded during a given proportion of the year are three times as much for maritime subtropical air-masses as for mid-latitude inland air-masses.

The frequency of intense rainfalls depends on location. Rainfalls in arid parts of Australia exceed 66 mm/h for only about five minutes each year on

Table 7.11. Relationship between the annual total precipitation (P mm/a) and the number of hours annually when a particular rainfall intensity is exceeded. The expressions were derived from measurements in south-east Asia and Panama

Rainfall intensity: mm/h	*Hours annually*	*Standard error*
0.25	0.45 P − 220 h/a	220 h/a
0.75	0.14 P − 7	32
2.5	0.087 P + 4[a]	17
6.4	0.039 P + 5	7
19.0	0.008 P + 3	5

Source: Atkinson 1971: 6.35.

Notes: [a] For example, the rainfall intensity exceeds 2.5 mm/h for 91 hours each year at a place with an annual total of 1000 mm/a, since (0.087 × 1000 + 4) equals 91. This figure is associated with a standard error of 17 h/a.

average, but 250 mm/h for that fraction of time in the monsoonal north of the continent. The figures are 25 mm/h and 150 mm/h, respectively, for 50 minutes annually. In Illinois, the precipitation exceeds 0.6 mm/h for 2 per cent of the time, 6 mm/h for 0.4 per cent and 60 mm/h for 0.02 per cent. The rainfall exceeds 0.25 mm/h for 1 per cent of the time in the south-west of the USA, but for 10 per cent in the north-east. Another way of describing the pattern of precipitation in the Americas is as follows: about half falls during 10 per cent of the time it is raining and 87 per cent of the precipitation occurs during the wetter half of the time of raining.

Return periods

Hourly rainfall data have been useful in assessing the intensity likely to be exceeded just once within a specified return period, as for extreme winds (see 'Estimating extreme winds', pp. 231–9). Such information is essential in designing flood-mitigation works. The assessment can be done by means of empirical relationships based on data from places where measurements have been taken for a long time. Various relationships are illustrated in Figure 7.15, in terms of what is called the 2-year/1-hour rainfall, denoted P (2,1). This is the intensity averaged over one hour with a return period of two years, and can be determined from only a few years of measurement. Subsequently, the diagram can be used to derive the rainfall over 2 hours occurring once in a century P (100,2), for instance. For that, the ratio on the left axis is 2.8, so the required rainfall is 2.8 times the known 2-year hourly amount.

A tentative map of the one-hour rainfall with a return period of two years is shown in Figure 7.16. It demonstrates the expected strong dependence on latitude.

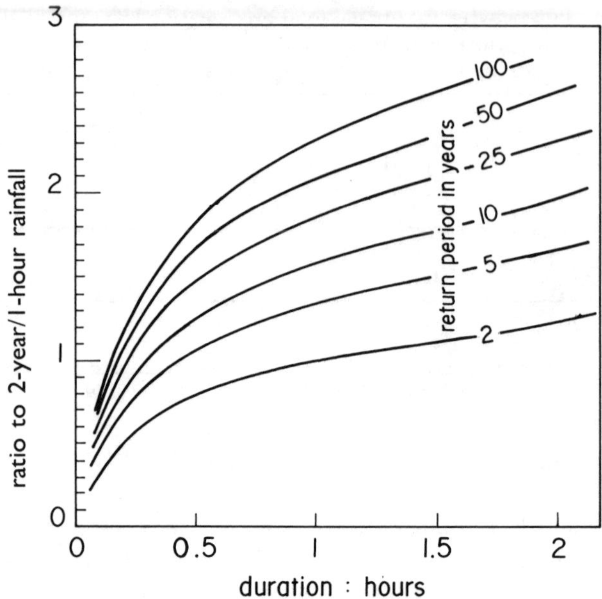

Figure 7.15 The dependence of the ratio of P(T,t) to P(2,1) on t, where the ratio is shown on the vertical axis and P(T,t) is the rainfall during t hours (shown on the horizontal axis) which is exceeded once each T years, the return period, shown by the various curves (Bell 1969: 324, by permission of the American Society of Civil Engineers)

Brooks (1950: 204) observed that P(T,t), i.e. the rainfall over t hours with a return period of T years, is approximately proportional to the fourth root of the product of T and t. In other words, increasing either the return period or the time of averaging by 16 times only doubles the appropriate rainfall. In more detail, Bell (1969: 323) reported that P(100,1) is twice P(10,1) and 2.9 times P(1,1) in the USA, Australia, South Africa and Puerto Rico. So a short record, sufficient to provide P(1,1), can be used to derive P(100,1), which is the rainfall in 1 hour, exceeded once each century.

Rains of brief duration

Rain may fall during storms which last less than an hour. For example, 61 per cent of the 1955 rain at a place in central Sudan was precipitated during storms lasting no more than 25 minutes. The intensity was up to 84 mm/h in the early part of the storms. Such rainfall, with an intensity above 60 mm/h for at least 5 minutes, is called a 'cloudburst'. About 31 cloudbursts occur in Jakarta each year, including five with moments when rain comes at a rate of more than 100 mm/h.

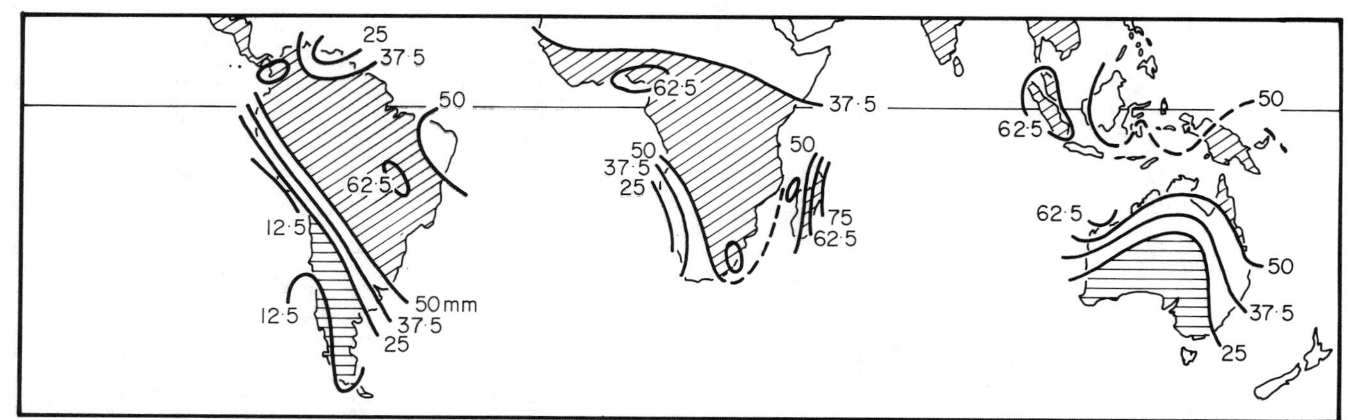

Figure 7.16 The rainfall (mm) over an hour, with a return period of two years (Reich 1963: 21, by permission of Elsevier Science Publishers, Physical Sciences and Engineering Division)

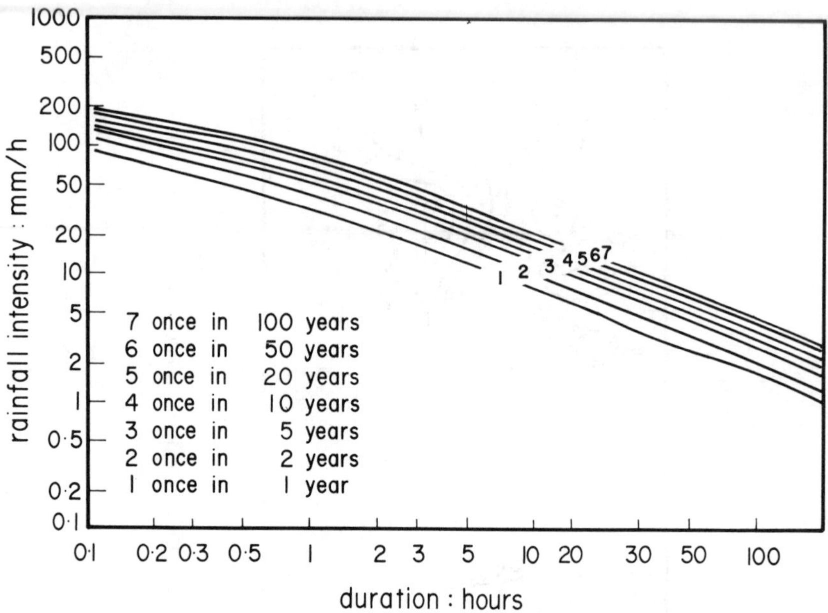

Figure 7.17 Frequency–intensity–duration diagram for Sydney (by permission of the Australian Bureau of Meteorology). As an example, the rainfall exceeds 20 mm over five hours about once each five years

The rainfall intensity over periods down to 6 minutes can be derived from a diagram like Figure 7.17. This is known as a 'frequency–intensity–duration' diagram, resulting from analysis of lengthy pluviograph and rain-gauge measurements at a place. The information is of great use in designing road bridges, dams, drainage culverts, etc.

HEAVY RAINFALLS

'Heavy rain' is defined variously as above 2.4 mm/h, above 4 mm/h or over 7.6 mm/h, for example. Even 4 mm/h is enough to create a roaring noise on roofs and misty spray on the roads. However, much higher intensities have been measured, as mentioned previously. Particularly heavy rains occur on steep windward slopes facing the sea, as the hills trigger thunderstorms and anchor them.

World-record measurements of precipitation averaged over various periods yield the following:

$$Pxt = 400\ t^{0.5} \quad \text{mm} \tag{7.9}$$

where Pxt is the world-record total rainfall over t hours. Thus, the highest rainfall within 120 minutes has been 570 mm, i.e. 285 mm/h. Or the record rate over 10 minutes is about three times that over 100 minutes, which is thrice that for 1000 minutes (i.e. 17 hours). An equivalent equation (but with 400 replaced by 56) relates maximum rainfalls in Norway.

Extreme rainfalls

It is of particular interest to those concerned with either dam design or flood control, for instance, to know what is called the Probable Maximum Precipitation or PMP. This is defined as follows: the theoretically greatest depth of precipitation for a given duration that is physically possible over a given size storm area at a particular location at a certain time of the year. The concept replaced that of the Maximum Possible Precipitation in about 1950. The PMP can bëestimated in at least four ways – either by statistics, by considering the meteorology of storms, by what is called a 'generalised' method, or by means of the 'Growth Factor'. These will now be described.

Statistical estimation

One procedure for estimating the PMP is to assume a purely random year-to-year variation of the annual maximum of the rainfall over a day, and then to use Gauss's formula – Eqn (4.6) – to calculate the intensity which is exceeded only once in T years, the 'return period'. That intensity **I**(T) is given by the following equation:

$$\mathbf{I}(T) = Pmx + k.s \qquad \text{mm/d} \tag{7.10}$$

where Pmx is the long-term mean of a series of annual maxima of daily rainfalls, s is the standard deviation of that series, and k is a factor which depends on the return period and on the rainfall. Values of k between 5–30 have been proposed, with an inverse dependence on the annual maximum storm, e.g. 17 if the average annual maximum 24-hour rainfall is 50 mm, 12 if 200 mm, 7 if 500 mm. Commonly the value is taken as 15.

The period t of totalling the rainfall also affects the value of k in Eqn (7.10). The value is 15 if the annual maximum *one-hour* rainfall is 22 mm, 10 if 50 mm, 5 if 100 mm. If the period is *30 minutes* (i.e. one considers a series of annual maximum half-hourly rainfalls), the appropriate value for k would be 10 instead of 15.

In practice, the choice of t and k is somewhat arbitrary and defines only an intensity which is rarely exceeded, rather than a maximum limit. In the case of storms in the USA, the method has given values which seriously underestimated the actuality, e.g. 343 mm instead of an observed 610 mm. The basic problem is that the complete set of data used in statistical estimation of the PMP provides little guidance in the crucial exceptional cases. The chief use of the procedure lies in providing a quick check on estimates obtained by other means.

Another procedure is similar to that used in estimating extreme winds (see Chapter 6). A Gumbel distribution is assumed instead of the less general Gaussian pattern. Then an extrapolation is made from available data to find the daily rainfall with an arbitrarily chosen return period. Annual maximum rainfalls from places as different as Tunisia, Madrid and Aberdeen fit the Gumbel equation. The extent of extrapolation is usually considerable, contrary to the general rule that one can extrapolate a rainfall series to about twice the record length with a reasonable degree of certainty, but only values for return periods of up to about a fifth of the record length can be regarded as being reliable.

All statistical procedures for deducing the PMP assume that the series of data is homogeneous, i.e. that extraordinary rainfalls arise from the same processes as those creating the general run of measurements, and that the record is long enough to include the rare circumstances which produce extreme rainfalls. But homogeneity is unlikely at low latitudes, where tropical cyclones intrude sporadically on the usual quite different pattern of convective rainfall. At Dakar in Senegal, for instance, there is no homogeneity but two distinct regimes of precipitation, resulting in a bimodal distribution of annual rainfalls. On the other hand, there may be homogeneity at high latitudes, and therefore statistical procedures are widely used in Europe.

Maximisation of storm data

A different method of deducing the PMP in a catchment is based on a consideration of the mechanics of storms. Observations of the water contents of the atmosphere during local storms are related to the rainfalls, so that the PMP can be calculated from an estimate of the maximum possible water content. This procedure is called 'maximisation' of the storms. For each storm, the atmospheric water content is inferred from the surface dewpoint (see Eqn (7.2)). The maximum possible water content (Wx) is reckoned as associated with the highest dewpoint ever recorded locally in the month of the storm. As a result, each storm with a rainfall Ps and a dewpoint Td, equalled or exceeded for 12 hours, yields a figure for the PMP:

$$\text{PMP} = \text{Ps.Wx/Ws} \qquad \text{mm/per storm} \qquad (7.11)$$

where Ws is the atmospheric water content during the storm, inferred from the sea-level equivalent of Td. The correction added to the measured Td to obtain the sea-level equivalent depends on the temperature. If the temperature at sea level is about 10°C, the correction consists of adding 0.57C° for each 100 m of the storm's elevation. At 20°C the increment is

0.49C°/100 m, and at 30°C 0.36C°/100 m. The ratio Wx/Ws in Eqn (7.11) is typically 1.3–1.8.

Transposition of storm data

Where there are no data available on major storms locally, figures may be taken from storms at other places, provided they are within 200 km horizontally (or less among mountains) and 300 m vertically, and are meteorologically similar. After maximisation of the storm data they are 'transposed' to the particular catchment, i.e. adjusted to allow for its different terrain. For example, the maximised sea-level value is increased by 10 per cent if the catchment slopes at 1 in 11, 25 per cent if 1 in 7, and 35 per cent if 1 in 5. If there is a mountain barrier between the site of the transposed storm and the catchment, the maximised sea-level value is reduced according to the estimated maximum dewpoint at the top of the barrier. But a barrier of more than 700 m above the storm area invalidates use of the data.

One difficulty in transposing storm data lies in the subjectivity of selecting transposable storms. Transposition is not possible in data-sparse countries, where it is necessary to use either a 'thunderstorm model' or else a 'generalised method' for estimating the PMP. The former involves assuming that thunderstorms anywhere in the world resemble the largest of the hundreds which have been intensively studied in the USA, after adjustment for local dewpoint and topographical conditions. The model is suited to thunderstorms contained within 500 km^2 extent and 6 hours' duration.

Generalised method

The generalised method of estimating the PMP is based on either Table 7.12 or equivalent diagrams, with subsequent allowance for geographical features of the particular catchment, e.g. distance inland and elevation. The basic information is taken from storms anywhere in the world, and the assumption is made that a storm's area is more important than its location. This method is increasingly widely used, e.g. in Australia, the USA and India, but not yet near the equator for lack of the required preparatory work.

It is a feature of the generalised method that the estimates it yields are higher than those obtained otherwise. For example, the PMP for one particular catchment of 2000 km^2 in Victoria in Australia is given as 560 mm/d, whereas transposition gives 410 mm/d, and in fact no more than 360 mm/d has been observed so far. The higher figures derived by the newer methods necessitate expensive modifications to dams.

Table 7.12. Relationship between the storm area and duration, and the maximum spatial-average rainfall measured at sea level

km² *Storm area*	*Storm duration:* *1 hour*	*6*	*12*	*24*	*48*
	Maximum spatially-averaged rain: mm				
26	236[a]	630	760	980	1000
260	160[a]	500	670	890	990
1300	150	320	630	830	910
5200		280	450	630	720
26,000		150	200	310	440
130,000		60	110	160	250

Source: Linsley *et al*. 1982: 83.

Note: [a]From Wiesner (1968), assuming a dewpoint of 25.5°C.

Growth-Factor method

An easy method favoured in the UK and Norway uses the empirical ratio (the 'Growth Factor') between the maximum daily rainfall (the PMP) and that with a return period of five years. The ratio for rainfalls of more than 80 mm/d is betweeen 3–4 (Forland and Kristoffersen 1989: 263). Comparisons of PMP estimates for nine basins in Norway obtained in this way show fair agreement with those derived either from maximisation with transposition or by the statistical method. For example, The respective estimates for one basin are 341 mm/d, 385 mm/d, and 352 mm/d, whereas the maximum observed over 49 years is 167 mm/d. In all nine basins the estimates were about twice the maximum so far observed.

However the PMP may be calculated, the figure has to be used with caution. There is no objective way of assessing its accuracy and there is an inherent contrast between the looseness of 'probable' and the precision of 'maximum'. Not all engineers accept the estimate, because of the high rainfall intensities which are calculated, necessitating what might appear to be unduly expensive civil engineering. And uncertainties about the mechanisms of storms and their efficiency in producing rain mean that any estimate of PMP is only an approximation. None the less, engineers need *some* guidance and the PMP is a means of providing it.

DEW

Dew provides a small supplement to the rain. For example, it can offset 6 per cent of the potential evaporation during autumn to spring in semi-arid Australia, and in British Columbia it equals about 13 per cent of the midsummer rainfall. About 8 litres were collected each night by a 28-metre fir tree in Washington State, i.e. 20 per cent of the daily evaporation. Dew provides moisture in arid areas for sheep, who lick it from the grass and fence-posts in the early morning. Dew also plays a part in the weathering of rocks in dry regions. But it is undesirable in humid areas, where it promotes diseases on the leaves of crops such as wheat and apples.

Dew is sometimes gathered in useful amounts. In ancient times sailors would collect it for drinking by hanging woollen fleeces at night on the side of the ship, not touching the sea. Then the dew would be squeezed out of the fleeces in the morning. This method of collecting dew was mentioned in the Old Testament (Judges 6: 36–40), and there are Hebrew prayers for dew in the summer.

'Dewponds' on the chalk hills of southern England date from Neolithic times and rarely dry out, providing water for sheep even in the most severe drought. They were made by lining pits with flints and stones, followed by a layer of straw and finally a thick coating of puddled clay. Even today, dew is collected on Lazarotte, a small dry island in the Atlantic, to nourish vines planted at the centres of dew 'traps', each consisting of a hole 3 m in diameter and 2 m deep, lined with volcanic ash. Similarly, dew is collected within gravel mounds 70 cm high in Israel and in the Sahara.

Measurement of dew

An instrument to measure the occurrence and amount of dew is called a 'drosometer'. There is no standard equipment, and a variety of methods have been used. For example, plastic tape which extends on being wetted. Or dew is measured by the increase of weight of exposed absorbent materials such as porous gypsum plates or blotting paper or expanded polystyrene. Unfortunately, these materials absorb moisture whenever the atmosphere has a high relative humidity, even though a film of dew is absent. It is better to measure dew by weighing initially dry filter paper before and after pressing it on to the natural dew-laden surface. A more complicated research instrument weighs a block of soil and vegetation to find the small increase caused by dew.

A common procedure involves 'Duvdevani blocks'. These are made of wood, 25 × 50 × 330 mm in size, painted with a special red paint and mounted a metre above the ground. The appearance of the collected dew is observed soon after sunrise and matched against standard photographs of eight different amounts of dew from 0.01 mm to 0.40 mm. A serious disadvantage is that the blocks do not resemble leaves of grass, for example, and may collect only a fifth as much dew. Also, the red paint deteriorates after a few months of sunshine, and the alteration affects the amount of dew collected.

The effect of the kind of surface is a problem. Balsa sticks a few millimetres wide can collect several times the amount on nearby grass, and over twice as much is deposited on a rubber sponge surface as on a bronze mesh. Up to 0.47 mm of dew per night was measured on a crop of sugar beet, but only 0.20 mm on to grass. Dew on to a wheat crop in South Africa is twice that measured above a lawn.

An instrument made by the German firm Thiess has been used at Marsfield. It continuously records the weight of a wire grid on which dew deposits. The grid is 17 cm above the ground.

The height of the collecting surface influences the quantity of dew. More has been measured at 8 cm than at 1.5 m above a lawn, presumably because the higher collector would be in a stronger wind and be too far from the ground for distillation, as explained below. However, Muller (1968) obtained 0.15 mm per night of dew on to a surface 5 cm above turf in Vienna, 0.30 mm at 1 m above, and zero at 5 m above. A similar maximum dewfall at an intermediate height (i.e. 50 cm) was also found in the arid north-west of India.

Amounts of dew

There have been many measurements of dew. Figures from Israel range from 25-200 mm/a. Amounts measured in Ohio, Copenhagen and Jamaica were 230 mm/a, 51 mm/a and 34 mm/a, respectively. Dew collection in the arid west of New South Wales is only 18 mm/a, and about 12 mm/a is collected in the interior of South Africa – mostly at the end of the rainy season. Less than 9 mm/a of dew is measured in almost rainless parts of north-west India. It is unfortunate that there is least dew in arid areas where it would be most useful.

Various amounts have been quoted for the *maximum* amount of dew in a single night. Severini *et al.* (1984) measured 20 mg/m^2.s, i.e. about 0.6 mm per night, and papers reviewed by Garratt and Segal (1988) also include rates up to 0.6 mm/night. At least six other authors reported maxima of 0.5 mm/d or more, another six 0.4–0.5 mm/d, eleven reported 0.3–0.4 mm/d, and two quoted less than 0.3 mm/d as the maximum. Data from Table 3.2 inserted into a formula of Yamamoto (1937: 95) indicate a dewfall of 0.32 mm per night.

Differences between climates do not seem to affect the maximum amounts in a consistent way, e.g. Jamaica, Germany, Israel and England yield the same values. On the other hand, Kessler (1985: 123) quoted 0.32 mm/d for the Belgian Congo, 0.30 mm/d for 11°N in India, but only 0.13 mm/d for 31°N, and 0.06 mm/d for Finland, i.e. less at higher latitudes.

The median of all the reports just mentioned is 0.35 mm/d. That amount of dew would be deposited if the dewpoint of a 100-metre depth of air fell from 18°C to 14°C, for instance. Also, such an amount of dew on to a roof with an area of 100 m^2, would yield 35 litres daily if it all ran off before evaporating.

Registrations of dew at Marsfield are only about 0.1 mm/d, like the 0.08 mm/d at Berlin. These typical amounts of dew are below the maximum quantities mentioned earlier on account of cloud, dry air or wind.

Dew formation

There are three sources of dew: dewfall, distillation and guttation. The last is found on the leaves of certain plants which continue to exude moisture from within, even at night. There is also an unusual absorption of moisture from the air in the case of a certain desert plant in Chile, which exudes a hygroscopic salt on to the leaf surface. This attracts about 0.1 mm of moisture each night.

Distillation occurs from the ground surface if it is wet and has been warmed by the previous day's sunshine. There is then a transfer of moisture by evaporation from the ground, followed by diffusion to exposed leaves cooled by long-wave radiation loss at night, and finally condensation there. Such distillation may be the main part of the dew formed in arid regions on calm nights. It does not represent additional water but recovery of some of the water lost by evaporation from the ground.

Dewfall is condensation from the air. It is precisely the reverse of surface evaporation and therefore obeys the same formulae, but with a negative sign, e.g. the Dalton equation mentioned in Table 3.1. A crop of wheat may receive two-thirds of the dew as dewfall on to that part of the canopy which is more than 60 cm from the soil and the rest as distillation from the ground on to the lower part of the foliage.

Considering the heat and moisture flows about a sward at night allows calculation of the amount of dew (see Note 2, pp. 338–9). This indicates that *distillation* would be about 0.024 mm/h (in the typical case of a heat flux at dawn of about 30 W/m^2, from within damp ground to its surface) and *dewfall* would be 0.025 mm/h, totalling 0.05 mm/h. This value is plausible, though higher than the 0.03 mm/h measured by Jennings and Monteith (1954), or the Marsfield measurement of around 0.01 mm/h discussed below.

Marsfield studies

Measurements by R. Nurse at Marsfield show that the screen dewpoint temperature is about 3C° less than the screen temperature at the onset of dew formation, i.e. the ground is then 3C° cooler than screen temperature. Dew begins to form early in the morning in January but at about 8 p.m. in July (i.e. winter). In all months, dew accumulates till sunrise, ie for about 4 hours in summer but 11 hours in winter. So, although rates of accumulation are faster in summer (i.e. 0.0125 mm/h instead of 0.0094 mm/h), the longer night leads to heavier falls in winter. The maximum in winter is 0.16 mm per night, and in summer 0.10 mm.

Other features of dew formation were shown in Table 3.2. The 9 a.m. screen dewpoint temperature is about 2.3C° above the screen daily minimum in both winter and summer, implying rapid evaporation after sunrise into the air near the surface.

Wind speeds during the formation of dew at Marsfield are sometimes less than the 0.5 m/s or thereabouts suggested in the literature as the lower limit for dew formation necessary to stir water vapour down to the surface. However, the threshold is uncertain because the standard anemometer is not accurate at such low speeds (see 'Measurement of wind', p. 49).

Whatever the lower limit, if any, there is certainly a reduced likelihood of dew formation at high speeds because stirring induced by the wind increases forced convection of heat down to the surface thus reducing the chance of cooling it to dewpoint. Severini *et al.* (1984) suggested an upper limit of 1 m/s for dew formation. But the evidence at Marsfield for rainless nights in 1981 is as follows: there was dew during 12 out of 18 completely still nights, on 19 out of 30 nights with a wind of about 0.5 m/s, on 11 out of 24 when there was 1.5 m/s, on 10 out of 27 when 2 m/s, and 14 out of 55 when the wind exceeded 2 m/s, showing a steady decline of dewfall probability from 67 per cent to 25 per cent. Extrapolation indicates an upper limit of 3 m/s for dew formation.

Other observations

The difference between dewfalls in summer and winter appears to vary with location. At Marsfield there is most in winter because of the longer night. But daylength varies less nearer the equator where there may be more dewfall in summer on account of the faster rate of formation, especially in humid, coastal climates. The amounts are similar in summer and winter at Pretoria, whereas in Israel there is more dew overall in summer than winter because of the different frequencies of clear nights. Any seasonal variation of cloudiness influences the incidence of dew.

There are heavier dewfalls in valley bottoms than on the sides of mountains, presumably because of the presence of rivers and wet ground and the ponding of cold air. However, at higher elevations there may be an increase of dewfall with height, despite the decrease of dewpoint implied in Figure 3.10. There was 0.10 mm/night for 37 nights per annum at 2900 m in Colorado, but only 0.03 mm/night for 66 nights lower down at 1300 m.

Dew lasts for several hours. In Ontario, it accumulates in August from about 10 p.m. until 9 a.m. (i.e. during 11 hours), as at Marsfield in winter. The dew can then remain on leaves several hours more. In Vienna, dew begins to form within an hour before sunset and lasts until 1–2 hours after sunrise. Sometimes the duration is reckoned as an hour more than the time that the screen relative humidity exceeds 90 per cent.

The time can be either calculated, or measured by means of a water-soluble pencil recording on a rotating aluminium disc on which the dew forms. The stretching of a wet lamb-gut membrane, or measuring the electrical resistance across an exposed surface, also can be used to detect the moisture.

Dew at Marsfield takes about 3 hours to dissipate after dawn. This represents an evaporation rate equivalent to 0.8 mm/d, which is low compared with values in Figure 3.1 as may be expected with the still air and weak irradiance just after dawn.

Fog-drip

Droplet deposition from fog is sometimes confused with dew. However, there is no phase-change from vapour to liquid with fog-drip since the cloud is already wet. Also, fog-drip can occur at any time of the day, whereas dew forms only at night.

The amount of water collected as fog-drip can be appreciable. Vertical gauze in the cloud on Table Mountain at Cape Town gathers many litres in a year. An ordinary radiation fog can provide 1 mm depth of water on any surface, i.e. one litre/m^2. The settling of such a fog into a rain gauge in calm conditions may register 0.2 mm/d. The fog drip collected in three months on a vertical harp of wires at the Californian coast (at a place with rainfall of 600 mm/a) amounted to 252 L/m^2 of vertical surface.

Schemenauer *et al.* (1987, 1988) described the collection of large quantities of fog on vertical nylon filaments across a frame which swings to face the wind. A frame of 0.25 m^2 can collect 50 L/a at 700 m elevation on a coastal mountain in Chile. Less is collected at other elevations. Presumably, shrubs would collect similar quantities.

Fog may deposit useful amounts of moisture on to vegetation. It is equivalent to 150 mm/a along the coastal regions of the Kalahari–Namib desert in southern Africa. Indeed, the moisture from fog is sufficient to support lush vegetation in high tropical islands in the Pacific and Atlantic oceans, even during months without rain. Moisture collected from cloud by trees in Hawaii effectively enhances the rainfall by 300 mm/a, and a redwood tree on the northern coast of California may collect the equivalent of 1.3 mm/d.

SNOW

Precipitation is likely to be snow wherever high latitude or elevation leads to a surface wet-bulb temperature below about 0.6°C. Half the precipitation is snow when air temperatures are 1.5°C in lowland Britain, 2.5°C in the USA, or 4°C in central Asia at 4000 m. In general, the percentage of a month's precipitation which lands as snow is approximately (50 − 5 T) per cent near sea level, where T is the monthly mean temperature, but

(75 – 8 T) per cent at 3100 m in the Alps, for instance. However, the percentage is greater in late winter and spring than in autumn and early winter, for the same screen temperature, because of steeper lapse rates in spring.

The effects of snow can be considerable. Three hundred millimetres depth prevents cattle from feeding and sheep from walking, and 600 mm necessitates people wearing skis or snow-shoes . There may be dangerous snow avalanches into valleys, whilst the melting of snow in the spring and summer provides water for the rivers.

Snowfall measurement

There are many methods of measuring snow depth. Light falls of snow or a mixture of rain and snow can be measured with a normal rain gauge, preferably Teflon-coated and fitted with a Nipher shield, as shown in Figure 2.10. The snow is melted after collection by bringing the gauge indoors, where the melt plus rainfall are measured together. Otherwise, a known amount of warm water can be added to the gauge to melt the snow. However, snowfall measurement with rain gauges is not recommended, because wind greatly affects the degree to which the amount of snow collected is representative of the local precipitation rate. Experiments in a wind tunnel suggest that a rain gauge seriously underestimates the snowfall.

A larger sample can be collected with a bucket on a weighing machine, with less vulnerability to the turbulence which occurs around the rim of a small rain gauge. The descent of the bucket is recorded on a paper chart. One problem is the imprecision due to oscillations caused by any winds over 7 m/s, and another is the risk of the mechanism icing up.

An alternative is a flat, rubber pillow of 1.8 m or 3 m diameter and 0.1 m depth, filled with antifreeze or alcohol. Or the pillow may be made of flexible stainless-steel, 1.2 m × 1.5 m × 13 mm. The weight of snow on the pillow raises the pressure on the liquid within, and the pressure is measured.

More commonly, snowfall is determined by measuring the depth on a previously cleared piece of ground, or, better, on a board laid on the snow after the previous measurement. The board should be at least 0.3 m × 0.3 m and have a specific gravity similar to that of the snow and be white and waterproof. A tube of about 40 mm diameter is used to cut a cylinder of snow vertically down to the board, and the tube is subsequently weighed to find the snowfall in terms of the equivalent depth of water.

To determine the average snowfall over an area it is necessary to find the snow depth at 10–20 points about 20 m apart. Fixed white stakes are marked off in centimetres and observed regularly. This tedious and crude procedure is made necessary by the considerable spatial variability of snowfall.

Density and albedo

The conversion of a value of snow depth to the equivalent depth of rain is complicated by the gradual increase of snow's density. Fresh powder-snow deposited in calm conditions may have a density which is only about 5 per cent that of water, whilst snow deposited in a gale may immediately have a density of 45 per cent. Generally, fresh snow is reckoned as having a density of 10 per cent that of water. However, the density increases to 14 per cent that of water, even within the first few hours. Then it compacts under its own weight and by the absorption of water that has melted at the surface, percolated down and refrozen.

An initial density of 10 per cent means that 3 mm of water would result from melting a 30 mm layer of snow. Thereafter, the density (D) increases according to the daily mean temperature (T) and the number of days with a mean wind above 6 m/s (N_6). Bruce and Clark (1966: 25) gave the following formula:

$$D = Do + 0.24\,N_6 + k.\Sigma(T) \qquad \text{per cent} \qquad (7.12)$$

where Do is the initial density (%), $\Sigma(T)$ is the sum of daily mean temperatures since the snow had a density Do, and k equals 0.7 until D exceeds 26 per cent, and then is 0.2. According to Eqn (7.12), 30 calm days with mean temperatures of 1°C would increase the snow's density from 10 per cent to 27 per cent, i.e. 0.6 per cent/day on average. This is four times what is observed in the Australian alps, i.e. a linear change from 30 per cent at the beginning of June to 62 per cent on 1 December (Kneen 1986).

After a month or two of spring, the snow has consolidated (i.e. 'ripened') to become 'firn', which consists of large grains saturated with water, all at 0°C, with an overall density of about 40 per cent. Any strong wind creates a smooth crust. Later snowfall causes further compaction, eventually forming glacier ice, which has a density about 90 per cent that of water.

The increase of the density of a snow layer is accompanied by a decrease of the surface albedo. In the USSR the albedo of initially 0.60–0.83 (being more for deeper snow layers) falls to about 0.50 when half the snow has melted, eventually reaching 0.30. Aguardo (1985: 205) found an albedo of 0.84 for a density of 10 per cent, 0.73 for 30 per cent and 0.46 for 50 per cent. Others have measured 0.88 for dry clean snow, 0.61 for wet clean snow and 0.29 for wet dirty snow. The albedo of a shallow layer of snow on a field falls from 0.80 to 0.20 within 10 days, whereas that of deep snow falls to 0.47 in that time. One concludes that a typical value for melting snow is 0.5.

Melting snow

Snow disappears either by melting and runoff or by sublimation from the solid state into water vapour in the atmosphere. The latter is relatively trivial, as is melting due to heat from the ground. Some melting is caused by rain, but this too is insignificant; the heat in a rainfall of P mm is sufficient to melt only 0.013 Tw.P mm of snow (in terms of the resulting run-off), where Tw is the wet-bulb temperature at the time of the rain. Whatever melting occurs on raindays is due less to the rain than to the accompanying winds and relative warmth.

The heat for melting comes chiefly from three sources: from solar irradiance, from the heat released when water vapour condenses on to the snow, and from warm winds. The relative importance of these depends on the circumstances. The example in Note 3, pp. 339–41 shows them as similar in magnitude. However, radiation melting is dominant where there is settled weather with no wind, and is generally the major factor in Canada. Likewise, 56 per cent of the energy used in melting at a place in Norway comes from the net irradiance. A figure of 52 per cent has been measured at a place in New Zealand, forced convection melting accounting for most of the rest. Convection melting dominates when there are warm winds. Condensation melting becomes important only in humid, cloudy and windy conditions.

It is traditional to regard the overall melting rate as proportional to the air temperature, the proportionality being called the 'melt factor'. Ten out of eleven methods of calculating snowmelt compared in a World Meteorological Organization study involved the melt-factor. Values given in 13 publications range from 1–9 mm/d.C°, because of fluctuations of irradiance, wind and humidity, with a median of about 3 mm/d.C°. The value for melting within a forest differs from that for snow in the open. In general, low values of the factor occur with high rates of melting.

The melt-factor concept has been claimed to be useful when daily mean temperatures exceed 2°C. However, Harding (1986: 14) quoted a case in Norway where there was no relationship whatsoever between daily mean temperatures and melting rates, and the melt-factor method of estimating snowmelt has also been found unsuccessful elsewhere in the case of snow on open grasslands.

Zuzel and Cox (1975) showed that the *partial* correlation coefficient (see Note 10, Chapter 4) of the melting rate at three sites in Idaho was 0.65 for net irradiance, 0.57 for wind, 0.56 for vapour pressure and, surprisingly, zero for air temperature. So the strong *simple* correlation coefficient between melting rate and temperature (e.g. 0.72 in Idaho) is due to the dependence of temperature on irradiance. In other words, air temperature is a good predictor but not itself the cause of melting. It is more useful to calculate the separate components of melting, as in Note 3, pp. 339–41.

Actual rates of melting (in terms of the equivalent depth of water that is formed) can be considerable. Maximum rates quoted in the literature include 12 mm/d in Vermont, 14 mm/d in New Zealand, 15 mm/d in the Great Lakes area, 22 mm/d in Austria, 25 mm/d in Ohio and the Himalayas, 30 mm/d in Australia, 31 mm/d in Scotland, up to 34 mm/d in New Guinea, 34 mm/d in the north-west USA, 38 mm/d in Canada, 40 mm/d in the USSR and 70 mm/d in Idaho. Hay and Fitzharris (1988: 205) quoted up to 70 mm/d in New Zealand, the average over 53 days being 38 mm/d. Such high rates imply considerable flooding of the rivers which collect the run-off if the catchment is large.

Amount of snow

The depth of snow depends on the balance of past rates of deposition and melting. Annual maximum depths at a snowfield of Australia exceeded 1250 mm in 5 per cent of years, 1000 mm in 50 per cent and 400 mm in 95 per cent. A like distribution occurs in the USA, but a Gumbel distribution is assumed in Canada. Such patterns allow calculation of the extreme load to be expected on a roof, for instance, since snowfall data can be treated like wind-speed records (see 'Estimating extreme winds', pp. 231–9).

Also, snow data can be treated like rainfalls to determine periodicities (see pp. 134–5, 278–83). Australian annual maximum snow depths indicate a slight alternation of heavy and light snowfalls (see Note 4, p. 341), reflecting the Quasi-Biennial Oscillation (see p. 279) and an associated fluctuation of the annual mean latitude at which anticyclones cross Australia's south-east coast. It appears that a year with deep snow is often followed by one of less than usual depth, especially if the latter coincides with an ENSO event (see pp. 281–2).

Duration of snow cover

Snow lasts longer at higher elevations because of heavier snowfalls and reduced air temperatures. Extra elevation by about 7 m, in areas above 400 m in Britain, increases the duration of snow cover by an extra day each year. In the western Carpathian mountains the change is about a day for each 11 m, and in Switzerland a day per 8 m. The altitude for a duration of 90 days, sufficient for a ski season, is 1700 m in south-east Australia, but only 930 m in Switzerland.

What is called the 'snowline' is the elevation above which there is permanent snow, even in midsummer. It varies with latitude and is higher in the northern hemisphere – at 60°N it is about 2500 m, whereas it is near sea level at 60°S. It rises to above 6500 m in the dry tropics of central Asia and South America and is around 5000 m at the equator. There is a glacier on the equator at 4400 m in Papua New Guinea. On the whole, the snowline corresponds to an annual warmest month with an average temperature of about −2°C.

The duration of snow is longer in places which are sheltered from the wind and facing away from the sun. For instance, it melts 15–20 days later on steep northern slopes in north-east Siberia than on slopes facing south. Also, slopes oriented towards the east hold snow longer than those facing west, since sunshine in the morning is offset by the chill of air cooled at night.

PRECIPITATION AND CLIMATE CHANGE

Global warming will change the pattern of rainfall around the world, though we cannot yet say exactly how. This will cause problems at a time when fresh-water supplies have already been strained by the rising population and increased *per capita* consumption (see 'Introduction', to this chapter pp. 250–1).

Past fluctuations of rainfall have been complex. Global cooling around AD 1200 led to a much reduced precipitation on the northern plains of the USA but increased rainfall on the southern plains, whilst the drying out of the Sahara was accompanied by filling of the Caspian Sea. The cooling that occurred in Europe in the summer months of the 16th century was accompanied by more *wet* years in Switzerland, contrary to the general rule. During the present century there has been an increase of summer rainfall in eastern Australia but a decrease of winter precipitation, i.e. a shift polewards of the monsoonal influence. On the other hand, there has been a steady increase since 1850 in the frequency of rainfalls of less than 12 mm/d in New Mexico, offset by fewer heavier rainfalls, which indicates a shift of the pattern of winds *towards* the equator at that longitude. These past vagaries point to similar complexity of the rainfall pattern in future.

In view of the importance and uncertainty of coming changes one has to be cautious in making predictions. What follow are simply conjectures based on the assumption of global warming by about 3C°.

Overall, there is likely to be more precipitation because of greater evaporation from warmer oceans. The Dalton equation (see Table 3.1) shows that an increase of ocean surface temperature from 15°C to 18°C, for instance, with a constant atmospheric relative humidity of 70 per cent, would enhance evaporation, and hence rainfall, by 21 per cent. The example discussed in 'Estimating and climate change' (pp. 106–8) indicates an increase of global rainfall by 27 per cent. Five elaborate computer models described by Mitchell (1989: 127) all showed increases around 10 per cent, as a result of doubling carbon dioxide concentrations.

However, the changes of precipitation will depend on location. Present calculations suggest that there will be higher summer rainfalls in many parts of the world (e.g. near the equator, in India and Australia, but only the southern half of the USA), alongside decreased annual rainfall over much of Africa and South America. The pattern will be affected by shifts of the general circulation as a result of a reduced difference between equator and polar temperatures, since global warming will affect polar regions especially. A poleward shift of the latitudes of wet and dry climates (see 'Spatial variations of rainfall', pp. 264–70) would bring occasional intense rainfalls, due to tropical cyclones, to latitudes now too high for that. Also, there would be an extension of monsoonal climates away from the equator, and perhaps a displacement of the belt of anticyclones which create zones of aridity.

Presumably, rainfalls will increase in coastal regions as a result of more oceanic evaporation, but this effect may not reach inland. There, the extra warmth will promote evaporation so that soils will be drier, causing poorer crops from many important areas at a time when the world's increased population needs more food.

Higher temperatures inland will increase the proportion of rainfall of the relatively intense convective kind. This will enhance soil erosion, and increase the risk of flooding, and the overtopping of dam-walls.

Estimating extreme rainfalls in the future will be made more difficult by the changes introduced into precipitation records by global warming. There will no longer be a homogeneous series of values which can be extrapolated statistically. In general, one would expect higher extreme values than the record so far would indicate.

Water supplies from snowfields will be reduced by global warming. Figures for the sea-level temperature and the height of the snowline at various latitudes (see 'Snow', pp. 305–10) imply that a 3C° warming will raise the snowline by about 400 metres, decreasing the amount of snow to be melted. Other figures given earlier for the average lapse rate (see Eqn (3.5)) and the duration of snow, as affected by elevation, indicate a reduction of snow duration by about 50 days or so.

In short, considerable research and enterprise will be needed to solve the water-supply problems of the future.

8

NOTES

CHAPTER 1

1 Documentation

For more details see the complete list of references of all the documents consulted during the writing of this book (over 2,000 of them) which is available from the author (see the Preface).

2 Simple method of temperature conversion

(i) Add 40 to the temperature you wish to convert; it does not matter whether it is °C or °F.
(ii) Multiply the sum by 5/9 if converting from °F to °C, and by 9/5 if converting from °C to °F.
(iii) Take away 40 from the result of the multiplication.
Example: Convert 50°F – 1 Add 40, hence 90; 2 Multiply by 5/9, hence 50; 3 Subtract 40, hence 10°C.

3 Converting a formula to SI units

The method consists of four steps:

(i) State the relationship between non-SI and SI units (see Table 1.4), with the former on the left side of each equation.
(ii) Replace each non-SI term in the given formula by its SI equivalent.
(iii) Simplify the new formula.
(iv) Validate it, to show that a convenient set of values gives the same answer in both the old and the new formulae.

The procedure is best explained by examples, as follows. Note that the SI version of an equation should contain figures with the same number of digits as in the original equation in order to indicate the same degree of exactness. Here we express non-SI units by capital letters and the same quantities in SI units by lower-case letters.

Example 1

A formula for the rate of the accretion of ice (I) on to a vertical surface is as follows:

$$I = 0.078\ U.P^{0.88} \qquad \text{inch/hour} \qquad (8.1)$$

where U is the wind speed (mph) and P is the rate of precipitation (inch/hour) in freezing conditions. We want to convert this to SI units.

(i) 1 inch = 25.4 mm   Hence 1 inch/hour = 25.4 mm/h
and 1 inch/hour   = 25.4 I mm/h = i mm/h
Thus   25.4 I = i   Hence I = i/25.4
Similarly   U = 2.24 u, where u is in m/s
Also,   P = p/25.4 = 0.0394 p, where p is in units of mm/h.
(ii) Hence $i/25.4 = 0.078\ (2.24\ u)\ (0.0394\ p)^{0.88}$
(iii) Thus $i = 25.4 \times 0.078 \times 2.24 \times 0.0394^{0.88}$

So $i = 0.26\ u.p^{0.88}$ mm/h   (8.2)

(iv) Take this case: U = 10 mph (i.e. u = U/2.24 = 10/2.24 = 4.47 m/s)
and   P = 2 inch/hour (i.e. p = 51 mm/h)
From Eqn (8.1) I $= 0.078 \times 10 \times 2^{0.88} = 1.46$ inch/hour
From Eqn (8.2) i $= 0.26 \times 4.47 \times 51^{0.88} = 37$ mm/h = 1.46 inch/h
The agreement of the numbers in italics validates Eqn (8.2).

Example 2

Consider an expression given by Curry *et al.* (1966: 128) for the error in reckoning the average rainfall over a catchment to be equal to the mean of measurements made by G rain gauges per square mile:

$$\log E = 0.711\ P^{0.5} + 0.663 \log G - 2.554 \qquad (8.3)$$

where P (inches) is the measured rainfall.
(i) 1 inch = 25.4 mm. Hence P inch = 25.4 P mm = p mm
Thus   P = p/25.4, where p is expressed in mm.
Likewise   G = 2.59 g, where g is the number of gauges per km^2,
per km^2,   (since 1 $mile^2$ equals 2.59 km^2)
and   E = e/25.4, where e is expressed in mm.
(ii) Hence, log (e/25.4) $= 0.711\ (p/25.4)^{0.5} + 0.663 \log (2.59\ g) - 2.554$
$= (\log e) - (\log 25.4) = \log e - 1.405$
(iii) Thus $\log e = 0.141\ p^{0.5} + 0.663 \log g + 0.663 \log 2.59 - 2.554 + 1.405$
Hence, $\log e = 0.141\ p^{0.5} + 0.663 \log g - 0.875$ (8.4)
(iv) Take the case of P = 1 inch
(i.e. p = 25.4 mm)
and G = 1 gauge/$mile^2$ (i.e. g = 0.386 gauges/km^2)
From Eqn (8.3),   E = *0.014 inches*
and from Eqn (8.4),   e = 0.365 mm = *0.014 inches*
So there is agreement, and therefore Eqn (8.4) is correct.

CHAPTER 2

1 A school climate station

Weather observation provides an introduction to science without the need for laboratory equipment, it is relevant to geography, and provides raw material for mathematics. It includes both theory and practical work and is relevant to everyday problems. For such reasons, a climate station can be a useful facility at a school.

However, the following practical matters should be considered before setting up a climate station for educational purposes:

1 It is important to write down at the outset the purposes of installing the climate station, the questions it is hoped to answer.

2 Efforts should be made to find local problems (such as the movements of air-pollution) to which the data can be applied, to make them relevant.
3 The observations could be part of the teaching of either science or geography – or preferably both – helping to link the subjects.
4 The commitment to run the weather station should not be unlimited. It should be reviewed each six months, for example. Derelict equipment teaches the wrong lesson.
5 Will it be necessary or possible to make measurements at the weekends or during holidays?
6 Can expert help be obtained to train the observers?
7 How will the roster of observers be organised?
8 Who will supervise the observers and collect the data?
9 The surface of the site should preferably consist of well-watered, mown grass. Who will do the required maintenance?
10 Arrangements have to be made for periodically recalibrating the equipment and for maintaining it.
11 Security of the equipment is a major consideration.

2 The effect of the environs on climate measurements

The dependence of readings on the microclimate is shown by the following anecdote from *Time* magazine (7 September 1981):

> *Some Like It Hot*
> Thanks to John Baudouine, of Bullhead City, Ariz. (pop. 15,000), his town is now officially hotter than ever before. Baudouine, a fireman, is no sorcerer: for four years he has been in charge of reporting daily weather statistics to the National Weather Service for his stretch of western Arizona desert. Last April a stickler from the Weather Service told him to move his thermometer from the firehouse's comparatively cool, sprinklered front lawn to more 'natural terrain'. Baudouine picked a dusty patch 100 yards away, and the high temperatures in Bullhead City were promptly four or five degrees higher. On eight days last month, in fact, television weathermen announced that the town – with temperatures as high as 115°F – was the warmest spot in the U.S.
>
> This has made some residents of Bullhead City hot under the collar. Dick Smith, who owns Dick and Lovella's Five Grand Cafe, is leading a petition drive to move the thermometer back. Says Smith of the new publicity glare: 'It's caused my business to fall off 20% to 30%.' But Baudouine is unapologetic. 'This kind of notoriety is good for the community,' he says. Cooler heads among Bullhead City businessmen seem to agree. According to a membership survey by the Chamber of Commerce, 90% think the summer superlative is 'good for business'.
>
>

3 The maintenance of pen recorders

The typical ink recorder used at a climate station requires careful maintenance. The recording pen may have either a fibre-tipped nib requiring renewal twice a year or so, or a metal nib. The latter needs regular cleaning with methylated spirits, using a small artist's brush to remove encrusted ink deposits. The slot between the points of the nib can be cleaned by drawing a sheet of paper through it.

Allow the pen to dry before adding ink and then transfer the ink from container to nib by means of a match or clean nail. Pens are replenished with an ink made of a mixture of a dye, glycerine and sugar in water. The glycerine reduces evaporation. Do not fill the nib completely, since the volume of ink may subsequently increase in humid conditions by absorption from the air. Then a piece of thin wire is drawn through the slit in the nib to induce the ink to flow to the points.

All surfaces of autographic instruments must be kept clean by regular wiping with a cloth. Spilt or splashed ink should be removed immediately.

Linkages may stick, so that sharp peaks or fluctuations in the trace are replaced by jerky steps. Clean the linkage with an oil-free cleaning fluid; never use oil or grease on any bearings, since they hold dust.

Care must be taken that the paper chart is accurately aligned on the rotating drum, else readings will be in error. A similar error can arise from expansion of the paper in humid climates.

4 Rounding numbers ending with 5

A problem arises in reducing the number of significant figures when the last digit is 5. It is clear that the digits 6, 7, 8, 9 raise the previous digit by unity, whereas 4, 3, 2, 1 are simply discarded. The question is whether to round up or down when the last digit is 5. Berry *et al.* (1945: 129) recommended rounding up, but that creates a bias, with five digits rounding up (i.e. 5, 6, 7, 8, 9) and only four rounding down (1, 2, 3, 4). An escape from that difficulty is provided by adopting some convention, such as rounding either up or down, to make the last retained digit either 0, 2, 4, 6 or 8, i.e. an even number (*ibid.*). However, the opposite convention (of rounding to make the last digit *odd*) was proposed by the Meteorological Office in the UK. An advantage of this is that it prevents zero being the last digit so there is no risk of an apparent rounding by an additional decade. This convention has been adopted in the present book.

5 Lag times

Eqn (2.2) can be developed to obtain expressions for obtaining the lag time L. For example, if t_2 is the time for the measurement error (e) to fall to half of a sudden change (ΔT) of the measured variable (i.e. $\Delta T/e$ becomes two), then taking natural logarithms of both sides of Eqn (2.2) gives L as $t_2/(\ln 2)$, i.e. 1.44 t_2. Alternatively, logarithms of values of the error, plotted on a graph against time, gives a straight line whose slope is 1/L.

6 Expressions for the saturation vapour pressure

There is a range of expressions to choose from, which are either simple or accurate. They include the equations of (i) Teten devised in 1930 with an error of less than 0.04 per cent over 0–35°C; (ii) Richards (1971), accurate within 0.1 per cent over −50°C to +140°C; (iii) Antoine (see McIntyre 1980: 41); and (iv) Lowe and Fricke (see Jensen 1983: 13), with an error of less than 0.01 per cent. A simplified version of the last (above 0°C) is as follows:

$$es = 6.1 + 0.27\,T + 0.034\,T^2 \qquad \text{hPa} \tag{8.5}$$

An alternative approximate equation is this (Linacre 1964b: 66):

$$es = 6.5\exp(T/16.1) \qquad \text{hPa} \tag{8.6}$$

where T is the temperature in degrees Celsius.

7 Saturation water-vapour pressure

Table 8.1 The effect of temperature on the saturation pressure of water vapour over a water surface. For example, it is 43.4 hPa at 30.4°C

Temperature: °C	*Saturation water-vapour pressure: hPa* 0	2	4	6	8
0	6.1	6.2	6.3	6.4	6.5
1	6.6	6.7	6.8	6.9	7.0
2	7.1	7.2	7.3	7.4	7.5
3	7.6	7.7	7.8	7.9	8.0
4	8.1	8.2	8.4	8.5	8.6
5	8.7	8.8	9.0	9.1	9.2
6	9.3	9.5	9.6	9.7	9.9
7	10.0	10.2	10.3	10.4	10.6
8	10.7	10.9	11.0	11.2	11.3
9	11.5	11.6	11.8	11.9	12.1
10	12.3	12.4	12.6	12.8	12.9
11	13.1	13.3	13.5	13.7	13.8
12	14.0	14.2	14.4	14.6	14.8
13	15.0	15.2	15.4	15.6	15.8
14	16.0	16.2	16.4	16.6	16.8
15	17.0	17.3	17.5	17.7	17.9
16	18.2	18.4	18.6	18.9	19.1
17	19.4	19.6	19.9	20.1	20.4
18	20.6	20.9	21.2	21.4	21.7
19	22.0	22.2	22.5	22.8	23.1
20	23.4	23.7	24.0	24.3	24.6
21	24.9	25.2	25.5	25.8	26.1
22	26.4	26.8	27.1	27.4	27.7
23	28.1	28.4	28.8	29.1	29.5
24	29.8	30.2	30.6	30.9	31.3
25	31.8	32.1	32.4	32.8	33.2
26	33.6	34.0	34.4	34.8	35.2
27	35.6	36.1	36.5	36.9	37.4
28	37.8	38.2	38.7	39.1	39.6
29	40.1	40.5	41.0	41.5	41.9
30	42.4	42.9	43.4	43.9	44.4
31	44.9	45.4	46.0	46.5	47.0
32	47.6	48.1	48.6	49.2	49.7
33	50.3	50.9	51.4	52.0	52.6
34	53.2	53.8	54.4	55.0	55.6
35	56.2	56.9	57.5	58.1	58.8
36	59.4	60.1	60.7	61.4	62.1
37	62.8	63.5	64.1	64.8	65.6
38	66.3	67.0	67.7	68.4	69.2
39	69.9	70.7	71.5	72.2	73.0
40	73.8	74.6	75.4	76.2	77.0
41	77.8	78.6	79.5	80.3	81.2
42	82.0	82.9	83.8	84.6	85.5
43	86.4	87.3	88.2	89.2	90.1
44	91.0	92.0	92.9	93.9	94.9

8 Underestimation by rainfall recorders

A siphoning gauge or a tipping-bucket pluviograph fails to record during the time of emptying. As a result, the instrument tends to under-estimate heavy rainfalls. An error of 8 per cent has been quoted for a tipping-bucket pluviograph when measuring 133 mm/h. It can be 12 per cent in the case of a siphoning gauge which takes 13.6 seconds to empty 10 mm of rainfall.

The error can be calculated for any circumstances as follows, reckoning on an emptying time of t seconds and a rainfall rate of P mm per second. If the bucket holds the equivalent of b mm of rain, the time between tippings is b/P seconds. Thus, a complete cycle takes (t + b/P) seconds, and the apparent rate of rainfall Pa is b/(t + b/P) mm/s. The percentage error 100 (P − Pa)/Pa is hence 100 P.t/b per cent. If b is 1 mm and t is 1 second, for example, the error is 7 per cent for 240 mm/h (i.e. for 0.067 mm/s). The underestimation would be 33 per cent if b is 0.2 mm.

9 Data editing

A typical editing programme checks the date, continuity of values with the previous three days' figures for the daily maximum and minimum temperatures, likewise for pan evaporation and for wet-bulb depression, that evaporation is below 20 mm/d, that any maximum is below 50°C and any minimum within −8°C to +32°C, that there are no excessive irradiance values, that any daily rainfall is below 500 mm, that the terrestrial minimum is within −12°C to + 32°C, the water-vapour pressure value is positive, the wet-bulb temperature is below that of the dry bulb, and so on. A test of the internal consistency of data from any single climate station would check the maximum and minimum temperatures from the thermograph against those from the maximum and minimum thermometers.

CHAPTER 3

1 Interpolation by the method of 'first differences'

Consider a series of numbers with one missing which has to be estimated. To do this, form another series consisting of the differences between adjacent pairs in the original series. The second set will have a double gap. Estimate the left-hand number in that gap by continuing rightwards the part of the series to the left of the gap. Add this estimated value to the number just to the left of the single gap in the original series. Likewise on the right, to obtain a second estimate for the original missing number. Then take the mean of the two estimates as the wanted figure in the original series.

2 The error in estimating one climatic element from measurements of others

For instance, we consider estimating the evaporation rate (E) by means of the Dalton equation:

$$E = 0.2\ u\ (es - e), \text{ approx} \qquad \text{mm/day} \qquad (8.7)$$

where u is the wind speed (a typical value being 4 m/s) measured with an assumed maximum error of 0.5 m/s, es is the saturation water-vapour pressure (svp) at the water's surface temperature (Ts, with a typical value of 20°C, measured with a maximum error of 0.5C°) and e is the air's water-vapour pressure (e.g. 12 hPa, with an error of 0.2 hPa). Table 8.1 shows that a 0.5C° error in Ts leads to an error of es equal to 0.7 hPa. So the error of (es − e) is $(0.7^2 + 0.2^2)^{0.5}$, i.e. 0.73 hPa. Hence, the fractional error is 0.73/(23 − 12), i.e. 6.6 per cent, where 23 hPa is the svp at 20°C. The fractional error of wind-speed measurement is 0.5/4, i.e. 12.5 per cent. Thus, the fractional error of the product [u (es − e)] is $(6.6^2 + 12.5^2)^{0.5}$, i.e. 14 per cent. This is the intrinsic error in estimating E, due to the specified errors in measuring u, Ts and e.

Another example involves Regnault's equation for the atmospheric water-vapour pressure (see Eqn (2.3)). If the error of measuring each psychrometer thermometer is 0.2C°, for instance, the gross error of the term (T − Tw) is $(0.2^2 + 0.2^2)^{0.5}$, i.e. 0.28C°. So the error of A (T − Tw) is 0.67 × 0.28, i.e. 0.19 hPa. An error of 0.2C° in measuring the wet-bulb temperature causes an error in determining the equivalent svp which depends on the temperature. Around 15°C, for example, a variation of 0.2C° alters the svp by 0.22 hPa. So the overall error in calculating the ambient vapour pressure would be $(0.22^2 + 0.19^2)^{0.5}$, i.e. 0.29 hPa. If the psychrometer temperatures were 20°C and 15°C respectively, the vapour pressure would be 13.7 hPa so the error would amount to 2.1 per cent.

Alternatively, the psychrometer error might be assessed by comparing the estimate of vapour pressure from psychrometer readings of 20.2°C and 14.8°C with that from 19.8°C and 15.2°C. (Here the readings have been varied by the specified error.) The respective vapour pressures are 13.2 and 14.2 hPa, the difference being 1.0 hPa. This is larger than the previous estimate of 0.29 hPa because we have now assumed coincidental extreme errors in measuring the thermometer temperatures, and such a coincidence is unlikely.

3 Comparison of the mean products with the product of means

Consider an element whose value changes from A to (A + a), and another which changes from B to (B + b). The product changes from A.B to (A.B + a.b + a.B + b.A). So the mean of the products is (2A.B + a.b + a.B + b.A)/2. On the other hand, the product of the mean values is [(2A + a)/2] [(2B + b)/2], i.e. (4A.B + a.b + 2a.B + 2b.A)/4. This is less than the mean of the products by a.b/4.

So the product of means is an underestimate, if a and b are either both positive or both negative. This is the case in Table 3.1, since both wind speed and (es − e) decrease together at night. It also follows that the product of means is an overestimate if a and b are opposite in sign.

4 The difference between dawn and afternoon lapse rates

There is a complicated dependence of the daily *minimum* temperature on elevation, on account of two processes. Firstly, there is the normal lapse condition of coldness at great height, because of the cooling associated with the expansion of gases rising to levels of lower pressure. Secondly, there is the tendency at night for cold air to slide downhill as a katabatic wind and to pond in the valleys, giving inversion conditions there. So after a cloudless night in a dry climate there may be an increase of daily minimum temperature in rising within the lowest one or two hundred metres of a valley bottom. Measurements on still, clear winter nights, showed a crest temperature up to 3C° warmer than at the bottom of a 90 m deep valley, and 7.6C° warmer in one 160 m deep. Further above, the daily minimum declines with greater height in the normal way.

The change of daily *maximum* with elevation also depends on the site and time. At one place in Wales that lapse rate varies between 2–11 C°/km and is least in summer. At another place it ranges 5–10 C°/km and is most in summer. However, the dependence of daily maximum on altitude is less variable than in the case of the daily minimum. The lapse rate of the daily maximum is 6.1–7.1 C°/km in Papua New Guinea, which is more than the lapse rate for the daily minimum of about 5.4 C°/km. In general, the rate over various kinds of terrain is about 7 C°/km, though the type of prevailing air-mass affects the precise value.

5 Variation inland of the annual mean temperature

Two factors particularly affect the temperature change of air blowing overland. Firstly, there is the temperature of the coastal water, which depends on the latitude, the north–south flow of coastal currents and the occurrence of upwelling nearby. Secondly, there is the net irradiance at the particular latitude, assuming a predominantly east or west wind. There is a net income of radiation at latitudes less than roughly 40 degrees, and hence a tendency there for warm land surfaces to heat the air. At higher latitudes, the loss of heat by a negative net radiation flux (especially in winter) leads to cooling of air as it travels overland.

Consider Figure 3.3. The top graph (Figure 3.3a) shows the expected increase at a latitude of 23°S, where easterly trade winds prevail. The latitudinal mean is 23.5°C (see Eqn (3.5)), and east coast temperatures average 22.5°C. So the difference (D) is about 1C°. The graph shows an increase of temperature from the east coast by about 1.2 C°/1000 km. The next graph, Figure 3.3b, for latitudes about 32°S, shows no regular variation of annual mean temperature across the continent, apart from a quite local effect at the coast, because the west coast temperature approximates the latitudinal mean.

There is a *cooling* of the air at higher latitudes, as Figure 3.3b shows for 40°N in the USA. The cooling is shown by the crosses for places lower than 300 m, after correcting for the altitude to obtain the sea-level equivalent (see Eqn (3.6)). The coastal temperature is about 16°C and the air cools to about 11°C over 4000 km, i.e. at a rate of 1.2 C°/1000 km. Figure 3.3c shows data from 52°N across Europe and Asia. There the Atlantic coastal temperature is about 10°C and the latitudinal mean 4.3°C. Cooling is by 10C° in about 8000 km, i.e. 1.2C°/1000 km, again.

A similar change is indicated by data from places within two degrees of the equator across Africa. In this case, there is a warming, as expected for low latitudes, from 24°C at the west coast towards the latitudinal mean of 26.5°C (see Eqn (3.5)). The rate of change is 4C° in 3300 km, i.e. 1.3C°/1000 km.

6 Continentality

This has been defined as a measure of the extent to which the climate at any place is subject to influences of the land, as opposed to the sea. The concept appears to have originated with H.W. Dove (1803–79).

Remoteness from the ocean has several consequences – increased annual and diurnal temperature ranges, reduced relative humidity and precipitation (but maybe a wetter summer due to increased thermal convection), and a shorter lag of temperature on seasonal changes of radiation. However, discussion of 'continentality' is usually restricted to a consideration of the annual range of temperature, which depends primarily on the distance inland and secondly on the latitude. So continentality is inferred by removing the latitude effect from the annual range. Various formulae have been proposed to achieve this, to obtain some *index* of continentality; they are all arbitrary and with little physical meaning. Another disadvantage is that people sometimes regard 'continentality' as a *cause* of the annual temperature swing, instead of a partial description of it. Also, the concept deflects attention away from the proper investigation of the physical processes involved and has no practical use. The ideas in Note 5 seem more worthwhile.

7 The dependence of temperature on irradiance

Concerning Figure 3.6:

Maximum or minimum solar irradiance occurs in June and December. But the consequent warming of the air is delayed by the time taken for the ground to heat up and cool down, as discussed in connection with Eqn (2.2). The lag is least at high altitude and most at low latitudes over the sea. Normally, it is about a month, corresponding to 30° of the annual cycle of 360° as the Earth circles the sun. Now, it happens that a loop caused by sinusoidal variations of perpendicular co-ordinates, in the case where the variations are always about 30° out of phase, has a particular tilted, oval shape, whose maximum vertical thickness is about 0.4 times the vertical range.

The parallelogram for Potsdam (see Figure 3.6) was calculated as follows. The latitude is 52°N, so Eqn (3.5) gives the annual temperature as 4.3°C. However, west European coastal temperatures in January and July are around 6°C and 14°C respectively, averaging 10°C (Linacre and Hobbs 1977: 30). This is much warmer than 4.3°C on account of the Gulf Stream in the Atlantic ocean. So the air is initially at about 10°C, subsequently cooling towards 4.3°C in travelling overland. If it is assumed that the temperature falls linearly at about 1C°/1000 km (see Note 5), the average in Ireland about 2000 km to the east of Valentia is about 2C° cooler. Hence, the annual average temperature at Potsdam is 8°C, except that it is further reduced by the altitude effect. The Potsdam climate station at Berlin is 81 m above sea level, so this effect is 0.5C° (see Eqn (3.6)). Therefore, the best estimate for the mean temperature is 7.5°C.

Eqn (3.6) gives the annual range (Ra) as 20.9C°, so the January temperature is −3.1°C [ie 7.5 − (20.9/2)], and the July value is 17.9°C. The vertical thickness of the parallelogram (TH) is 0.4 times Ra, i.e. 8.4C°.

The extreme monthly irradiance averages are obtained from the empirical expressions in the section 'Estimating screen temperature' (p. 78). Hence Figure 3.6.

8 The daylength

See Table 8.2 p. 321

9 The difference between screen and surface minimum temperatures

There can be substantial differences between screen and surface minimum temperatures (Tmin and Tgn, respectively). Table 3.2 shows that the surface of the ground at Marsfield reaches a minimum temperature which is up to 6.4C° colder than the screen minimum. Various authors have reported differences of about 2C° for Australia as a whole, more than 5.5C° when there is frost in Papua and New Guinea, up to 6.5C° in Germany, 6–12C° in Canada, and typically 6C° in January in India. The difference in Japan equals (4.4 + 0.2 Tmin), where Tmin is the screen minimum temperature, e.g. the difference is about 6.4C° when Tmin is 10°C. Thus differences of 6C° are representative.

The difference varies on account of wind and cloud. It is as follows in Canada in autumn:

$$\text{Tmin} - \text{Tgn} = 7.3 - 0.32\ \text{C} - 0.6\ \text{u} \qquad \text{C}^\circ \tag{8.8}$$

where C is the amount of cloud in tenths (not oktas), and u is the mean wind speed at night (m/s). But if the effect of wind speed is ignored, the best empirical equation for Canadian data is the following (Bootsma 1976: 438):

$$\text{Tmin} - \text{Tgn} = 8.6 - 0.67\ \text{C} \qquad \text{C}^\circ \tag{8.9}$$

Table 8.2. The daylength in each half-month of the year at various latitudes. The definition of daylength here is the period between moments when the upper edge of the sun's disc appears exactly on the horizon, if this is unobstructed and refraction is normal

Latitude	*Daylength for various half-months in the* **Northern** *hemisphere*																							
	January		*February*		*March*		*April*		*May*		*June*		*July*		*August*		*September*		*October*		*November*		*December*	
60°	6.3	7.3	8.5	9.9	11.0	12.5	13.8	15.0	16.5	17.7	18.6	18.8	18.5	17.5	16.4	15.0	13.6	12.2	10.9	9.5	8.2	7.0	6.2	5.9
50	8.3	8.9	9.7	10.6	11.3	12.4	13.3	14.2	15.0	15.8	16.2	16.4	16.2	15.6	14.9	14.1	13.2	12.2	11.3	10.3	9.4	8.7	8.3	8.1
40	9.5	9.9	10.4	11.0	11.5	12.3	13.0	13.6	14.2	14.6	14.9	15.0	14.9	14.5	14.1	13.5	12.8	12.1	11.5	10.9	10.3	9.8	9.5	9.3
30	10.3	10.6	10.9	11.4	11.7	12.2	12.7	13.1	13.5	13.8	14.0	14.1	14.0	13.8	13.4	13.0	12.6	12.1	11.7	11.3	10.8	10.5	10.3	10.2
20	11.0	11.2	11.4	11.7	11.9	12.2	12.5	12.7	13.0	13.2	13.3	13.3	13.3	13.1	12.9	12.7	12.4	12.1	11.9	11.6	11.3	11.1	11.0	10.9
10	11.6	11.7	11.8	11.9	12.0	12.1	12.3	12.4	12.5	12.6	12.7	12.7	12.7	12.6	12.5	12.4	12.3	12.1	12.1	11.8	11.7	11.6	11.6	11.5
0	12.1	12.1	12.1	12.1	12.1	12.1	12.1	12.1	12.1	12.1	12.1	12.1	12.1	12.1	12.1	12.1	12.1	12.1	12.1	12.1	12.1	12.1	12.1	12.1
	July		*August*		*September*		*October*		*November*		*December*		*January*		*February*		*March*		*April*		*May*		*June*	
	Daylength for various half-months in the **Southern** *hemisphere*																							

Source: Linacre and Hobbs 1977: 241.

That implies a maximum difference of 8.6C°, once again confirming that screen and surface minimum temperatures can differ appreciably.

The surface minimum in places in England with sandy soil may be about 3C° cooler than where the soil is a better conductor of heat. Also in England, the grass minimum is not so much less than the screen minimum when temperatures are low.

10 Measuring net irradiance

Several instruments have been described for measuring net irradiance. An American instrument was developed by J.T.Gier and R.V. Dunkle, and an Australian device by Funk (1959). A Russian instrument due to S.B. Khvoles was described by Ioffe and Revut (1966: 45), and home-made instruments have been discussed by several authors. Most instruments are for research; there are few climate stations where net irradiance is measured on a routine basis.

In use, the net radiometer should be placed 1–2 metres above the appropriate surface. If it is closer, the instrument appreciably shades the surface being examined. But it averages conditions over too large an area if the instrument is high, e.g. over a circle of 18 m diameter if the radiometer is at 2 metres.

Birds or animals may damage the fragile plastic dome of a net radiometer.

The overall error of measuring net irradiance is commonly regarded as within 5 W/m^2, though much larger measurement errors have been revealed by instruments used side by side. Even the comparatively accurate Funk instrument may have errors of 20 W/m^2, which are of the order of 25 per cent of typical measurements. Other authors have quoted errors of 10 per cent and 10–20 per cent of the daily value. This is worse than the errors involved in measuring solar irradiance (see Chapter 5).

11 Radiation nomenclature

The term 'radiation' is often used loosely to refer either to amounts of radiant energy, or to its flux density, or to the subject in general. It is more accurate to distinguish between: (a) the radiant *energy* (measured in joules); (b) the flow or *flux* of such energy (in joules per second, i.e. watts); (c) the flux through unit area (called the radiant *flux density* W/m^2); (d) the flux *from* unit area, called the '*exitance*', in units of W/m^2; and (e) the radiant flux density *on to* unit area, called the '*irradiance*'. Usually we are concerned with the average irradiance during an hour, a day or a month, for instance.

What is called the 'exposure' (or 'irradiation') is the amount of radiant energy received by unit area during a specified period (joules/metre^2). This is a different use of the word 'exposure' from that in Chapter 2, 'Climate stations'.

Sometimes net irradiance is termed the 'radiation balance'. This seems undesirable because of an ambiguity about the English word 'balance'. It can mean either an equality (as in 'energy balance') or quite the opposite, a departure from equality, a remainder after inputs and outputs have been cancelled as in calculating a bank balance. Using the term 'net irradiance' as the label for the radiation-flux remainder avoids any uncertainty and emphasises that it relates to some particular surface.

12 Extra-terrestrial radiation

Table 8.3 Monthly mean values of the extra-terrestrial irradiance (Qa) in units of W/m^2

Latitude	*Jan*	*Feb*	*Mar*	*Apr*	*May*	*Jun*	*Jly*	*Aug*	*Sep*	*Oct*	*Nov*	*Dec*	*Average*
60°N	40	101	194	319	17	68	450	362	248	133	60	27	235
50	105	171	259	364	436	475	463	397	304	201	126	89	283
40	175	239	316	396	453	476	469	423	350	264	195	160	327
30	245	301	363	423	456	470	466	437	388	321	265	229	364
20	310	354	400	436	450	453	452	438	413	366	323	296	391
10	367	398	424	433	431	424	426	429	425	405	376	360	408
0	416	430	434	421	400	387	390	407	425	429	419	410	414
10°S	454	450	432	395	357	336	341	375	414	441	450	452	408
20	480	457	417	357	307	277	286	329	385	440	469	483	391
30	494	452	388	311	248	215	227	276	350	425	475	502	364
40	496	432	350	253	185	149	159	216	301	398	468	508	327
50	487	403	303	189	117	83	94	149	246	361	452	505	283
60	471	363	242	124	54	26	34	84	181	312	429	497	235

Source: Linacre and Hobbs 1977: 241.

13 Atmospheric irradiance

The atmospheric irradiance of the ground (Qld, i.e. the flux density of long-wave radiation downwards from the atmosphere) comes primarily from tri-atomic gases (i.e. water vapour, ozone and carbon dioxide) and from cloud base. With a clear sky, Qld varies little each day, because the radiant energy comes, in effect, from a layer of the atmosphere well above the ground, remote from daily changes of ground temperature.

This is shown by comparing the screen temperature T, the cloud-base temperature and the effective sky temperature, in terms of their emission of radiation. In each case, we assume emission according to the Stefan-Boltzmann equation (Eqn (3.16)), with an emittance of unity. Also, we assume that T equals 15°C, for example. Then Eqn (3.14) gives 298 W/m^2 for the clear-sky radiation, and that value in Eqn (3.16) implies a sky temperature of minus 4 °C, i.e. 19C° colder than the screen temperature, if the sky's emittance is regarded as unity. Hence, the normal free-air lapse rate of 6.5C°/km (see 'Estimating screen temperature', this chapter) implies that the downwards long-wave exitance (Qld) comes, in effect, from the atmosphere at about 3000 m above the ground.

If the sky is overcast there is an increase of Qld by about 72 W/m^2 (see Eqn (3.18)), to 370 W/m^2 in the above example. So the sky's effective temperature becomes 11°C (from Eqn (3.16)), which is only 4C° less than the screen temperature, and hence relates to 670 m elevation. This is the implied height of cloud base.

14 Surface emittance

The emittance (**e**) is about 0.97 for vegetation, for soils, for water and for various other surfaces. Paltridge and Platt (1976: 135) quoted 0.99 for water, 0.97 for concrete, 0.96 for gravel and asphalt, 0.93 for sandstone, 0.91 for sand, all averaging 0.95. Ioffe and Revut (1966: 42) mentioned 0.95–0.97 for the emittance of snow. The value is close to unity in each case.

CHAPTER 4

1 The percentile number

1 The percentile in Eqn (4.1) is 100 m/(N + 1), not 100 m/N. The former avoids the final value in the ranked set being regarded as the highest possible. Also, it makes the last percentile differ from 100 per cent as much as the first differs from zero, thus achieving symmetry.

Several alternative expressions have been published for obtaining the percentile, including those of (i) Hazen – i.e. 100 (m – 0.5)/N; (ii) Gringorten – i.e. 100 (m – 0.44)/(N + 0.2); (iii) Cunnane – i.e. 100 (m – 0.4)/(N + 0.2); and (iv) that used within the UK Meteorological Office – i.e. 100 (m – 0.31)/(N + 0.38). However, the expression in Eqn (4.1), popularized by Gumbel (1954: 14), is probably the most commonly used.

2 The percentile of an annual rainfall P is 100 C(P), where C(P) is the cumulative distribution. In other words, C(P) is the fractional likelihood that the selected annual rainfall P exceeds any value drawn at random from a set of annual values. It is exemplified by Figure 4.10(b) and can be related to the 'return period' (T, the average time between occurrences of the selected rainfall or more) as follows:

The chance of P or less occurring in any particular year equals 100 C(P) per cent, i.e. 100 m/(N + 1), from Eqn (4.1). So the chance of P being *exceeded* is [1 – C(P)], called the 'exceedance', E(P). The inverse of E(P) is the return period T. Therefore,

$$T = 1/E(P) = 1/[1 - C(P)] \quad \text{years} \qquad (8.10)$$

$$= (N + 1)/(N + 1 - m) \quad \text{years} \qquad (8.11)$$

Other authors have given different expressions for T, e.g. (N + 1)/(N – m), or (N + 0.12)/(N – 0.44 – m).

3 If an event's annual probability (p) is 0.1 (i.e. 10 per cent), then the return period is 1/p, i.e. 10 years. However, this does not mean that the event will happen precisely once each decade. The likelihood of the event happening exactly once in a particular decade would be $10 \times p \times (1 - p)^9$, i.e. 39 per cent in this example. Similarly, a return period of 50 years means a 33 per cent chance of occurrence at least once in 20 years and a 1 per cent chance of at least three times in that period.

4 Table 8.4 concerns weekly precipitation at Grenfell in New South Wales, Australia. It illustrates the way to analyse weekly rainfall data in terms of return periods. For example, the weekly rainfall exceeded 6 mm each 2.6 weeks on average.

2 Ambiguous terminology

The following are quoted from climatology papers by respected authors. The reader is invited to try to guess the various meanings each quotation might have:

- 'annual hourly wind velocity'
- 'two-year average mean monthly daily maxima'
- 'mean minimum of the coldest month'
- 'mean monthly maximum' temperature
- 'annual daily maximum rainfall'
- 'annual mean maximum' temperature
- 'mean maximum temperature'
- 'mean weekly maximum screen'

Table 8.4 Weekly rainfalls at Grenfell during 1953–72, i.e. 748 weeks

Weekly rain (Pw): mm	*No. of such weeks*	*Rank m*	*No. of weeks when Pw exceeded*	*% of weeks when Pw exceeded*	*Return period in weeks*
0	297	297	451	60.3	1.7
1–6	167	464[a]	284[b]	38.0[c]	2.6[d]
7–12	95	559	189	25.3	3.9
13–18	55	614	134	17.9	5.6
19–25	46	660	88	11.8	8.5
26–31	28	688	60	8.5	12.5
32–37	14	702	46	6.1	16.4
38–43	14	716	32	4.3	23.3
44–50	5	721	27	3.6	27.8
51–56	8	729	19	2.5	40.0
57–62	1	730	18	2.4	41.6
63–68	3	733	15	2.0	50.0
69–75	3	736	12	1.6	62.5
76–81	5	741	7	0.9	111
over 81	7	748	0	0	–

Notes: [a] i.e. 297 + 167; [b] i.e. 748 – 464; [c] i.e. (284/748) × 100; [d] i.e. 100/38.0, in other words, the weekly rainfall at Grenfell exceeded 6 mm each 2.6 weeks, on average.

Difficulty of interpretation arises because, for example, 'mean monthly maximum temperature' is not the same as 'maximum monthly mean temperature'. The order of the words in each case implies the procedure for deriving the value. In the first case, the highest temperature reached in a month has been averaged with the equivalents from other months, whereas in the second case one selects the highest of a set of monthly average temperatures.

If no period is stated before 'mean', 'average', 'maximum' etc, the word 'long-term' is implied (e.g. over 30 years or so), as in 'mean temperature'. So 'mean monthly temperature' makes no sense, because 'monthly temperature' has no significance. Likewise, 'maximum mean monthly temperature' should be expressed more clearly as 'monthly mean daily maximum temperature', if that is what is intended.

The rule is as follows: Each word such as 'mean', 'maximum', or 'total', must be *immediately preceded* by specification of the period or area of application.

3 The error in averaging relative humidity values

As an example, consider a volume of air at 10°C with a relative humidity of 50 per cent (i.e. the vapour pressure is 6.2 hPa), mixed with an equal volume at 30°C and 75 per cent (i.e. 31.8 hPa). The mixture would have a temperature of 20°C, a vapour pressure of 19.0 hPa (this being the average), and thus a relative humidity of 85 per cent. That is quite unlike the mean of 50 per cent and 75 per cent, i.e. 63 per cent. So relative humidity cannot be averaged by simple arithmetic. The error comes from treating ratios as though they are scalar quantities.

This is confirmed by the studies of Kalma (1968: 252). He found that hourly relative-humidity readings, each accurate within 5 per cent, give a 24-hour average quite different from the mean of the daily maximum and minimum values. The answer is to average vapour-pressure values instead.

4 Using the Runs Test to test for homogeneity

Consider the following series of 50 atmospheric pressures measured at Marsfield at 9 a.m. on 50 Thursdays in 1976 (i.e. the first value refers to 15 January 1976 and the last to 23 December 1976): 999 hPa, 1000, *1016, 1015*, 1011, 1010, 1011, 1006, *1016*, 1008, *1013, 1018, 1019, 1020, 1018*, 1006, *1023, 1027, 1022, 1022, 1023, 1025, 1009 1016, 1013, 1025, 1017, 1026*, 1011, *1013*, 1010, *1022, 1022*, 1006, 998, *1020, 1019, 1027*, 1005, 1003, 1009, 1009, 1002, 1005, 1005, 1010, 1009, 1007, 1006, 1010 hectopascals. Is the series homogeneous?

The median is 1012 hPa and higher values have been indicated in italic. A 'run' comprises a sequence of values either all in italic or all not. It can be seen that there are 17 runs in the whole series of 50 figures. So the number of runs does not fall between 22 and 30, given in Table 4.6 as the limits for a homogeneous series of that length. Therefore it is *not* homogeneous. (In fact, the series shows an annual cycle, with a tendency towards higher pressures in autumn and winter.)

5 Vector addition

An alternative to the graphical method of vector addition shown in Figure 4.9, is to do it numerically. This involves decomposing each vector into its eastward and northward parts. For example, the second period of travel (BC) in Figure 4.9 involves 108 km in a NNE wind, i.e. an angle of 202.5° clockwise from the north. So there is a negative, eastwards movement of 108 sin 202.5, i.e. −41 km, and a northward travel of 108 cos 202.5, i.e. −100 km. The total eastwards motion over the three periods is thus [86 − 41 + 0,] i.e. +45 km, and the total northward travel is [0 − 100 + 86], i.e. −14 km. Combining these components of the overall travel gives a distance (AD) equal to $[45^2 + (-14)^2]^{0.5}$, i.e. 47 km. The direction of AD is found by calculating the angle whose tangent is +45/(−14), i.e. −73° (or 73° anticlockwise from north). This approximates west-north-west, which is −67.5°.

6 Annual mean temperatures at Sydney

Table 8.5 Annual mean temperatures at Sydney, 1901–84

Year	*T:°C*	*Year*	*T:°C*	*Year*	*T:°C*	*Year*	*T:°C*	*Year*	*T:°C*	*Year*	*T:°C*	*Year*	*T:°C*
1901	17.1	1913	17.7	1925	17.3	1937	17.7	1949	17.4	1961	17.8	1973	18.7
1902	17.3	1914	18.3	1926	18.1	1938	17.9	1950	17.8	1962	17.8	1974	17.4
1903	17.2	1915	17.7	1927	17.0	1939	17.8	1951	17.7	1963	17.8	1975	18.4
1904	17.2	1916	17.4	1928	18.2	1940	18.1	1952	17.6	1964	18.0	1976	18.1
1905	16.9	1917	17.2	1929	17.4	1941	17.4	1953	17.6	1965	17.8	1977	18.4
1906	17.6	1918	17.4	1930	17.6	1942	17.8	1954	17.7	1966	17.4	1978	17.9
1907	17.4	1919	18.3	1931	17.5	1943	16.8	1955	17.8	1967	17.7	1979	18.4
1908	17.2	1920	17.3	1932	17.4	1944	17.4	1956	17.6	1968	18.2	1980	18.9
1909	17.0	1921	18.0	1933	17.1	1945	17.4	1957	17.8	1969	17.9	1981	18.4
1910	17.4	1922	18.1	1934	17.3	1946	17.6	1958	18.3	1970	17.6	1982	18.2
1911	17.4	1923	18.0	1935	17.2	1947	17.6	1959	18.1	1971	17.8	1983	18.3
1912	17.5	1924	17.4	1936	17.5	1948	17.2	1960	17.6	1972	17.9	1984	17.9

7 Similarity of samples

As an example of a problem involving scatter, consider the case of adjacent wind meters, one being compared with the other. The question is whether the readings are significantly different?

One way of answering this is to find the difference between n pairs of simultaneous readings, and treat the n differences as a sample of a possible infinity of comparisons. Find the mean difference (**d**) and the standard deviation (s) of the differences. Then the standard error, involved in assuming that the sample mean is the true difference, is given by $s/n^{0.5}$. If this standard error is less than **d**/2, we can say, with a 95 per cent level of confidence, that there is indeed a significant difference between the instruments' readings.

8 Periodicity in rainfall figures

Examination of a long series of annual totals is occasionally thought to reveal regular rhythms of rainfall which offer guidance to conditions in the coming year. As an example, there is a belief in north-west New South Wales that four years of 'good' years alternate with seven 'bad', making an eleven-year cycle related to the well-known sunspot periodicity. To check this we can use 97 years' data from Tibooburra.

The method involves working out the likelihood of one wet year being followed by another. *If there is any cycle greater than four years, the chance of a second year being wet is greater when the first is wet than when it is dry*. For instance, the chance of one wet year following another is 75 per cent (i.e. three out of four) if there is a cycle of four wet and seven dry years, whereas the chance of a wet year after a dry one is only 14 per cent (i.e. one out of seven). Likewise, if four wet years alternate with four dry (i.e. there is an eight-year cycle), the chance of wet after wet is 75 per cent, which once more is greater than the chance of wet after dry, i.e. 25 per cent. Inspection shows that this rule is quite general for any cycle, except in the case of two wet and two dry years alternating.

Now let us consider the data from Tibooburra. Each year can be categorised as 'wet' if the rainfall exceeds the long-term average of 224 mm. Examination of the data then shows that the chance of one wet year after another is 13 out of 36 wet years, i.e. *0.36*. On the other hand, the chance of a wet after a dry year is 23 out of 61 dry years, i.e. *0.38*. So there is almost no difference, ie the chance of a wet year is not increased if the previous year was wet. This means that there are no cycles in the weather.

This is confirmed by calculating the likelihood of a run of wet years. The expected number of runs of *four* wet years during 97 years is given by the following $97 \times 0.63 \times 0.38 \times 0.36 \times 0.36 \times 0.36 \times 0.64$, where 0.63 is the chance of a dry year (i.e. 61 out of 97), and 0.64 is the chance of a dry year after a wet one (i.e. $1 - 0.36$). The product shows 0.7 such runs, i.e. about one. In fact, there was one, in good agreement. The actual and calculated numbers of runs of other lengths are also similar. In other words, the assumption of no cycles leads to a reasonable explanation of the actual measurements.

9 The frequencies of runs of dull days at Bangkok

On average there are 72.2 dull days each year in Bangkok, so the initial probability of a dull day p(d) is 72.2/365, i.e. 0.198 (Exell 1982: 6). Therefore, the chance of a bright day p(b) is $[1 - p(d)]$, i.e. 0.802. Hence, the chance each year of a clear day followed by n dull days and then a clear day, assuming no persistence (i.e. assuming that each day's condition is independent of what happened the day before), is 0.802 $\times$ 0.198^n $\times$ 0.802 $\times$ 365, i.e. 235 $\times$ 0.198^n. Call this *p1*.

However, if there is persistence, the likelihood of such a run of dull days depends on the transition probabilities p(d/b), p(d/d) and p(b/d), where these mean the chance of a dull day after a bright day, etc. The Markov model for the chance of n dull days in a run is p(b) $\times$ p(d/b) $\times$ p(d/d) $\times$ p(b/d) $\times$ 365. At Bangkok, 24.8 of the 72.2 dull days annually are preceded by a dull day, so p(d/d) is 24.8/72.2, i.e. 0.343. Hence, p(b/d), ie [1 − p(d/d)], is 0.657. Also, p(d/b), the chance of a dull day after a bright one, is (72.2−24.8)/292.7, i.e. 0.162. Thus, the Markov model estimate is 0.162 $\times$ 0.343^{n-1} $\times$ 0.657 $\times$ 365, i.e. 31.2 $\times$ 0.343^{n-1}. Call this p2.

Values of *p1* and *p2* for various length of run (n) are given in Table 4.14, in comparison with observed run frequencies.

10 The correlation coefficient (r)

The link between two variables, x and y, is defined by the following expression, where Σ [x − **x**]. for instance, is the sum of differences from the average **x** of the various values of x

$$r = \Sigma\,[(x - \mathbf{x}).(y - \mathbf{y})]/\{\,\Sigma\,[x - \mathbf{x}]^2.\Sigma\,[y - \mathbf{y}]^2\}^{0.5} \qquad (8.12)$$

where $\Sigma\,[x - \mathbf{x}]^2$ is the sum of the squares of differences from the mean value of x, and **y** is the mean value of y.

The coefficient has a value between +1 and −1. It approaches one of these extreme values if all the points on a graph lie precisely on a straight 'regression' line. In other words, a value near +1 or −1 is regarded as showing a 'strong' association. The number is negative if an increase of x is associated with a decrease of y.

The coefficient is necessarily zero if the variables x and y are independent of one another, i.e. a variation of one is not associated with a consistently positive or negative change of the other. But the converse may not be true. A zero coefficient does not guarantee independence, for the following reason: there is a possibility of a real connection between two variables, even when the correlation coefficient is small, because of an important difference between, on the one hand, the 'simple correlation coefficient' (see Eqn (8.12)) relating a dependent variable to a single one of several variables which control it, and, on the other hand, what is called the 'partial correlation coefficient'. The latter quantifies the association between two variables *when all other variables are held constant*, whereas in simple correlation one ignores the possibility of confounding factors.

If there are confounding factors, a simple correlation of x and y may yield a low coefficient even when they are strongly connected in fact. Such is the situation in assessing the effect of wind speed on the evaporation rate of water, for example. Wind speeds tend to be low on fine days when higher temperatures increase evaporation, and the considerable effect of the low wind speeds in reducing evaporation is offset by the higher temperatures. For that reason, a misleadingly low coefficient is found for the effect of wind on evaporation, e.g. only 0.35. This is less than for the simple relationships between evaporation and the number of hours of either bright sunshine (i.e. 0.60), temperature (i.e. 0.63) or relative humidity, (i.e. −0.70). Simple correlations of daily evaporation have been found with coefficients of only 0.232 for wind, but 0.601 for irradiance, 0.304 for temperature and 0.457 for water-vapour pressure-deficit.

On the other hand, a 'partial correlation coefficient' shows the effect of one particular independent variable alone, free of the effects of confounding factors. In the case of water evaporation, there is a *high* positive partial correlation coefficient for the effect of wind, in conformity with the Dalton equation discussed in Table 3.1. In fact, there is a larger partial correlation coefficient for wind (i.e. 0.54) than for daily maximum temperature (0.43), hours of bright sunshine (0.42) or relative humidity (−0.41). Likewise, there is a simple correlation between evaporation and wind involving a coefficient of only 0.19, whereas the partial correlation coefficient is 0.50.

The error of relying on simple correlations was discussed by Pokorny *et al.* (1974: 350), using data on suicide rates in Houston (Texas) related to weather variables. Simple correlation suggested high correlation coefficients for the association of suicide with wind direction or fogginess, which hardly makes sense. But the partial correlation coefficients are so low that these factors are clearly negligible, as one would expect.

The correlation coefficient is a measure of the usefulness of an equation, not a direct indication of the likelihood that the apparent association is real rather than occurring by chance. The degree of reality (or the 'statistical significance') depends also on the number of pairs of x and y values. Table 8.6 shows how the correlation coefficient and the number of pairs of observations affect the significance level, i.e. the risk of achieving the coefficient by chance.

Table 8.6 Effects of the number of pairs of values and of the stipulated significance level, on the correlation required to establish a relationship

Number of pairs of values	*Significance level* 10%	5%	2%	1%	0.1%
	Values of the correlation coefficient				
4	0.900	0.950	0.980	0.990	0.999
6	0.729	0.811	0.882	0.917	0.974
8	0.621	0.707	0.789	0.834	0.925
10	0.549	0.632	0.715	0.765	0.872
15	0.441	0.514	0.592	0.641	0.780
20	0.378	0.444	0.516	0.501	0.679
27	0.323	0.381	0.445	0.487	0.597
37	0.275	0.325	0.381	0.418	0.519
47	0.243	0.287	0.338	0.372	0.465
62	0.211	0.250	0.295	0.325	0.408
82	0.183	0.217	0.256	0.283	0.357
102	0.164	0.195	0.230	0.254	0.321

11 The parts of a scientific report

1 The *Summary* or Abstract should be brief enough for adoption by an abstracting journal (such as *GeoAbstracts*). It should answer two questions – what was done and what was discovered. The latter should preferably be stated numerically. The Summary should not repeat the title. It must not include specialist terms, nor require explanation.

2 The *Introduction* defines the terms used, explains the problem to be solved and why it is worth solving, and outlines the scope of what is to follow. The problem

should preferably be expressed in the form of a question, with a question mark. (The answer should be in the Conclusions).

3 The *Previous Work* section consists of a literature review to ascertain the extent to which the question has been answered already. This defines the gaps still to be filled. References should not be introduced in any later section of the report.

4 The *Theory* section is the place for conjecture about the possible answer to the question, based on Previous Work and creative insight. The hypothesis to be tested is stated here.

5 The *Method* section should give enough detail to allow anyone else to confirm the results by repeating the measurements. Any special circumstances or materials should be described

6 *Results* consist of measured values in tables, graphs and equations, with just enough comment to highlight the main features.

7 In the *Discussion* is a consideration of the implications of the results, especially in terms of answering the initial question. Comparisons with Previous Work mentioned already and any inadequacies of the measurements would be considered. Also, fresh questions that arise from the results. Most Discussions are too long and verbose.

8 The *Conclusions* must be separate and brief, consisting chiefly of a bald answer to the initial question. It is entirely factual, stating what new facts have been discovered. Its starkness contrasts with the relative discursiveness of the more free-ranging Discussion. That's why they should not be mixed together, as is often done, unfortunately. Where possible, the Conclusions should be quantitative rather than adjectival.

9 The *References* are most conveniently given in terms of the Harvard system, i.e. author's name and the year of publication. However, it needs more space than a serial numbering of the references, and therefore is more costly to print and declining in use. A system gaining ascendancy has references given in the text by the sequential number of the reference in an alphabetical series at the end. Unfortunately, it is easy to create mistakes of referencing that way, and the numbers in the text are less helpful to the reader than authors' names and dates.

12 Assessing a report

Teachers and reviewers often have to grade reports, which is commonly done rather subjectively. Also, if the assessment is non-numerical (e.g. in terms of Greek letters), more arbitrariness is involved in adding it to other marks to achieve an overall grade. Better is a numerical assessment based on explicit criteria, such as the following ABCDE aspects:

Accuracy – the truth of the content and numeracy of presentation,
Breadth – the adequacy of covering the topic,
Clarity – fluency within a lucid structure,
Depth – the extent to which matters are explained rather than simply mentioned;
Enterprise – the element of ceative thought.

Equal weight might be given to each aspect. Thus, a total mark out of 20, say, would include a mark out of 4 for accuracy, and so on. A mark of 2 out of 4 would be satisfactory, 3 would be good, 4 perfect. Such marks show the writer where improvement is needed.

CHAPTER 5

1 Calculation of the hour angle, solar elevation and azimuth

Example 1

What is the hour angle (H) at 10.30 a.m. on 1 March at 153°E?

We assume that there is no need to allow for summer daylight-saving-time. The standard meridian is 150°E, the nearest multiple of 15°. The Local Standard Time is seen to be 10.5 hours and the difference between local and standard longitudes (L − Lo) is 3 degrees, representing 0.2 hours difference in solar noon, i.e. (L − Lo)/15. This time-difference will be negative, since these longitudes are to the east. The Equation of Time is seen in Figure 5.1 to be −8 minutes, i.e. −0.1 hours. Then one can calculate the True Solar Time by combining Eqns (5.3) and (5.4) to avoid the Universal Time term. In other words, Tt equals [Tls − (L − Lo)/15 + E/60], or [10.5 + 0.2 − 0.1], i.e. 10.6 hours. Thus, the hour angle is 15 (10.6 − 12), i.e. −21°, where the negative sign shows that the sun lies to the east, the time being before noon.

Example 2

What is the hour angle at 11 a.m. on 1 December at 106.5°W?

Once again, it is assumed that there is no daylight-saving correction needed. The standard meridian this time is 105°W. Hence, Tlm is [11.0 − (106.5 − 105)/15], i.e. 10.9 hours. The Equation of Time is 11.3 minutes (see Figure 5.1), i.e. 0.2 h. Therefore Tt, which equals (Tlm + E/60) is (10.9 + 0.2), i.e. 11.1 hours, or 0.9 hours before solar noon. So the hour angle is 15 × (−0.9), i.e. −13 degrees.

Example 3

What is the solar elevation when the declination is 10° and the hour angle 20° (i.e. it is afternoon, so the sun is to the west) and the latitude is 30°S (i.e. A equals −30)?

Eqn (5.5) gives the elevation as 46 degrees above the horizon.

Example 4

In the case above, what is the azimuth of the sun?

Eqn (5.6) gives (sin az) equal to (− cos 10 × sin 20/cos 46), equivalent to −29° for the azimuth, i.e. 29 degrees west of north.

Example 5

What is the sun's azimuth at 11 a.m. on 1 December at 50°N, 106.5°W?

From Example 2, the hour angle is −13 degrees, and from Figure 5.1 the declination is −21 degrees. Eqn (5.5) gives the solar altitude as 18 degrees. Thus, Eqns (5.6) and (5.7) give the azimuth as +13 degrees, i.e. 13 degrees east of south.

2 Calculation of the solar irradiance

The case of a transmission coefficient of 0.75 and an air-mass of 2 is shown in the section 'Nature of solar radiation' (this to imply direct irradiance of the ground (Qd.sin sa) of 385 W/m^2, with a clear sky). However, the global irradiance Qs includes a diffuse component also (see Eqn (5.9)):

$$Qs = D + Qd.\sin sa \qquad W/m^2 \qquad (8.13)$$

where sa is the angle of solar elevation. The latter is equal to 30° if the air-mass is 2, i.e. [sin sa] equals 0.5. We can deduce D from Eqn (5.14) since the extra-terrestrial irradiance (Qa) equals I.sin sa, where I approximates the solar constant, i.e. 1367 W/m^2.

$$\text{Hence: } 3\ D/Qs = 1367 \times 0.5/Qs - 1 \qquad (8.14)$$

$$\text{i.e. } 683 - 3D = Qs = D + 385 \qquad \text{i.e. } D = 75 \quad W/m^2 \qquad (8.15)$$

$$\text{Hence} \quad Qs = 385 + 75 = 460 \quad W/m^2 \qquad (8.16)$$

3 Attenuation of the solar beam

Definitions

The bracketed term in Eqn (5.20) (i.e. − **a**.m) is an index of atmospheric opacity. However, in place of the extinction coefficient **a** and relative air-mass m (given by Eqn (5.15)) some authors prefer to use a 'volume absorption coefficient' (bv, in units of $metre^{-1}$) and 'optical path' (**x**, in units of a metre, the *distance* traversed by the beam through a layer of the atmosphere). The distance **x** is the product of the layer's depth and the relative air-mass. So (−**a**.m) in Eqn (5.20) is replaced by (−bv.**x**).

Alternatively, one may use the 'mass absorption coefficient' (bm, in units of m^2/kg) and the 'optical mass' (in units of kg/m^2), the *mass* of atmosphere traversed by a beam of unit cross-sectional area. Then the bracketed term in Eqn (5.20) becomes (−bm.d.*x*) for a layer of density d. This bracketed term is called the 'optical depth', integrated for all layers of the atmosphere.

The extinction coefficient **a** has the advantage of being independent of units. However, it depends on atmospheric pressure and temperature, and on radiation wavelength. So it is an approximation to take a single value as representing all sunlight and the whole atmosphere.

Older literature refers to the extinction coefficient in decadal rather than exponential terms, i.e. the right-hand side of Eqn (5.20) is replaced by 10^{-z}, where z equals (0.434 **a**.m), since $10^{0.434}$ is equal to the exponential constant 2.718.

Calculation

The fraction of the direct radiant energy transmitted by the atmosphere is given by the following (Davies and Hay 1980: 35):

$$t = ts\ (to.tg + tw - 1) = Qd/I \qquad (8.17)$$

where ts is the transmittance of the aerosols in the atmosphere, to is that of the ozone, tg that of the dry, clean air, and tw that of the water vapour. Qd is the direct irradiance of the ground and I that of the same beam, but outside the Earth's atmosphere – roughly equal to the solar constant.

$$\text{Hence} \qquad ts = Qd/I\ (to.tg - [1 - tw]) \qquad (8.18)$$

Qd is measured and I is inferred by measuring Qd for various values of the optical air-mass (i.e. at different times of the day) and then extrapolating to zero air-mass. The transmittances are given as follows (Hay and Darby 1984: 356):

$$to = 1 - 0.00212\ X/(1 + 0.0042\ X) - 0.013\ X^{0.195} \qquad (8.19)$$

where X is [3.5 m], and m is the relative optical air-mass (see Eqn (5.15)).

$$tg = 0.986 - 0.103\ m + 0.0173\ m^2 + 0.00198\ m^3 \qquad (8.20)$$

$$tw = 1 - 1/(2.03 + 8.5/\mathbf{w}^{0.365}) \qquad (8.21)$$

where $$\mathbf{w} = m\ (1013\ p)^{0.75}\ w \quad \text{millimetres} \qquad (8.22)$$

and w on the right in Eqn. (8.22) is the sea-level atmospheric water content (in millimetres not centimetres), given by [exp (2.257 + 0.0545 Td)], where Td is the surface dewpoint temperature. This expression for w was quoted by Hay and Darby (1984: 356), but others are reported in chapter 7. The pressure p in Eqn (8.22) is derived from Eqn (5.16) for a given elevation. The pressure is about 900 hPa at 1 km, 800 hPa at 2 km, and 630 hPa at 4 km.

Examples

Examples in Table 8.7 show that the ozone transmission hardly varies from 0.98, that the clean, dry air's transmission is within 0.87–0.93, i.e. about 0.90, and that the water-vapour absorption is within 0.09–0.15, i.e. about 0.12 in a wide range of conditions. As a result, the overall transmission of clean air is about 0.76.

Table 8.7 Calculations showing the effect of altitude, elevation and dewpoint, on the total optical transmission of clean, moist air

Solar altitude (sa): degrees	90	90	90	90	90	90	50	50	50	50	50	50	30	30	30
Elevation (h): km	0	0	0	2	2	4	0	0	0	2	2	4	0	2	2
Dewpoint (Td): °C	25	15	5	15	5	5	25	15	5	15	5	5	15	15	5
Air-mass (m) [a]	1.0	1.0	1.0	0.79	0.79	0.62	1.30	1.30	1.30	1.02	1.02	1.02	2.00	1.57	1.5
Ozone transmission (to) [b]	0.98	0.98	0.98	0.98	0.98	0.98	0.97	0.97	0.97	0.98	0.98	0.98	0.97	0.97	0.97
Air transmission (tg) [c]	0.90	0.90	0.90	0.92	0.92	0.93	0.89	0.89	0.89	0.90	0.90	0.90	0.87	0.88	0.88
Water vapour absorption (− tw) [d]	0.15	0.13	0.11	0.11	0.10	0.09	0.15	0.13	0.11	0.11	0.10	0.09	0.13	0.11	0.10
Total transmission (t) [e]	0.73	0.75	0.77	0.79	0.80	0.82	0.71	0.73	0.75	0.77	0.78	0.79	0.71	0.74	0.75

Notes: [a] From Eqn (5.15); [b] From Eqn (8.19); [c] From Eqn (8.20); [d] From Eqn (8.21); [e] From Eqn (8.17)

4 Components of the diffuse irradiance

The ratio Qd/I in Eqn (5.17) is the 'transmittance', whose value might be 0.56, for example. It follows from Eqn (5.10) that the ratio of circumsolar to isotropic diffuse irradiance on to a horizontal surface would be (D.Qd/I)/(D − D.Qd/I), i.e. 1/(I/Qd − 1), or 1.3 in this example. However, if cloud over half the sky reduces Qd/I to 0.28, for instance, the ratio of circumsolar to isotropic diffuse irradiance would decrease to 0.39. In fact, 0.2 has been observed in the case of skies at Melbourne, which implies appreciable cloudiness there.

5 Deriving the sunshine-duration time from a Campbell-Stokes card

The sun's track on the card may consist of holes burnt by intermittent sunshine, but overlapping, to form a continuous slot. This is confusingly like that created by continuous sunshine. Or the periods of bright sunshine may each be too brief to allow the burning of holes so that it seems that there has been continuous cloud. As a result, errors can arise in the interpretation of the record.

It is good practice to reduce the errors by adopting a standard procedure in order to approach the desired error of no more than 0.1 hours for the daily total of sunshine duration:

1 When there is a clear burn with round ends, reduce the length at each end by an amount equal to half the radius of the curvature at the end of the burn; this normally shortens each burn by 0.1 h.
2 When burns are circular, consider 2 or 3 burns together as 0.1 h; 4, 5 or 6 together as 0.2 h, 7, 8 or 9 as 0.3 h and so on.
3 When the mark is only a narrow line, the whole length is measured, even when the card is only slightly discoloured.
4 When a clear burn is temporarily reduced in width by at least a third, subtract 0.1 h for each such reduction, though the total subtracted is not to exceed half the total burn time.

CHAPTER 6

1 Winds over a city and over an airfield nearby

It is instructive to calculate a city wind from measurements at an airfield nearby. Consider Eqn (6.2) for the case of the airfield and take appropriate values for the zero-plane displacement (0.07 m), the surface roughness (3 cm, i.e. 0.03 m), and the height of the gradient wind (250 m). Assume, for instance, a measurement of 6 m/s at 10 m above the ground, and neutral conditions. So the gradient wind speed (ug) may be calculated as follows:

$$ug = 6 \ln [(250 - 0.07)/0.03]/\ln [(10 - 0.07)/0.033] = 9.3 \text{ m/s} \qquad (8.23)$$

The wind at a standard height of 10 m from the ground in a city is only indirectly related to the wind above the city, because the urban zero-plane displacement (which is about two-thirds of the heights of the buildings) exceeds 10 m. In these circumstances it seems more useful to specify the wind environment of the city in terms of the speed (ut) at the top of the city's canopy at the level of the roofs of the buildings. If that height is 60 m the plane at which Eqn (6.2) gives a zero wind is 40 m above the ground, i.e. 20 m below roof level and 460 m below the level of the urban gradient wind, reckoned as occurring at 500 m (see p. 202). The roughness is 3 m (see Table 6.3). Now we may use Eqn (6.2) again, this time taking the gradient wind as the reference value, in order to determine the roof-level wind:

$$ut = 9.3 \ln [20/3]/\ln [460/3] = 3.5 \quad \text{m/s} \qquad (8.24)$$

This is the driving force of the winds at street level, and of those impacting on to the walls of buildings. It is notably less than the airfield measurement of 6 m/s – unless the streets channel the winds.

2 A simple form of the Fisher-Tippett type 1 equation

The Fisher-Tippett type 1 equation (see Eqn (6.10)) is as follows:

$$C(U) = \exp[-\exp\{-g(U - Um)\}] = 1 - E(U) = 1 - 1/T \quad (8.25)$$

where E(U) is the exceedance, and T is the return period (see Note 1, Chapter 4).

Hence, $$U = Um - [\ln\{-\ln(1 - 1/T)\}]/g \quad (8.26)$$

Now, ln (1−1/T) equals 1/T, approximately, provided T exceeds about 10. Hence,

$$Ut = Um - [\ln\{1/T\}]/g = Um + (\ln T)/g \quad (8.27)$$

3 Examples of the estimation of extreme winds

Example 1

Consider the values in Table 8.8, based on data published by WMO (1966: XIII.10). Plotting graphs of (ln U) against (ln ln T), and of (− ln [− ln C(U)]) against U, shows that the latter gives the more linear diagram. So the Gumbel distribution is more appropriate than the Weibull in this case. For the *100-year* value, C(U) is 0.99 and so (− ln [− ln C(U)]) is 4.60. This value in the Gumbel graph corresponds to 77 mph, i.e. 34 m/s.

For a *50-year* return period, the reduced variate is 3.90, corresponding to 71 mph, i.e. 32 m/s. This is slightly less than the 35 m/s derived by means of an assumed Weibull distribution.

Table 8.8 An analysis of values of the extreme winds in nine years

Wind (U): mph	*38*	*38*	*41*	*44*	*47*	*50*	*64*
Rank	*1*	*2*	*3*	*6*	*7*	*8*	*9*
C(U) %	–	20	50	60	70	80	90
E(U) %	–	80	50	40	30	20	10
1/T	–	1.13	2.00	2.50	3.3	5.0	10
ln ln T	–	−2.1	−0.37	−0.09	0.18	0.48	0.83
ln V	–	3.64	3.71	3.78	3.85	3.91	4.16
−ln [–[nC(U)]	–	−0.48	+0.37	0.67	1.03	1.50	2.25

Example 2

A summary of data on all the three-hour mean wind speeds at Des Moines during 1965–74 is in Table 8.9, in the form of a cumulative frequency distribution C(U). There would be 292,000 such three-hour periods in 100 years, so the required return period (T) would be 292,000 units. Trial and error show that the data fit a Weibull distribution. Graphing (ln ln T) against (ln U) gives a line with slope 1.92, so the distribution almost fits the Rayleigh equation also. The 100-year reduced variate (i.e. ln ln 292,000) equals 2.53 and the value of (ln U) equivalent to that is 2.85 in this example. The exponential of 2.85 gives the 100-year extreme three-hour mean wind as 17 m/s.

Table 8.9 Cumulative values [C(U) percent] of three-hour mean winds [U: m/s] at Des Moines during 1965–74

Wind (U):	1.5	2.0	2.5	3.1	3.6	4.2	4.7	5.1	6.1	7.1	8.2	10	13
Percentile C(U)[a]:	7	12	21	30	39	50	57	66	79	87	95	99	99.9

Source: Takle and Brown 1978: 557.

Note: [a]i.e. the percentage of values less than or equal to the value of U above.

Example 3

The fastest wind over a mile in a month may be estimated from some empirical relationship to the *monthly* average (Um). At seven places in New England the maximum is given by (2.9 Um + 2.4) m/s. Thus, the Gust Factor is about 3, which is between values of less than 2 (when gusts are compared with *hourly* mean winds, as in Table 6.6) and values of more than 5, when the comparison is with *annual* mean winds (see 'Gusts', pp. 227–8).

4 Selecting a site for a wind-energy collector

Great care is needed in choosing a place to install a wind-energy collector in view of the spatial variation of winds. One method of selecting a site is to rely on expert subjective judgement, and this appears about as reliable as selecting a site after measurements of winds at many different places. Experts in Hawaii picked eight of the ten sites which in fact had the most winds. However, a more systematic procedure is as follows. It involves identifying promising regions, then localities within the regions, sites within the localities, and position at a site:

1 Examine any maps of wind measurements which are available, in order to identify windy regions.
2 Scrutinise contour maps of the selected regions on the scale of 1 in 50,000 (i.e. 1 centimetre on the map is equivalent to 500 m, or 1 km is shown by 2 cm) to identify localities where strong winds are possible, given the shape of the coast, plains and hills, and the directions from which the stronger winds come. One would look for long sloping valleys aligned with the prevailing gradient winds, exposed ridges across the winds, isolated hills which are conical, exposed coastlines and high plateaux. Localities vulnerable to either salt spray, lightning, large hail or sandstorms would be ruled out.
3 Visit the selected localities to interview local residents in order to learn about sites with the strongest winds. Collect climatological data for the selected localities to determine possible hazards. For instance, the frequencies of icing conditions or lightning should preferably be low.
4 Use a mobile anemometer to sample wind speeds for comparison with a central benchmark anemometer which is recording continuously to find sites in each locality with apparently high windiness ratios. Avoid places with tall vegetation or rough terrain, or with high land within a few kilometres – even to leeward. Places near airports, places remote from users or near residential areas, are also weeded out. Good sites do not have woods or rocky ground near the summit. The hilltop area should

preferably be small, surrounded by regular and smooth slopes of about 16 degrees for a few hundred metres. Easy access is important. Look for deformation of the vegetation and bare terrain.

5 Install a recording anemometer at 2 m at each of the several most promising sites to run for as long as practicable, preferably for a month or two, before being shifted to other possible sites in order to find the mean wind speeds, gustiness, frequency of hours with a mean above limits such as 5 m/s, 10 m/s etc., and to find the directions of the winds. A wind survey of a valley in Germany, for example, involved seven anemometers in fixed positions during two months with 13 more put at various places for 2–4 weeks at a time. The aim is to find places with winds in the range 4.5–27 m/s. It is reckoned in the UK that a really good site has winds averaging above 11 m/s at 3 m.

6 Examine winds above the ground by means of kites or balloons. Simultaneously carry out experiments at the various sites to establish the degree of turbulence within the lowest 50 m by watching the flapping of 2 m lengths of ribbon at intervals along the string to a kite. Also, undesirable windshear should be checked, using rising balloons or else smoke-trails from rising rockets. All this should be done in various conditions of solar radiation, wind speed and direction, in order to sort out sites subject to undesirable gusting, which is particularly common in mountains.

7 Install recording anemometers at 20 m at the selected best few sites, for a year or two, to identify the best in terms of the continuity of winds above the rated speed of the envisaged wind-energy collector (WEC). Often this stage has to be curtailed to avoid delays in construction.

8 Choose the best site, bearing in mind such practical considerations as land ownership, access, security, effect on the environment, firm ground for tower foundations, and proximity to the power demand. To install an array of WECs, sites must be found which are separated by at least ten times the height of each WEC, to avoid mutual interference. In a study of possible sites in the Orkney Islands of Scotland, five places had to be ruled out because of the environmental impact, another three on account of the effect on TV reception, two for infringement of land rights, three because of poor exposure and four where there was poor access to the power grid.

9 Install a close array of anemometers at the best site to find the exact position for the WEC, since many of the best sites are found in complex terrain where wind conditions vary spatially to a considerable degree.

CHAPTER 7

1 The error in using measurements by a single rain gauge to determine the average rainfall on to an area

The following expressions have been published for the error E, the difference between the true catchment rainfall (assessed from a large number of gauges) and the average of measurements with a smaller, more practicable number. The error is expressed in terms of the rainfall in each storm or rainday (P mm), and either the number (N) of gauges in area A (km^2), or the area sampled by each gauge (G: km^2), i.e. A/N.

(i) Linsley and Kohler (1951: 246):

$$E = 0.023\ P^{0.47}\ G^{0.6} \quad \text{mm} \qquad (8.28)$$

This relates to a basin in Ohio. By taking logarithms of the terms in such an expression, one obtains equations like those following:

(ii) Curry *et al.* (1966: 128, and see Note 3 (Chapter 1) and Eqn (8.4)) for an area in Michigan:

$$\log E = 0.14\ P^{0.5} + 0.66 \log G - 0.87 \qquad (8.29)$$

(iii) Corbett (1967: 114), based on measurements in Illinois:

$$\log E = 0.24\ P^{0.5} + 0.29 \log G - 0.73 \qquad (8.30)$$

(iv) Huff (1970: 42), where D is the storm duration in hours:

$$\log E = 0.68 \log P + 0.94 \log G - 0.01 \log D - 0.75 \log A - 9.50 \qquad (8.31)$$

(v) Schaake (1978: 55), based on data from central Illinois:

$$\log E = 0.65 \log P + 0.82 \log G - 0.22\ D - 0.45 \log A - 0.50 \qquad (8.32)$$

It may be noticed that all these equations come from measurements in the northern USA. However, work in Nigeria tends to confirm the form of Eqns (8.31) and (8.32). The error there is proportional to the power 0.58 of the rainfall (like 0.68 or 0.65 in the equations), and to the power 0.76 of the area per gauge, like 0.94 or 0.82.

All the equations involve the engineer's trick of reckoning that a dependent variable (in this case the error) can be regarded as the *product* of factors each representing controlling variables and that the factors can be assumed to be independent of each other. These two assumptions are convenient but hard to justify.

2 The heat flux in the ground near the surface at dawn, and dew formation

Six authors have given examples of around 80 W/m^2 for the heat flux upwards to a *bare ground* surface before dawn. However, only about 50 W/m^2 was measured in Oregon, 56 W/m^2 in China, and merely 14–23 W/m^2 at Pune in India. Presumably, the lower values reflect differences of soil conductivity, of heating during the previous day, of cloudiness, and maybe weeds.

Another dozen or so papers give 5–50 W/m^2 as the dawn flux *under crops* of grass or the like. Other reported values include about 40 W/m^2 between midnight and dawn in a Nebraska field, 35–56 W/m^2 under irrigated lucerne, 42 W/m^2 in the USSR, and 20 W/m^2 under grass in the UK in summer. Thus, 30 W/m^2 may be taken as representative. This is less than half the typical value for bare ground.

The range of observed nocturnal heat fluxes (G) under grass presumably reflects various amounts of vegetation, described by the leaf area index (LAI). The latter is the area of leaf on unit area of ground. Thompson (1981: 5) reckoned G as 19 W/m^2 when the net exitance is 97 W/m^2 and LAI is five, but 28 W/m^2 if the LAI is only two.

As regards the ground-heat flux during the *daytime*, this varies with the height of the vegetation. As a fraction of the nett irradiance (Qn), it falls from 0.3 to 0.1 as lucerne stubble grows from 100 cm to 450 cm and thereafter remains at 0.1. For bare soil in Tunisia the average G/Qn is 22 per cent.

Dew formation: distillation

At night-time, the ground heat flux G beneath a sward energises (i) heat flux and (ii) distillation of water from the soil up through the crop. The latter involves a flux of latent heat L.Di through the canopy, which can be estimated from Eqn (3.28). In this case, the input of energy is not the net irradiance Qn, but G. Also, the resistance to fluxes through the crop is so large that the term d.c.S/r is zero. So the distillation rate Di is given by the following:

$$L.Di = \Delta.G/(\Delta + Ks) \qquad W/m^2 \qquad (8.33)$$

where L is the latent heat of evaporation, Δ is the change of saturated vapour pressure with a change of 1C°, with a value of 0.8 hPa/C° near 10°C (see Note 7, Chapter 2), and Ks is the psychrometric constant, i.e. 0.67 hPa/C° at sea level (see Chapter 3, 'Estimating evaporation'). If G is 30 W/m^2, the latent-heat flux is 16.3 W/m^2 (i.e. 30 × 0.8/[0.8 + 0.67]), equivalent to distillation at the rate of 0.024 mm/h, since 28 W/m^2 evaporates 1 mm/d or 0.042 mm/h.

Dew formation: dewfall

On a still clear night at about 10°C when dew may form, the radiative loss of heat from the surface to the sky is given by Eqn (3.17), i.e. 97 W/m^2. About 30 W/m^2 is offset by G, leaving 67 W/m^2, provided by dewfall and heat flux from the air. Dewfall is estimated from Eqn (3.28) again, but now the irradiance Qn is negative, since upwards, and the moisture flow Df also is negative, since downwards. The positive term d.c.S/r becomes 4 Δ.u (T − Td) as in deriving Eqn (3.30). We will assume a dewpoint Td of 6°C, say, and a wind speed u of 1 m/s. Hence the following:

$$-L.Df = 0.8\ \{-67 + 4\ (10 - 6)\}/(0.8 + 0.67) \qquad W/m^2 \qquad (8.34)$$

That means a dewfall of 28 W/m^2, i.e. 0.041 mm/h. The total rate of dew formation in the present case is thus estimated as about 0.06 mm/h.

In the same way, one may calculate the effects of temperature, wind speed and dewpoint on dew formation. There is no dewfall if 4 u (T − Td) exceeds (Qnl − G), i.e. it is less if either the wind speed is high, humidity low, sky cloudy or the ground bare.

3 The melting of snow

The melting of snow results from the forced convection of sensible heat from the wind (Hc), the liberation of latent heat (Hd) when water vapour condenses on the snow, and the absorption of net irradiance (Qn). A flux of 1 W/m^2 will melt 0.25 mm/d of snow (water equivalent) since the latent heat of melting is about 340 kJ/kg. Rain also causes a small amount of melting.

It is interesting to consider conditions like those at Cooma (at 841 m in New South Wales) in October, which is the month when most melting occurs. Then the monthly mean temperature is 12°C, the dewpoint 3°C, and the wind 2 m/s. These values will be assumed in what follows.

Melting by warm wind

The convective transfer of heat (Hc) to snow is limited by the smoothness of the surface and the stability of the lowest part of the atmosphere. The following equation applies to the case of a surface at temperature Ts (i.e. 0°C in the case of snow) and a roughness of 0.05 cm:

$$Hc = h\ (T - Ts) = 2.3\ u.T \qquad W/m^2 \qquad (8.35)$$

Here the heat transfer coefficient h was derived from values for convective transfer from water, which has about the same roughness (see Table 6.3) and was mentioned in Chapter 3 (see 'Estimating evaporation').

Eqn (8.35) gives a convective melting rate (Mc) as follows:

$$\mathrm{Mc} = 0.6\ \mathrm{u.T} \qquad \mathrm{mm/d} \tag{8.36}$$

Hence, there would be 14 mm/d at 12°C and 2 m/s.

Condensation melting

The surface of snow is also heated when the water-vapour pressure of the atmospheric moisture (e hPa) exceeds that of the snow's surface at 0°C, i.e. 6.1 hPa. Then there is either sublimation of the vapour directly into the solid state or condensation into liquid at 0°C. We will consider the second case, which applies when melting is occurring. The condensation liberates about 2490 kJ/kg, which melts some of the snow.

The latent-heat flux (Hd) depends on the difference between the vapour pressures of the air (e) and snow, the density of the air d, its specific heat c, and the surface diffusion resistance r:

$$\begin{aligned}\mathrm{Hd} &= \mathrm{d.c.(e - 6.1)/Ks\ r} = \mathrm{h.\Delta.(Td - 0)/Ks} \\ &= \mathrm{h.\Delta.Td/Ks} = \mathrm{2.3.u.\Delta.Td/Ks} \qquad \mathrm{W/m^2}\end{aligned} \tag{8.37}$$

where $\triangle$ and Ks are as in Eqn (3.28), u is the wind speed and Td is the dewpoint. The value of Ks at 841 m elevation is 0.60, and $\triangle$ at 0°C is 0.5 hPa/C°, so the flux of condensation heat in a wind of 2 m/s with a dewpoint of 3°C, is 12 mm/d. At higher elevations the rate would be increased by a smaller value of Ks.

Radiation melting

The third factor in melting snow is the daytime sunshine. Its importance in causing melting depends greatly on the snow's albedo (a).

Eqns (3.13), (3.18), (5.35) and (5.37) yield the following:

$$\mathrm{Qn} = (0.25 + 0.5\ \mathrm{n/N})\ (1 - \mathrm{a})\ \mathrm{Qa} + \mathrm{T} - 75\ \mathrm{n/N} - 27,\ \text{approx.} \qquad \mathrm{W/m^2} \tag{8.38}$$

where n is the number of hours of bright sunshine in a daylength of N hours, T is the screen temperature and Qa is the extra-terrestrial irradiance (see Table 8.3). Evaluation of Eqn (8.38) for cloudless conditions in October at 40°S when temperatures are 12°C, for instance, gives a net irradiance which varies from −65 W/m^2 to +85 W/m^2 as the albedo decreases from 0.9 to 0.4. In other words, radiation fluxes lead to *cooling* of the snow until other processes have melted enough snow to leave a dirty surface. The critical value for the albedo is about 0.7. A typical albedo of 0.5 leads to a net irradiance of +54 W/m^2 in the example, which would create melting at a rate of 13 mm/d.

Overall melting rate

The total melting rate is the sum of Mc, Md and the equivalent for irradiance melting. For the example given, the total equals 14 + 12 + 13 mm/d, i.e. 39 mm/d. This resembles the pattern of (11 + 6 + 20), i.e. 37 mm/d, quoted by Hay and Fitzharris (1988) in New Zealand, and is within the range of rates mentioned in the section on 'Snow', p. 309. In practice, the rate may be reduced by cloud, by air dryness, winds less than 2 m/s, air temperature less than 12°C, extra-terrestrial irradiance below 400 W/m^2 or by a higher albedo. In addition, radiation heating may be diminished by orientation of the snow surface away from the sun.

It used to be thought that radiation melting was less important than heating by warm air. This applied in Wisconsin, for instance, except when it occurred late in the snow season. However, Barnaby (1980: 128) found a much stronger correlation between melting rate and solar irradiance than with air temperature. The importance of irradiance is also corroborated by studies in Russia and Japan, showing that blackening the snow's surface with coal dust (to increase the absorption of radiant energy) appreciably accelerates melting. Other evidence for the significant influence of irradiance heating of snow consists of measurements of daytime energy fluxes in California, where the net irradiance in April and May is about three times the convective energy flux.

One concludes that none of the three factors causing melting can be ignored, but condensation is important only when the air is warm and moist, convection when there is a warm, strong wind, and irradiance when skies are clear on a summer's day. Irradiance accounted for over half the melting in 24 out of 32 cases, and condensation the least in 31 of the cases (Paterson 1969: 62).

Various formulae have been published for the overall melting rate. The first sort allows for each of the various melting processes, including empirical factors to allow for forest cover and for slope and orientation of the land. However, formulae which are even simpler relate the total melting rate merely to the daily mean temperature on the assumption that net irradiance heating and condensation heating, as well as convective heating, are proportional to the air temperature. Hence, the example above suggests 3.3 mm/d.C°. But this procedure is inaccurate (see 'Snow', p. 308).

4 The annual alternation of snow depth

The annual maximum depth of snow at Spencer's Creek (at 1830 m in the Snowy Mountains in New South Wales) is shown in Table 8.10. The chance of an abnormal year after a sub-median year, or the reverse, is seen to be 0.73 (i.e. 19 years out of 26), which is appreciably above the 0.50 expected from random events. In fact, the Runs Test (see Table 4.6) shows that 20 runs in 27 values can occur by chance on only 6 per cent of occasions. In other words, there has indeed been some tendency for abnormal and subnormal values to alternate.

Table 8.10 Maximum depths of snow at Spencer's Creek in different years

Year	*Depth m*	*Year*	*Depth m*	*Year*	*Depth m*
1954	1.13	1963	1.92	1972	*2.67*
1955	*2.30*	1964	*3.56*	1973	1.23
1956	*3.18*	1965	1.46	1974	*2.87*
1957	*1.54*	1966	1.76	1975	1.80
1958	*2.33*	1967	1.85	1976	1.61
1959	1.47	1968	*3.07*	1977	*2.27*
1960	*2.93*	1969	1.15	1978	1.87
1961	1.92	1970	*2.81*	1979	1.71
1962	*2.12*	1971	*2.07*	1980	*1.98*

Source: Data from the Snowy Mountain Hydro-electricity Commission.

Note: The values in italic exceed the median of 1.92 m.

BIBLIOGRAPHY

The following are the articles whose authors' names are mentioned in the previous chapters. The many more which have been quoted tacitly are listed in the complete documentation which is available from the author (see the Preface).

Aguardo, E. (1985) 'Snowmelt energy in southern and east-central Wisconsin', *Annals of the Association of American Geographers* 75, 203–11.

Alcock, R.K. and D.G. Morgan (1978) 'Investigations of wind and sea state with respect to the Beaufort scale', *Weather* 33, 271–7.

American Meteorological Society (1981) *Bulletin of the American Meteorological Society* 62, 87.

Andersson T. (1963) 'On the accuracy of rain measurements and statistical results from rain studies with dense networks'. *Archiv und Geophysik* 4, 307–32.

Angstrom, A. (1924) 'Solar and terrestrial radiation', *Quarterly Journal of the Royal Meteorological Society* 50, 121–6.

Aslyng, H.C. and B.F. Nielsen (1962) 'The radiation balance at Copenhagen', *Archiv für Meteorologie, Geophysik und Bioklimatologie/Archives for Meteorology, Geophysics and Bioclimatology* 10B, 342–58.

Asuncion, M.T. (1971) 'An analysis of potential evapotranspiration and evaporation records at Los Banos', M.Sc. thesis, University of Philippines.

Atkinson, B.W. (1981) *Mesoscale Atmospheric Circulations* (Academic).

Atkinson, G.D. (1971) *Forecasters' Guide to Tropical Meteorology*, Technical Report 240, (Air Weather Service), US Air Force 347pp.

Atwater, M.A. and J.T. Ball (1978) 'A numerical solar radiation model based on standard meteorological observations', *Solar Energy* 21, 163–70.

Bagnall, D. (1982) 'The effect of temperature and radiation on crop growth in Australia', *Proceedings Seminar on Farmers and the Weatherman, Sydney*, (Australian Institute Agricultural Science) 47–59.

Bailey, R.G. (1985) 'The factor of scale in ecosystem mapping', *Environmental Management* 9, 271–75.

Ballantyne, C.K. (1983) 'Precipitation gradients in Wester Ross, northwest Scotland'. *Weather* 38, 379–87.

Barnaby, I.E. (1980) 'Snowmelt observations in Alberta', *Proc. Western Snow Conference*, (Colorado State University) 128–37.

Barrett, E.C. (1976) 'Cloud and thunder' in Chandler and Gregory (1976) 199–210.

Barry, R.G. (1970) 'A framework for climatological research with particular reference to scale concepts', *Transactions of Institute of British Geographers* 49, 61–70.

—— (1981) *Mountain Weather and Climate*, (Methuen).

BC Grape Growers (1984) *Atlas of Suitable Grape Growing Locations* (Association of British Columbia Grape Growers; Agriculture Canada)

Beckinsale, R.P. (1957) 'The nature of tropical rainfall'. *Tropical Agriculture* 34, 76–98.

Bell, F.C. (1969) 'Generalized rainfall-duration-frequency relationships'. *Journal of Hydraulics Division* (American Society of Civil Engineers) 95, 311–27.

Berliand, T.A. (ed.) (1970) *Solar Radiation and Radiation Balance Data* (Leningrad: Hydrometeorology Publishing House).

Berry, F.A., E. Bollay and N.R. Beers (eds) (1945) *Handbook of Meteorology* (McGraw Hill).

Black, J.N. (1956) 'The distribution of solar radiation over the Earth's surface', *Archiv für Meteorologie, Geophysik und Bioklimatologie/Archives for Meteorology, Geophysics and Bioclimatology* B7, 165–89.

Bolsenga, S.J. (1965) 'The relationship between total precipitable water and surface dewpoint on a mean daily and hourly basis', *Journal of Applied Meteorology* 4, 430–2.

Bootsma, A. (1976) 'Estimating minimum temperature and climatological freeze risk in hilly terrain', *Agricultural Meteorology* 16, 425–43.

Bridgman, H.A. (1978) 'Wind and wind-with-rain conditions at the site of the NEWMED extension to the Royal Newcastle Hospital', *Part I: Climatological Analysis*. Research Paper in Geography 18 (University of Newcastle, New South Wales).

Brodie, H.W. (1964) 'Instruments for measuring solar radiation', *Hawaiian Planters' Record* 57, 159–97.

Brooks, C.E.P. (1950) *Climate in Everyday Life* (Benn).

Brooks F.A. (1959) *An Introduction to Physical Microclimatology Syllabus 397* (Univeristy of California: Davis)

Bruce, J.P. and R.H. Clark (1966) *Introduction to Hydrometeorology* (Pergamon).

Brunt, A.T. (1966) 'Forecasting rainfall for agriculture', *Proc. WMO Seminar on Agricultural Meteorology, Melbourne* (Australian Bureau of Meteorology) 525–34.

Bryson, R.A. and J.F. Lahey (1958) *The march of the seasons*. Rept AF-19(604)-992 (Department of Meteorology, University of Wisconsin).

Budyko, M.I. (1974) *Climate and Life* (Academic).

Burroughs, W.J. (1980) 'Quasi-cycles in meteorology', *Weather* 35, 156–61.

Butler, D.M. *et al.* (1984) *Earth Observing System*. Working Group Rept (National Space Aviation Administration, Goddard Space Flight Center, Greenbelt, Maryland).

Cengiz, H.S., J.M. Gregory and J.L. Sebaugh (1981). 'Solar radiation prediction from other climatic variables', *Transactions of American Society of Agricultural Engineers* 24, 1269–72.

Chaine, P.M. and P. Skeates (1974) *Wind and Ice Loading Criteria Selection* (Environment Canada).

Chandler, T.J. (1976) Urban climatology and its relevance to urban design. WMO Tech. Note 149 (World Meteorological Organization).

—— and S. Gregory (eds) (1976) *The Climate of the British Isles* (Longman).

Chang, J.H. (1971) *Problems and Methods in Agricultural Meteorology* (Taiwan: Oriental Publishing Co.).

Changnon S.A. (1984) 'Climate fluctuations and impacts', *Bulletin of the American Meteorological Society* 66, 142–51.

Choudhury, N.K.D. (1963) 'Solar radiation at New Delhi', *Solar Energy* 7, 44–52.

Clarke, R.H., A.J. Dyer, R.R. Brooke, D.G. Reid and A.J. Troup (1971). The Wangara experiment: boundary layer data. Tech. Paper 19 (Commonwealth Scientific Industrial Research Organization, Australia)

Clothier, B.E., J.P. Kerr, J.S. Talbot and D.R. Scotter (1982). 'Measured and estimated evapotranspiration from well-watered crops', *New Zealand Journal of*

Agricultural Research 25, 301–7.

Cogley, J.G. (1979) 'Albedo of water as a function of latitude', *Monthly Weather Review* 107, 775–81.

Collares-Pereira, M. and A. Rabl (1979) 'The average distribution of solar radiation-correlations between diffuse and hemispherical and between daily and hourly insolation values', *Solar Energy* 22, 155–64.

Conrad, V. (1946) *Methods in Climatology* (Harvard University Press).

Corbett, E.S. (1967) 'Measurement and estimation of precipitation on experimental watersheds', in W.E. Sopper and H.W. Lull (eds), (1967) *International Symposium on Forest Hydrology* (Pergamon). 107–29.

Court, A. (1960) 'Reliability of hourly precipitation data', *Journal of Geophysics Research* 65, 4017–24.

—— (1987) 'Ten proposed rules for numerical diagrams', *Eos* 68, 1642–3.

—— (1989) 'Climagram origins and Landsberg's improvements', in D. Driscoll and E.O. Box (eds), (1989) *Proceedings of 11th ISB Congress, West Lafayette* (The Hague: SPB Academic Pub.) pp.3–7.

Critchfield, H.J. (1987) 'Sources of climatic data', in J.E. Oliver and R.W. Fairbridge (eds), *Encyclopedia of Climatology* (Van Nostrand Reinhold) 272–76.

Curry, H.A., R.Z. Wheaton and E.H. Kidder (1966). 'Errors resulting from mean rainfall estimates on small watersheds'. *Transactions of American Society of Agricultural Engineers* 9, 126–8.

Dahlstrom, B. (1973) Investigation of errors in rainfall observations. Rept 34 (Department of Meteorology, University of Uppsala).

Davidson, B., N. Gerbier, S.D. Papagianakis and P.J. Rijkoort (1964) *Sites for Wind Power Installations*. WMO Tech. Note 63 (World Meteorological Organization).

Davies, J.A. (1967) 'A note on the relationship between net radiation and solar radiation', *Quarterly Journal of the Royal Meteorological Society* 93, 109–15.

—— and J.E. Hay (1980) 'Calculation of the solar radiation incident on a horizontal surface', in J.E. Hay and T.K. Won, *Proceedings of First Canadian Solar Radiation Data Workshop* (Canada: Ministry of Supply and Services; also New Jersey: Rowan & Allanshield), 32–58.

—— and D.C. McKay 1982. 'Estimating solar irradiance and its components', *Solar Energy* 29, 55–64.

Davis, F.K. and H. Newstein (1968) 'The variation of gust factors with mean windspeed and with height', *Journal of Applied Meteorology* 7, 372–8.

Day, J.A. and G.L. Sternes (1970) *Climate and Weather*, (Addison-Wesley).

Day, R.A. (1979) *How to Write and Publish a Scientific Paper*. (Philadelphia: I.S.I. Press).

Deacon, E.L. (1949) 'Vertical diffusion in the lowest layers of the atmosphere', *Quarterly Journal of the Royal Meteorological Society* 75, 89–103.

de Boer, H.J. (1950) 'On the relation between rainfall and altitude in Java, Indonesia', *Chronica Naturae, Djakarta* 106, 424–7.

de Lisle, J.F. (1966) 'Mean daily insolation in New Zealand', *New Zealand Journal of Science* 9, 992–1005.

Dennett, M.D., J. Rodgers and R.D. Stern (1983) 'Independence of rainfalls through the rainy season and the implications for the estimation of rainfall probabilities', *Journal of Climatology* 3, 375–84.

de Vries, D.A. (1958) 'Two years of solar radiation measurements at Deniliquin', *Australian Meteorology Magazine* 22, 36–49.

Dickson, R.R. (1967) 'The climatological relationship between temperatures of successive months in the United States', *Journal of Applied Meteorology* 6, 31–8.

Dixon, K.W. and M.D. Shulman (1984) 'A statistical evaluation of the predictive abilities of climatic averages', *Journal of Climate and Applied Meteorology* 23, 1542–52.

Druyan, L.M. (1985) 'Wind climate studies for WECs siting', *Journal of Climatology* 5, 95–105.

Dury, G. (1980). 'Step-functional changes in precipitation at Sydney', *Australian Geographical Studies* 18, 62–78.

Dutton, J.A. (1976) *The Ceaseless Wind: An Introduction to the Theory of Atmospheric Motion* (McGraw-Hill).

Edwards, K. (1979) *Rainfall in New South Wales. Technical Handbook 3* (New South Wales Soil Conservation Service).

Elliott, W.P. (1958) 'The growth of atmospheric internal boundary layer', *Transactions of American Geophysics Union* 39, 1048–54.

Elsom, D.M. and G.T. Meaden (1984) 'Spatial and temporal distributions of tornadoes in the United Kingdom 1960–1982'. *Weather* 39, 317–23.

Evans, M. (1980) *Housing, Climate and Comfort.* (London: Architectural Press).

Exell, R.H.B. (1982) 'The availability of solar and wind energy in Thailand', 5th Nat. Conf., Canberra (Australian Institute of Physics).

Fairbridge, R.W. (ed.) (1967) *The Encyclopedia of Atmospheric Sciences and Astrogeology.* (van Nostrand Reinhold).

Fiedler, F. and H.A. Panofsky (1970) 'Atmospheric scales and gaps', *Bulletin of the American Meteorological Society* 51, 1114–19.

Filipov, V.V. (1968) *Quality Control Procedures for Meteorological Data*, Planning Report 26, World Weather Watch (World Meteorological Organization).

Fitzpatrick, E.A. and W.R. Stern (1973) 'Net radiation estimated from global solar radiation', *Proceedings of Symposium for Plant Response to Climatic Factors*, Uppsala, 1970 (United Nations Education and Science Cultural Organization), 403–10.

Flavin, R.K. (1981) *Rain Attenuation Considerations for Satellite Paths*, Research Laboratory Report 7505 (Telecom Australia).

Folland, C.K. (1988) 'Numerical models of the raingauge exposure problem, field experiments and an improved collector design', *Quarterly Journal of Royal Meteorological Society* 114, 1488–1516.

Forland, E.J. and D. Kristoffersen (1989) 'Estimation of extreme precipitation in Norway', *Nordic Hydrology* 20, 257–76.

Fritschen, L.J. (1967) 'Net and solar radiation relations over irrigated field crops', *Agricultural Meteorology* 4, 55–62.

Fukui, E. (1962) 'Meaning of climate and three-dimensional division of climatology', *Journal of Geography (Tokyo)* 71, 232–6.

Funk, J.P. (1959) 'Improved polythene shielded net radiometer', *Journal of Scientific Instruments* 36, 267–70.

Gaffney, D.O. (1975) 'Global radiation in the Australian region with reference to some climatic variations', Paper to Australian Conference on Climate and Climate Change, Monash University, Melbourne.

Gani, J. (1975) 'The use of statistics in climatological research', *Search* 6, 504–8.

Garratt, J.R. and M. Segal (1988) 'On the contribution of atmospheric moisture to dew formation', *Boundary-Layer Meteorology* 45, 209–36.

Gates, D.M. (1962) *Energy Exchange in the Biosphere* (Harper & Row).

Geiger, R. (1966) *The Climate Near the Ground* (Harvard University Press).

Gentilli, J. (ed.) (1971) 'Climates of Australia and New Zealand', in *World Survey of Climatology* (Elsevier) Vol. 13.

Gibbs, W.J., J.V. Maher and M.J. Coughlan (1978) 'Climatic variability and extremes', in A.B. Pittock, L.A. Frakes, D. Jensen, J.A. Peterson and J.W. Zillman, *Climatic Change and Variability* (Cambridge University Press) 135–50.

Gloyne, R.W. and J. Lomas (1980) *Lecture Notes for Training Class II and Class III*

Agricultural Meteorological Personnel. WMO Publ. 551 (World Meteorological Organization).

Godske, C.L. (1966) 'Methods of statistics and some applications to climatology', in WMO 1966, *Statistical Analysis and Prognosis in Meteorology*, WMO Tech. Note 71 (World Meteorological Organization) 9–86.

—— (1969) 'The future of meteorological data analysis', in WMO, *Data Processing for Climatological Purposes*, WMO Tech. Note 100 (World Meteorological Organization), 52–63.

Goel, S.M. and A.S. Aldabagh (1979) 'A distance weighted method for computing average precipitation', *Journal of the Institute of Water Engineers and Science* 33, 451–4.

Gois, C.Q. (1974) 'Some criteria used in hydrologic studies with inadequate data', *Design of Water Resources Projects with Inadequate Data, Proceedings of Symposium Madrid* (Association of International Science and Hydrology) 241–52.

Golding, E.W. (1955) *The Generation of Electricity by Wind Power* (London: Span; also published by John Wiley, 1976).

Goltsberg, I.A. (ed.) (1969) *Microclimate of the USSR* (Washington: Israel Program for Scientific Translations, US National Science Federation).

Gomes, L. and B.J. Vickery (1976) 'On thunderstorm wind gusts in Australia', *Civil Engineering Transactions (Institute of Engineers, Australia)* CE18, 33–9.

Graham, A.E. (1982) 'Winds estimated by the Voluntary Observing Fleet compared with instrumental measurements at fixed positions', *Meteorological Magazine* 111, 312–27.

Greene, G. and F. Nelson (1983) 'Performance of a frost hollow as a hemispherical thermal radiator', *Archiv für Meteorologie, Geophysik und Bioklimatologie/Archives for Meteorology, Geophysics and Bioclimatology* B32, 263–78.

Gumbel, E.J. (1954) *Statistical Theory of Extreme Values and Some Practical Applications*, Applied Mathematics Series 33 (US National Bureau Standards).

Guttman, N.B. and M.S. Plantico (1987) 'Climate temperature normals', *Journal of Climate and Applied Meteorology* 26, 1428–35.

Hann, J. (1897) *Handbuch der Klimatologie* (2nd edn), translated by R. dec. Ward (1903) J. Engelhorn, Stuttgart).

Hanna, S.R. (1987) 'An empirical formula for the height of the coastal internal boundary layer', *Boundary-Layer Meteorology* 40, 205–7.

Hanson, K., G.A. Maul and W. McLeish (1987) 'Precipitation and the lunar synodic cycle', *Journal of Climate and Applied Meteorology* 26, 1358–62.

Harbeck, G.E. and E.W. Coffey (1959) 'A comparison of rainfall data obtained from rain-gauge measurements and changes in lake levels', *Bulletin of the American Meteorological Society* 40, 348–51.

Harding, R.J. (1986) 'Exchanges of energy and mass associated with a melting snowpack', in F.W. Morris (ed.), *Modelling Snowmelt-induced Processes,* Symposium at Budapest, IAHS Publ. 155 (International Association of Hydrological Science), 3–15.

Hargreaves, G.H. (1981) 'Simplified method for rainfall intensities, *Journal of the Irrigation Drainage Division* (American Society of Civil Engineers) 107, 281–8.

Hawke, G.S., R. Hyde, E.T. Linacre and C.A. McGrath (1975) 'Sydney's winds and air pollution', Environment 75 Conference, Sydney 1975, (New South Wales State Pollution Control Commission), 19–34.

Hay, J.E. (1976) 'A revised method for determining the direct and diffuse components of the total short-wave radiation', *Atmosphere* 14, 278–87.

—— (1984) 'An assessment of the mesoscale variability of solar radiation at the Earth's surface', *Solar Energy* 32, 425–34.

—— (1986) 'Calculation of solar irradiances for inclined surfaces', *Atmosphere-Ocean* 24, 16–41.

—— and R. Darby (1984) 'El Chichon-influence on aerosol optical depth and direct, diffuse and total solar irradiances at Vancouver, B.C.', *Atmosphere-Ocean* 22, 354–68.

—— and J.A. Davies (1980) 'Calculation of the solar radiation incident on an inclined surface', in J.E Hay and T.K. Won, *Proceedings of First-Canadian Solar Radiation Data Workshop* (Canada: Ministry of Supply and Services; also New Jersey: Rowan & Allanshield), 59–72.

—— and B.B. Fitzharris (1988) 'The synoptic climatology of ablation on a New Zealand glacier', *Journal of Climatology* 8, 201–15.

—— and K.J. Hanson (1985) 'Evaluating the solar resource: a review of problems resulting from temporal, spatial and angular variations', *Solar Energy* 34, 151–61.

—— and T.K. Won (1980) *Proceedings of First Canadian Solar Radiation Data Workshop* (Canada: Ministry of Supply and Services; also New Jersey: Rowan & Allanshield).

Heermann, D.F., G.J. Harrington and K.M. Stahl (1985) 'Empirical estimation of daily clear sky solar radiation', *Journal of Climate and Applied Meteorology* 24, 206–14.

Henry, A.J. (1919) 'Increase of precipitation with altitude', *Mon. Weather Review* 47, 33–41.

Hines, C.O. and I. Halevy (1977) 'On the reality and nature of a certain sun–weather correlation', *Journal of Atmospheric Science* 34, 382–404.

Hirst, A. and E.T. Linacre (1978) 'Associations between coastal sea-surface temperatures, onshore winds and rainfalls in the Sydney area', *Search* 9, 325–7.

Hogstrom, A.S. and U. Hogstrom (1978) 'A practical method for determining wind frequency distributions for the lowest 200 m from routine meteorological data', *Journal of Applied Meteorology* 17, 942–54.

Hookey, P.G. (1965) 'A home-made rainfall recorder', *Weather* 20, 193–6.

Hopkins, J.S. (1977) 'The spatial variability of daily temperatures and sunshine over uniform terrain', *Meteorology Magazine* 106, 278–92.

Houghton, D.D. (ed.) (1985) *Handbook of Applied Meteorology* (Wiley).

Houghton, J.T. (1977) *The Physics of Atmospheres* (Cambridge University Press)

Hounam, C.E. (1945) 'The sea breeze at Perth', *Weather Development and Research Bulletin* 3 (Royal Australian Air Force Meteorology Service) 20–56.

—— and J.J. Burgos, M.S. Kalik, W.C. Palmer and J. Rodda (1975) *Drought and Agriculture*, WMO Tech. Note 138 (World Meteorological Organization).

Hu, H.C. and J.T. Lim (1983) 'Solar and net radiation in peninsular Malaysia', *Journal of Climatology* 3, 271–83.

Huff, F.A. (1970) 'Sampling errors in measurement of mean precipitation', *Journal of Applied Meteorology* 9, 35–44.

Idso, S.B., D.G. Baker and B.L. Blad (1969) 'Relations of radiation fluxes over natural surfaces', *Quarterly Journal of Royal Meteorological Society* 95, 244–57.

Impens, I. and R. Lemeur (1969) 'The radiation balance of several field crops', *Archiv für Meteorologie, Geophysik und Bioklimatologie/Archives for Meteorology, Geophysics and Bioclimatology* B17, 261–8.

Ingram, W.J. (1989) 'Modelling cloud feedbacks in climate change', *Weather* 44, 303–11.

Ioffe, A.F. and I.B. Revut (eds) (1966) *Fundamentals of Agrophysics* (Israel Program of Scientific Translations).

Iqbal, M. (1979) 'Correlation of average diffuse and beam radiation with hours of bright sunshine', *Solar Energy* 23, 169–73.

—— (1983) *An Introduction to Solar Radiation* (Academic).

Jagannathan, P., R. Arlery, H.T. Kate and M.V. Zavarina (1967) 'A note on climatological normals', WMO Tech. Note 84 (World Meteorological Organization).

James, C. (1969) 'Analysis of rapid methods for fitting straight line graphs', *Physics Education* 4, 101–5.

Jennings, E.G. and J.L. Monteith (1954) 'A sensitive recording dew balance', *Quarterly Journal of the Royal Meteorological Society* 80, 222–6.

Jensen, D. (1983) 'Computer simulation of an aerological diagram', *Meteorology Australia* 3, 13–16.

Jobson, H.E. (1972) 'Effect of using averaged data on the computed evaporation', *Water Resources Research* 8, 513–18.

Johnson, G.T. (1979) 'Evaluation of schemes for estimating surface-wind strength', *Atmosphere and Environment* 13, 437–42.

—— (1982) 'Climatological interpolation functions for mesoscale wind fields', *Journal of Applied Meteorology* 21, 1130–6.

—— and E.T. Linacre (1978) 'Estimation of winds at particular places', in E.T. White, P. Hetherington and B.R. Thiele (eds), *Proceedings of International Clean Air Conference, Brisbane* (Ann Arbor Scientific), 751–62.

Johnson, W.B. and R.E. Ruff (1975) 'Observational systems and techniques in air pollution meteorology', in D.A. Haugen (ed), *Lectures on Air Pollution and Environmental Impact Analysis* (American Meteorological Society), 243–74.

Jones, D.J. and T.L. Wigley (1990) 'Global warming trends', *Scientific American* 263, 84–91.

Jones, D.M.A. and A.L. Sims (1978) 'Climatology of instantaneous rainfall rates', *Journal of Applied Meteorology* 17, 1135–40.

Jones, H.G. (1983) *Plants and Microclimate.* (Cambridge University Press) 323pp.

Jones, K.R., O. Berney, D.P. Carr, and E.C. Barrett (1981) Arid zone hydrology for agricultural development. FAO Irrig. & Drainage Paper 37 (United Nations, Food and Agriculture Organisation).

Jones, P.D., T.M.L. Wigley, C.K. Folland, D.E. Parker, J.K. Angell, S. Lebedeff and J.E. Hansen (1988) 'Evidence for global warming in the past decade', *Nature* 332, 790.

Kalma, J.D. (1968) 'A comparison of methods for computing daily mean air temperature and humidity', *Weather* 23, 248–52, 259.

—— (1972) 'The radiation balance of a tropical pasture: net all-wave radiation', *Agricultural Meteorology* 10, 261–75.

—— and R. Badham (1972) 'The radiation balance of a tropical pasture', *Agricultural Meteorology* 10, 251–75.

—— and P.M. Fleming (1972) 'A note on estimating the direct and diffuse components of global radiation', *Archiv für Meteorologie, Geophysik und Bioklimatologie/Archives for Meteorology, Geophysics and Bioclimatology* B20, 191–205.

Katsaros, K.B., L.A. McMurdie, R.J. Lind and J.E. Devault (1985) 'Albedo of a water surface, spectral variation effects of atmospheric transmittance, sun angle and wind speed', *Journal of Geophysical Research* 90, 7313–21.

—— S.D. Smith and W.A. Oost (1987) 'HEXOS-humidity exchange over the sea: a program for research on water-vapor and droplet fluxes from sea to air at moderate to high wind speeds'. *Bulletin of the American Meteorological Society* 68, 466–76.

Keene, P. (1981) 'Winds in Sydney', M.Sc. diss., Macquarie University, Sydney.

Kennedy, R.E. (1940) 'Average daily air mass', *Monthly Weather Review* 68, 301–3.

Kerr, R.A. (1984) 'The moon influences western US drought', *Science* 224, 587.

Kessler, A. (1985) 'Heat balance climatology', *World Survey of Climatology* (Elsevier) 1A, 224 pp.

Klein, W.H. (1948) 'Calculation of solar radiation and the solar heat load on man', *Journal of Meteorology* 5, 119–29.

Kneen, T. (1982) 'Snow piled high', *Meteorology Australia* 2, 4–7.

Kotoda, K. (1986) Estimation of river basin evapotranspiration. Environmental Resources Center Paper 8 (University of Tsukuba, Japan).

Lacy, R.E. (1977) *Climate and Building in Britain* (Building Resources Establishment, UK).

Lamb, H.H. (1972) *Climate: Present, Past and Future.* Vol 1 – *Fundamentals and Climate Now* (Methuen).

Landsberg, H.E. (1957) 'Review of climatology, 1951–1955', *Meteorology Monographs.* (American Meteorological Society) 3, 1–43.

—— (1958) *Pysical Climatology* (Gray Printing Co., DuBois, Penn.) 446pp.

—— (1982) 'Climatology – now and henceforth', (Department of Meteorology, University of Maryland) or *WMO Bulletin* 31, 361–8.

Langley, R. (1979) *Practical Statistics Simply Explained.* (Pan).

Lee, R. (1966) 'An organizational solution of the weather forecasting problem', *Bulletin of American Meteorological Society* 47, 438–44.

Lethbridge, M.D.V. (1970) 'Relationship between thunderstorm frequency and lunar phase and declination', *Journal of Geophysics Resources* 75, 5149–54.

Linacre, E.T. (1964a) 'Determinations of the heat-transfer coefficient of a leaf', *Plant Physiology* 39, 687–90.

—— (1964b) 'A note on a feature of leaf and air temperatures', *Agricultural Meteorology* 1, 66–72.

—— (1968) 'Estimating the net-radiation flux', *Agricultural Meteorology* 5, 49–63.

—— (1969) 'Empirical relationships involving the global radiation intensity and ambient temperature at various latitudes and altitudes', *Archiv für Meteorologie, Geophysik und Bioklimatologie/Archives for Meteorology, Geophysics and Bioclimatology* B12, 1–20.

—— (1969b) 'Net radiation to various surfaces', *Journal of Applied Ecology* 6, 61–75.

—— (1981) 'Caracteristicas das varias escalas de clima', *Boletin da Sociedade de Meteorologia* 5, 11–14.

—— (1982) 'Effect of altitude on the daily temperature range', *Journal of Climatology* 2, 375–82.

—— and J.A. Barrero (1974) 'Surveys of surface winds in the Sydney region', *Proceedings of International Geographical Union Regional Conference, and 8th New Zealand Geographical Conference, Palmerston North* (New Zealand Geographical Society) 247–60.

—— and J.E. Hobbs (1977) *The Australian Climatic Environment* (Jacaranda Wiley).

Lindner G. and K. Nyberg (eds) (1973) *Environmental Engineering* (Reidel).

Linsley, R.K. and M.A. Kohler (1951) 'Variations in storm rainfall over small areas', *Transactions of American Geophysics Union* 32, 245–50.

—— M.A. Kohler and J.L.H. Paulhus (1982) *Hydrology for Engineers*, 3rd edn (McGraw-Hill).

Liu, B.Y.H. and R.C. Jordan (1960) 'The interrelationship and characteristic distribution of direct, diffuse & total solar radiation', *Solar Energy* 4, 1–19.

Lloyd, E. (ed.) (1984) *Handbook of Applicable Mathematics* (John Wiley).

Lockwood, J.G. (1974) *World Climatology: An Environmental Approach.* (Arnold).

Lowry, W.P. (1972a) *Compendium of Lecture Notes on Climatology for Class IV Meteorological Personnel*, WMO Publ. 327 (World Meteorological Organization).

—— (1972b) *Compendium of Lecture Notes on Climatology for Class III Meteorological Personnel.* WMO Publ. 335 (World Meteorological Organization).

Ludlam, F.E. (1980) *Clouds and Storms* (Pennsylvania State University).

Lund, I.A. (1968) 'Relationships between insolation and other surface weather observations at Blue Hill, Massachusetts', *Solar Energy* 12, 95–106.

Lunde, P.J. (1980) *Solar Thermal Engineering* (John Wiley).

McConnell, D (1988) 'Making a simple thermometer screen', *Weather* 43, 198–203.

McGee, O.S. (1984) 'Preliminary analysis of the regression equations linking precipitable water and surface dewpoint in South Africa', *South African Journal of Science* 80, 282–5.

McGrath, C.A. (1972) 'The development of the sea-breeze over Sydney and its effect on climate and air pollution', M.Sc. thesis for Macquarie University.

McIlroy, I.C. and D.E. Angus (1964) 'Grass, water and soil evaporation at Aspendale', *Agricultural Meteorology* 1, 201–24.

McIntyre, D.A. (1980) *Indoor Climate* (London: Applied Science Publisher).

McWhorter, J.C. and B.P. Brooks (1965) 'Climatological and solar radiation relationships', Bull. 715 (Mississippi Agricultural Experiment Station).

Mani, A. and S. Rangarajan (1982) *Solar Radiation Over India*. (New Delhi: Allied Publishers Private Ltd.).

Manning, H.L. (1956) 'The statistical assessment of rainfall probability and its application in Ugandan agriculture', *Proceedings of the Royal Society* B144, 460–80.

Markham, C.G. (1970) 'Seasonality of precipitation in the United States', *Association of American Geographers Annals* 60, 593–7.

Martin, K.G. (1976) 'Australian climatic data and weathering', *Australian Oil and Colour Chemistry Association Proceedings and News* 13, 9–15.

Mason, B.J. (1970) 'Future developments in meteorology: an outlook to the year 2000', *Quarterly Journal of the Royal Meteorological Society* 96, 349–68.

—— (1976) 'Towards an understanding and prediction of climatic variations', *Quarterly Journal of the Royal Meteorological Society* 102, 473–98.

Meteorological Office (1956) *Handbook of Meteorological Instruments: Part I Instruments for Surface Observations* (London: H.M. Stationery Office).

Mihara, Y. (ed.) (1974) *Agricultural Meteorology of Japan* (University of Tokyo Press).

—— and T. Ando (1974) 'Background and history of agricultural meteorology in Japan', in Y. Mihara (ed.), *Agricultural Meteorology of Japan* (University of Tokyo Press), 1–10.

Mitchell, J.F.B. (1989) 'The greenhouse effect and climate change', *Review of Geophysics* 27, 115–40.

Mitchell, J.M., C.W. Stockton and D.M. Meko (1979) 'Evidence of a 22-year rhythm of drought in the western United States related to the solar cycle since the 17th century', in B.M. McCormac and T.A. Seliga (eds.), *Solar–Terrestrial Influences on Weather and Climate* (Academic), 125–43.

Monteith, J.L. (1965) 'Radiation and crops', *Experimental Agriculture* 1, 241–51.

—— (1973) *Principles of Environmental Physics* (Arnold).

—— and G. Szeicz (1960) 'Performance of a Gunn-Bellani radiation integrator', *Quarterly Journal of the Royal Meteorological Society* 86, 91–4.

—— and G. Szeicz (1961) 'The radiation balance of bare soil and vegetation', *Quaterly Journal of the Meterological Society*, 87, 159–70.

Mueller, C.C. and E.H. Kidder (1972) 'Raingauge catch variations due to airflow disturbances around a standard raingauge', *Water Resources Res.* 8, 1077–82.

Muffatti, A.H.J. (1966) 'Statistical methods in agrometeorology', Proceedings of WMO Seminar Agricultural Meteorology, Melbourne (Australian Bureau of Meteorology) *2*, 291–342.

Muller, W. (1968) 'The role and measurement of dew', Proceedings of Symposium on Agricultural Methods, *Natural Resources Research* 7, 303–7.

Munn, R.E. (1966) *Descriptive Micrometeorology* (New York: Academic Press).

—— (1970) *Biometeorlogical Methods*. (Academic Press) 336pp.

Murtagh, G.J. (1976) 'Relations between net radiation, global solar radiation and sunshine on the north coast of New South Wales', *Australian Meteorology Magazine* 24, 111–14.

Neuwirth, F. (1980) 'The estimation of global and sky radiation in Austria', *Solar Energy* 24, 421–6.

Nielsen, L.B., L.P. Prahm, R. Berkowicz and K. Conradsen (1981) 'Net incoming radiation estimated from hourly global radiation and/or cloud observations', *Journal of Climatology* 1, 255–72.

Nieuwolt, S. (1974) 'The influence of aspect and elevation on daily rainfall: some examples from Tanzania', WMO (1974c) 16–29.

Nkemdirim, L.C. (1970) 'Empirical methods for the assessment of insolation and evaporation in southern Scotland', *Archiv für Meteorologie, Geophysik und Bioklimatologie/Archives for Meteorology, Geophysics and Bioclimatology* B18, 131–41.

—— (1972) 'Relation of radiation fluxes over prairie grass', *Archiv für Meteorologie, Geophysik und Bioklimatologie/Archives for Meteorology, Geophysics and Bioclimatology* B20, 23–40.

Nunez, M. (1988) 'A comparison of three approaches to estimate daily totals of global solar radiation in Australia, using GMS data', *Australian Meteorology Magazine* 36, 25–33.

Obasi, G.O.P. and N.N.P. Rao (1977) 'A detailed study of solar radiation and potential photosynthesis distribution in Kenya', in WMO, *Solar Energy, Proceedings of WMO/UNESCO Symposium Geneva, 1976*, WMO Publ. 477 (World Meteorological Organization), 114–24.

Ojo, O. (1970) 'The distribution of mean monthly precipitable water vapour and annual precipitation efficiency in Nigeria', *Archiv für Meteorologie, Geophysik und Bioklimatologie/Archives for Meteorology, Geophysics and Bioclimatology* B18, 221–8.

Oke, T.R. (1978) *Boundary-layer Climates* (Methuen).

Okolowicz, W. (1976) *General Climatology* (Warsaw: Polish Scientific Publishers).

Oldeman, L.R. (1987) 'Characterisation of main experimental sites and sub-sites and questions of instrumentation', in A.H. Bunting (ed.), *Agricultural Environments* (Commonwealth Agriculture Bureau International), 101–12.

—— and D. Suardi (1976) 'Climatic determinants in relation to cropping patterns', Private communication.

Oliver, J. (1959) 'Rainfall variations over a small area', *Meteorology Magazine* 88, 289–93.

Orlanski, I. (1975) 'A rational subdivision of scales for atmospheric processes', *Bulletin of the American Meteorological Society* 56, 527–30.

Osborn, H.B. (1984) 'Estimating precipitation in mountainous regions', *Journal of Hydraulic Engineering* (American Society of Civil Engineers) 110, 1859–63.

Padmanabhamurty, B. and V.P. Subrahmanyam (1964) Some studies on radiation at Waltair', *Indian Journal of Pure and Applied Physics* 2, 293–5.

Painter, H.E. (1981) 'The performance of a Campbell-Stokes sunshine recorder compared with a simultaneous record of the normal incidence irradiance', *Meteorology Magazine* 110, 102–9.

Paltridge, G.W. (1975) 'Net radiation over the surface of Australia', *Search* 6, 37–9.

—— and C.M.R. Platt (1976) *Radiative Processes in Meteorology and Climatology* (Elsevier).

Parker, D.E. (1989) 'Observed climatic change and the greenhouse effect', *Meteorology Magazine* 118, 128–31.

Pasquill, F. (1971) 'Effects of buildings on the local wind', *Philosophical Transactions of the Royal Society* A 269, 439–56.

Paterson, W.S.B. (1969) *The Physics of Glaciers* (Pergamon).

Penman, H.L. (1948) 'Natural evaporation from open water, bare soil and grass',

Proceedings of the Royal Society A 193, 120–45.

Penney, C.L. (1981) 'A study of evapotranspiration from irrigated crops in the south east of South Australia', *Research Report 38* (Institute of Atmosphere and Marine Science, Flinders University S. Australia) 252pp.

Peterson, W.A. and I. Dirmhirn (1981) 'The ratio of diffuse to direct solar irradiance (perpendicular to the sun's rays) with clear skies – a conserved quantity throughout the day', *Journal of Applied Meteorology 20*, 826–8.

Pinker, R.T., O.E. Thompson and T.F. Peck (1980) 'The albedo of a tropical evergreen forest', *Quarterly Journal of the Royal Meteorological Society* 106, 551–8.

Pittock, A.B. (1978) 'A critical look at long-term sun-weather relationships', *Review of Geophysics and Space Physics* 16, 400–20.

—— (1983) 'Solar variability, weather and climate: an update', *Quarterly Journal of the Royal Meteorological Society* 109, 23–55.

Pokorny, A.D., F. Davies and W. Harberson (1974) 'Suicide, suicide attempts and weather', in J.H. Sims and D.D. Baumann (eds.), *Human Behaviour and the Environment* (Chicago: Maaroufa Press), 345–54.

Prescott, J.A. (1940) 'Evaporation from a water surface in relation to solar radiation', *Transactions of Royal Society of South Australia* 64, 114–25.

Quenouille, M.H. (1959) *Rapid Statistical Calculations* (London: Charles Griffin).

Raju, A.S.N. and K.K. Kumar (1982) 'Comparison of point cloudiness and sunshine derived cloud cover in India', *Pure and Applied Geophysics* 120, 495–501.

Rao, C.R.N. and W.A. Bradley (1983) 'Estimation of the daily global solar irradiation at Corvallis (Oregon, USA) from the hours of bright sunshine, the daily temperature range and relative humidity', *Journal of Climatology* 3, 179–85.

Rao, N.J., Y. Viswamadham and G.S.S. Nunes (1980) 'Moisture relationships for the southern hemisphere', *Pure and Applied Geophysics* 118, 1076–89.

Reddy, K.K. (1987) 'Studies on albedo at a tropical station', in AMS, *Proceedings of 18th Conference Agricultural Meteorology, Lafayette* (American Meteorological Society) 43–4.

Reich, B.M. (1963) 'Short-duration rainfall-intensity estimates', *Journal of Hydrology* 1, 3–28.

Reitan, C.H. (1963) 'Surface dew point and water vapour aloft', *Journal of Applied Meteorology* 2, 776–9.

Reynolds, G. (1965) 'A history of raingauges', *Weather* 20, 106–14.

Richards, J.M. (1971) 'Simple expression for the saturation vapour pressure of water in the range −50°C to 140°C', *British Journal of Applied Physics* 4, L15–18.

Rietveld, M.R. (1978) 'A new method for estimating the regression coefficients in the formula relating solar radiation to sunshine', *Agricultural Meteorology* 19, 243–52.

Rijks, D.A. (1967) 'Water use by irrigated cotton in Sudan: reflection of shortwave radiation', *Journal of Applied Ecology* 4, 561–8.

Robertson, G.W. (1970) 'Rainfall and soil-water variability with reference to land-use planning', *Technical Series 1*, WMO/UNDP Project on Meteorological Training and Research, (Manila: Philippines Weather Bureau and Department of Meteorology, University of Philippines).

Robinson, D.A. (1990) 'The US co-operative climate-observing systems', *Bulletin of the American Meteorological Society* 71, 826–31.

Rodskjer, N. (1978) 'Net and solar radiation over bare soil, short grass, winter wheat and barley', *Swedish Journal of Agricultural Research* 8, 195–206.

Rosenberg, N.J. (1974) *Microclimate: the Biological Environment* (John Wiley).

—— B.L. Blad and S.B. Verma (1983) *Microclimate: the Biological Environment* (John Wiley).

Rouse, W.R., P.F. Mills and R.M. Stewart (1977) 'Evaporation in high latitudes', *Water Resources Research* 13, 909–14.

Rusin, N.P. (1970) 'Precipitation development', in S. Teweles and J. Giraytys (eds.), *Meteorological Observations and Instrumentation* (American Meteorological Society) 283–6.

Russell, E.W. (1960) 'The agricultural uses of meteorological data', in D.J. Bargman (ed), *Tropical Meteorology in Africa* (Nairobi: Munitalp Foundation), 390–401

Satterlund, D.R. and J.E. Means (1978) 'Estimating solar radiation under variable cloud conditions', *Forest Science* 24, 363–73.

Schaake, J.C. (1978) 'Accuracy of point and mean areal precipitation estimates from point precipitation data', Tech. Memo. NWS-HYDRO (US National Oceanic and Atmospheric Administration, Washington).

Schemenauer, R.S., P. Cereceda and N. Carvajal (1987) 'Surface-based measurements of cloud liquid water contents and their relationship to terrain features', *Journal of Climate and Applied Meteorology* 26, 1285–91.

—— H. Fuenzalida and P. Cereceda (1988) 'A neglected water resource: the Camanchaca of South America', *Bulletin of the American Meteorological Society* 69, 138–47.

Schneider, S.H. (1990) 'The global warming debate heats up', *Bulletin of the American Meteorological Society. 71*, 1292–1304.

Schulze, R.E. (1980) 'Potential flood producing rainfall of medium and long duration in southern Africa', (Water Resources Commission, Pretoria).

Schumaker, J., B. Bajusz, L. Pochop and J. Borrelli (1984) 'Predicting temperature and precipitation at ungaged sites in the Upper Green River basin', in AMS, Proceedings of 3rd Conference of Mountain Meteorology, Portland (American Meteorological Society), 152–4.

Schwerdtfeger, P., U. Radok, J. Bennett, B. van Meurs, I. Piggin, A. Ussher and A. Wu (1975) 'The microenvironment of plants', in L.P. Smith (1975), 3–50.

Schwerdtfeger. W. (1984) *Weather and Climate of the Antarctic* (Elsevier).

Severini, M., M.L. Moriconi, G. Tonna, and B. Olivieri (1984) 'Dewfall and evapotranspiration determination during day and night-time on an irrigated lawn', *Journal of Climatology and Applied Meteorology* 23, 1241–6.

Sevruk, B. (1982) Methods of correction for systematic error in point precipitation measurement for operational use, WMO Operational Hydrol. Rept 21, WMO Publ. 589 (World Meteorological Organization).

Shaw, E.M. (ed.) (1967) *Ground-level Climatology*, Publ. 86 (American Association of Advanced Science).

Shaw, R.H. (1956) 'A comparison of solar and net radiation', *Bulletin of American Meteorological Society* 37, 205–6.

Schulze, R.E. (1975) 'Incoming radiation fluxes on sloping terrain: a general model for use in southern Africa' *Agrochemophysica* 7, 55–60.

Showalter, A.K. (1962) 'Surface weather measurements', *Bulletin of American Meteorological Society* 43, 454–6.

Singer, S.F. (ed.) (1989) *Global Climate Change* (Paragon House).

Smith, K. (1975) *Principles of Applied Climatology* (McGraw-Hill).

Smith, L.P. (ed.) (1975) *Progress in Plant Biometeorology 1.* (Swets & Zeitlinger, Amsterdam) 474pp.

Smith, R.B. (1979) 'The influence of mountains on the atmosphere', *Advanced Geophysics* 21, 87–230; also in K. Browning (convenor) (1981) *Nowcasting: Proceedings of the International IAMAP Symposium, Hamburg* (European Space Agency), 37–41.

Smith, W.L. (1966) 'Note on the relationship between total precipitable water and surface dewpoint', *Journal of Applied Meteorology* 5, 726–7.

Snijders, T.A.B. (1986) 'Interstation correlations and non-stationarity of Burkina Faso rainfall', *Journal of Climatology and Applied Meteorology* 25, 524–31.

Snow, J.T., M.E. Akridge and S.B. Harley (1989) 'Basic meteorological observations for schools: surface winds', *Bulletin of American Meteorological Society* 70, 493–508.

Stanhill, G. (1958) 'Rainfall measurements at ground level', *Weather* 13, 33–4.

—— (1962) 'Solar radiation in Israel', *Bulletin of Research Council of Israel* 11G, 34–41.

—— (1965) 'A comparison of four methods of estimating solar radiation', in F.E. Eckardt (ed.), *Methodology of Plant Eco-Physiology* (UNESCO), 55–61.

Stewart, R. and V.A. Mohnin (1977) 'Solar energy development and implementation', in WMO, *Solar Energy: Proc. WMO/UNESCO Symp. Geneva, 1976,* WMO Publ. 477 (World Meteorological Organisation), 357–71.

Stringer, E.T. (1972) *Techniques of Climatology* (Freeman).

Stull, R.B. (1985) 'Predictability and scales of motion', *Bulletin of American Meteorological Society* 66, 432–6.

Sumner, C.J. (1966) 'A sunshine sensing device for long-period recording', *Quarterly Journal of Royal Meteorological Society* 92, 567–9.

Szeicz, G. (1968) 'Measurement of radiant energy', in R.M. Wadsworth (ed.), *The Measurement of Environmental Factors in Terrestrial Ecology* (Oxford: Blackwell), 109–30.

Tagg, J.R. (1957) Wind data related to the generation of electricity by wind power. Report C/T 115. (Electrical Res. Assoc., UK).

Takahashi, L. (ed.) (1985) *World-wide Weather* (Balkema).

Takle, E.S. and J.M. Brown (1978) 'Note on the use of Weibull statistics to characterize windspeed data', *Journal of Applied Meteorology* 17, 556–9.

Templeman, R.F., H.R. Oliver, M.R. Stroud, M.E. Walker, T. Altay, and S.L. Pike, (1988) 'Some wind speed and temperature observations during the storm of 15–16 October 1987', *Weather* 43, 118–22.

Tepper, M. (1959) 'Mesometeorology – the link between macroscale atmospheric motions and local weather', *Bulletin of the American Meteorological Society* 40, 56–72.

Thompson, N. (1981) 'The duration of leaf wetness', *Meteorology Magazine* 110, 1–12.

Tomlinson, A.I. (1987) 'Wet and dry years – seven years on', *Soil and Water (N.Z.),* winter, 8–9.

Trenberth, K.E. (1983) 'What are the seasons?', *Bulletin of the American Meteorological Society* 64, 1276–81.

Tucker, G.B. (1964) 'Solar influences on the weather', *Weather* 19, 302–11.

Tuller, S.E. (1976) 'The relationship between diffuse, total and extraterrestrial solar radiation', *Solar Energy* 18, 259–63.

—— (1977) 'The relationship between precipitable water vapour and surface humidity in New Zealand', *Archiv für Meteorologie, Geophysik und Bioklimatologie/ Archives for Meteorology, Geophysics and Bioclimatology* A 26, 197–212.

Tyson, P.D. (1981) 'Atmospheric circulation variations and the occurrence of extended wet and dry spells over Southern Africa', *Journal of Climatology* 1, 115–30.

van der Hoven, I. (1957) 'Power spectrum of horizontal wind speed. *Journal of Meteorology* 14, 160–4.

Verral, K.A. and R.L. Williams (1982) 'The use of surface-wind flow-patterns to estimate the transport of air pollutants in the Gladstone region', Paper to Ann. Conf., Sydney (Australia and New Zealand Association of Advanced Science).

Viswanadham, Y. (1981) 'The relationship between total precipitable water and surface dewpoint', *Journal of Applied Meteorology* 20, 3–8.

Wallen, C.C. (1968) 'Definitions and scales in climatology as applied to agriculture', *Proceedings of Regional Seminar on Agricultural Meteorology, Wageningen.* (World Meteor. Organ.), 207–12.

Wallen, C.C. (ed.) (1977) 'Climate of Central & Southern Europe', *World Survey of Climatology* (Elsevier) 6, 248pp.

Walter, A. (1967) 'Notes on the utilization of records from third order climatological stations for agricultural purposes', *Agricultural Meteorology* 4, 137–43.

Ward, R. de C. (1903) *Handbook of Climatology* (Macmillan, New York).

Warne, D.F. and P.G. Calnan (1977) 'Generation of electricity from the wind', *Proceedings of Institute of Electrical Engineers* 124, 963–85.

Webb, E.K. (1975) Evaporation from catchments. Conference on Prediction in Catchment Hydrology (Australian Acadamy of Science) 203–36.

Wendland, W.M. (1982) 'Wind power as an electric energy source in Illinois', *Journal of Applied Meteorology*, 21, 423–8.

Wendler, G. and N. Ishikawa (1973) 'Heat balance investigations in an arctic mountainous area in northern Alaska', *Journal of Applied Meteorogy* 12, 955–62.

Wieringa, J. (1977) 'Wind representativity increase due to an exposure correction, obtainable from past analog station wind records', in WMO, *Technical Conference on Instruments and Methods of Observation, Hamburg,* WMO Publ. 480 (World Meteorological Organization), 39–44.

—— (1980) 'Representativeness of wind observations at airports', *Bulletin of American Meteorology Society* 61, 962–71.

Wiesner, C.J. (1968) 'Estimating the Probable Maximum Precipitation in remote areas', Paper to Ann. Conf., Christchurch (Australia and New Zealand Association of Advanced Science).

Williams, M. (1982) 'Issues relating to the development of solar water heater ratings', Paper to Workshop, Sydney (Australian and New Zealand Section, International Society for Solar Energy).

Williams, R.J. (1985) 'Lillooet climatological network report', Priv. comm.

WMO (1960) *Guide to Climatological Practices,* WMO Publ. 100 (World Meteorology Organization).

—— (1966) *Statistical Analysis and Prognosis in Meteorology,* WMO Technical Note 71 (World Meteorological Organization).

—— (1974a) *Guide to Hydrological Practices,* WMO Publ. 168 (World Meteorological Organization).

—— (1974b) *Physical and Dynamical Climatology: Proceedings of Symposium, Leningrad, 1971,* WMO Publ. 347 (World Meteorological Organization).

—— (1981a) *Meteorological Aspects of the Utilization of Wind as an Energy Source,* WMO Technical Note 175 (World Meteorological Organization).

—— (1981b) *Meteorological Aspects of the Utilization of Solar Radiation as an Energy Source,* WMO Tech. Note 172 (World Meteorological Organization).

—— (1983a) *Guide to Meteorological Instruments and Methods of Observation,* 5th edn, WMO Publ. 8 (World Meteorological Organization).

—— (1983b) *Guide to Climatological Practices,* 2nd edn. WMO Publ. 100 (World Meteorological Organization).

Won, T.K. and E.J. Truhlar (eds) (1980) *An Introduction to Meteorological Measurements and Data Handling for Solar-Energy Applications* (International Energy Agency, Organization for Economic Cooperation and Development, US Dept Energy).

Woodhead, T. (1972) 'Mapping potential evaporation for tropical East Africa: the accuracy of Penman estimates derived from indirect assessments of radiation and windspeed', Proceedings of Symposium of World Water Balance, Reading, 1970

(World Meteorological Organization), 232–41.

Yamamoto, G. (1937) 'On the rate of condensation of dew', *Geophysical Magazine (Tokyo)* 11, 91–6.

Yamamoto, R. (1980) 'Variability of northern-hemisphere mean surface-air temperature during the recent two hundred years', in S. Ikeda, E. Suzuki, E. Uchida and M.M. Yoshino (eds), *Statistical Climatology* (Elsevier) 307–25.

Ye, Y. (1986) 'The relationship between temperature variation and low temperature in the growing season during the recent 70 years in Hunan province', *Geographical Research (China)* 5, 81–9.

Yoshino, M.M. (1975) *Climate in a Small Area* (University of Tokyo Press).

Zuzel, J. and L. Cox (1975) 'Relative importance of meteorological variables in snowmelt', *Water Resources Research* 11, 174–6.

INDEX

Note: Page numbers in italic refer to figures or tables.